S0-ANO-404

Coastal and Estuarine Studies

Coastal and Estuarine Studies

41

A.I. Robertson and D.M. Alongi (Eds.)

Tropical Mangrove Ecosystems

American Geophysical Union

Washington, DC

Editors

Alistar I. Robertson
Australian Institute of Marine Science
PMB No 3
Townsville MC
Queensland 4810
AUSTRALIA

Daniel M. Alongi
Australian Institute of Marine Science
PMB No 3
Townsville MC
Queensland 4810
AUSTRALIA

Library of Congress Cataloging-in-Publication Data

Tropical mangrove ecosystems / A.I. Robertson and D.M. Alongi (eds.).
p. cm. — (Coastal and estuarine series ; 41)
Includes bibliographical references.
ISBN 0-87590-255-3
1. Mangrove swamp ecology. I. Robertson, A. I. (Alistar I.)
II. Alongi, D. M. (Daniel M.) III. Series.
QH541.5.M27T76 1992
574.5′26325—dc20 92-44357
CIP

ISSN 0733-9569
ISBN 0-87590-255-3

Printed in the United States of America.

Preface

Mangrove forests are a dominant feature of tropical coasts. Like their terrestrial counterparts these forests are under threat worldwide through a variety of destructive human practices. As is also the case with tropical terrestrial forests, management decisions about mangrove ecosystems are currently being made often without adequate fundamental knowledge of the processes controlling natural ecosystem function.

Since the mid 1970's, the Australian Institute of Marine Science (AIMS) has had in place a research program investigating the structure and dynamics of mangrove ecosystems in tropical Australia. Early results of this research were summarized in 1982 in the book entitled *Mangrove ecosystems in Australia: structure, function and management*, edited by B.F. Clough. We felt that the time was ripe for a review of work during the decade 1981-1991, and so approached all past and present AIMS mangrove research workers to contribute to this volume. With the exception of Steve Blaber and Colin Woodroffe, all authors have worked at AIMS as staff members, or visiting scientists. We thought it appropriate that since John Bunt instigated much of the early work on mangrove ecosystems in northeastern Australia, he should introduce this volume.

While we are sure that this volume will serve as a useful source book for managers of mangrove wetlands, most chapters identify the often substantial gaps in our knowledge of these systems. Given the rate of loss of mangrove forests worldwide, the challenge for future workers on tropical mangrove systems will be to fill these important gaps in our knowledge, while at the same time publicizing their research results and making them available to managers (see Chapter 11, this volume).

All major chapters were reviewed by one external and one AIMS scientist. We thank Marylin Ball, Betsy Jackes, Chari Pattiaratchi, Richard Pearson, Gordon Thayer, Bruce Thom, Ivan Valiela and Bill Wiebe for review of chapters. All word processing was performed by Frances Conn, Steve Clarke, Rhonda Lyons, Sue Smith and Kim Wicks. Marty Eden produced all of the final figures and Christine Cansfield-Smith produced the camera ready text and did the copy editing. We also acknowledge the scientific support sections at AIMS for their professional assistance with much of the research reviewed in this volume.

Alistar Robertson and Daniel Alongi
Townsville, June 1992.

Contents

List of Contributors

Dr D.M. Alongi
Australian Institute of Marine Science
PMB No 3
Townsville MC
Queensland 4810
AUSTRALIA

Dr S.J.M. Blaber
CSIRO Marine Laboratories
PO Box 120
Cleveland
Queensland 4163
AUSTRALIA

Dr K.G. Boto
Australian Institute of Marine Science
PMB No 3
Townsville MC
Queensland 4810
AUSTRALIA

Dr J.S. Bunt
4/6 McDonald Street
Potts Point
N.S.W 2011
AUSTRALIA

Dr B.F. Clough
Australian Institute of Marine Science
PMB No 3
Townsville MC
Queensland 4810
AUSTRALIA

Dr N.C. Duke
Smithsonian Tropical Research Institute
Box 2072
Balboa
REPUBLIC OF PANAMA

Prof. Y. Mazda
School of Marine Science and Technology
Tokai University
Orido
Shimizu
Shizuoka
JAPAN 424

Dr P. Ridd
Department of Physics
James Cook University of North Queensland
Townsville
Queensland 4811
AUSTRALIA

Dr A.I. Robertson
Australian Institute of Marine Science
PMB No 3
Townsville MC
Queensland 4810
AUSTRALIA

Dr A. Sasekumar
Department of Zoology
University of Malaya
Kuala Lumpur
MALAYSIA

Dr T.J. Smith III
Rookery Bay National Estuarine Research Reserve
Florida Department of Natural Resources
10 Shell Island Road
Naples
Florida 33962
U.S.A.

Dr E Wolanski
Australian Institute of Marine Science
PMB No 3
Townsville MC
Queensland 4810
AUSTRALIA

Dr C. Woodroffe
Department of Geography
University of Wollongong
PO Box 1144
Wollongong
N.S.W. 2500
AUSTRALIA

1

Introduction

J.S. Bunt

There is presently wide public and scientific concern to achieve a fuller and more effective understanding of all of Earth's natural ecosystems. For that objective to be realised with tropical mangrove systems will require undiminished effort throughout the present decade and probably well into the next century, because our understanding of mangrove forests lags behind that of many other ecosystems. Nevertheless, as will be seen from the present book, great advancements have been made during the last decade, particularly in understanding the structure and function of mangrove ecosystems in the Indo-Pacific region. In this brief introductory chapter I will outline the history of system-level research on mangroves, with particular emphasis on work in tropical Australia, by the Australian Institute of Marine Science.

Many human communities have a traditional dependence on mangroves for their subsistence, and a wide range of natural products from mangroves are presently utilized, as documented by Saenger *et al.*, (1983). For some time, however, there has also been exploitation of mangrove resources for broader economic objectives, e.g. forestry. Harvests of that kind, when well-managed, are probably sustainable in the long term as evidenced by the continued high yield of timber products and charcoal from the Matang mangrove forest in Malaysia (Ong, 1982). On the other hand, some other forms of exploitation, often for short-term gain, can be irreversibly destructive. Clear felling of virgin forests for wood chips and paper pulp is a notable example. The conversion of mangrove forests for mariculture ponds at large scales, notwithstanding the recovery of fishery products, is also, of course destructive of the native forests and, at least in practical terms, irreversible (Ong, 1982). Reclamation of mangrove forests for agricultural, industrial, urban and other forms of land-based development is also increasing and causing irreversible damage in coastal regions throughout the tropics (Hatcher *et al.*, 1989). Such impacts, not to mention the additional burdens of pollution and changes in hydrological regimes caused by freshwater diversion projects, have adverse effects not only on subsistence dwellers in and near mangrove habitats, but also on other members of society, who depend indirectly on intact coastal resources. A consequence of this has been greater public concern for coastal lands, including mangrove forests, and increased research and management effort throughout the tropics.

There is a vast scientific literature on mangroves reflecting a wide spectrum of interest (e.g. see Rollet, 1981). However, prior to about 1970 the impetus for most research was enquiry for its own sake rather than research in relation to application for such goals as management and conservation. Indeed, a good deal of the systems-oriented research which became so popular in marine biological science from the mid 1950's on, and which so readily lends itself to application, was at first stimulated as much out of basic as of practical interest.

Currently, however, virtually all system-scale research based in the mangroves is linked to the application of sustained-yield management of the natural resource. Of course, it should be noted that since the values associated with mangroves vary with socio-economic setting, research priorities themselves tend to be variable. Nevertheless, a goal of most research on mangroves is an understanding of processes that control mangrove ecosystem function.

To place the present book in context, it is useful to consider some important stages in past research on mangrove ecosystems. Studies in the south-east of the United States in the 1960's provided much of the impetus for mangrove system research worldwide. Research undertaken concurrently by Heald (1969) and Odum (1970) on mangrove productivity in Florida attracted wide scientific and public attention to the general significance of mangrove communities in the coastal zone. A substantial body of work in the Americas has been built on these foundations. Reviews of the state of knowledge in the mid 1970's, based mainly on work in Florida were produced by Lugo and Snedaker (1974) and Walsh *et al.,* (1975). The programs of research based in Florida are continuing and have resulted in the beginning of similar research elsewhere in the New World (eg. Panama, Ecuador, Venezuela, Costa Rico, Puerto Rico, Mexico, Brazil, Columbia). Much of this research has been supported by the United Nations as well as national governments, indicating the perceived importance of mangroves to tropical countries.

More than 50% of the 10^5km^2 of mangrove forest worldwide occurs in the Old World (Saenger *et al.*, 1983). As with the situation in the New World, prior to the 1970's there was published a large literature, mainly on the floristics, distribution, autecology and silviculture of mangrove tree species, for the Old World tropics (see Chapman, 1976; Rollet, 1981). More process-oriented work in South-east Asia, aimed at solving conservation and management questions, began in the 1970's and Soepadmo *et al.*, (1984) is a very useful compendium of work from Indonesia, Malaysia, India and Thailand during that period.

Australia has 11,500 km^2 of mangrove forest, extending over approximately 7000 km of the Australian mainland coast and island shorelines (Galloway, 1982) between the latitudes of 10°S and 38°45'S. Prior to European settlement (and presently in some regions of Australia), coastal aboriginal communities used mangroves as sources of food and other basic resources (e.g. Meehan, 1982) but had little if any impact on the forests themselves. Because of their remoteness from most centres of population mangrove forests in Australia still remain close to pristine, in contrast to other locations in the Indo-Pacific region where a variety of developments have led to massive losses in mangrove forest areas (Hatcher *et al.,* 1989). Mangrove forests in tropical Australia therefore offer excellent opportunities for basic research and for gaining the types of insight needed to underpin effective protective management.

Even so, prior to about 1960 research interest in the Australian mangroves was sporadic and driven by curiosity about a little known vegetation and habitat. It was not until the South African researcher, Macnae (1966) published the results of quite coarse-scaled ecological surveys along the eastern seaboard of Australia that interest in mangrove research blossomed. There followed a marked expansion of activity in this field among researchers in the life and earth sciences already resident in Australia. Advances made to about the early 70's have been well-documented by Saenger *et al.*, (1977). The successes achieved are noteworthy since so few facilities existed at the time to provide direct support for research focussed in tropical Australian marine environments.

A decision by the Commonwealth Government of Australia in 1972 to create the Australian Institute of Marine Science (AIMS) as an independent entity and to locate it in or near Townsville in northern Queensland, already the site of James Cook University of North Queensland, suddenly and significantly expanded the national capacity in tropical marine research. When the Institute began research in 1974, one of the first programs to be initiated was directed to studies of "inshore productivity" with a focus of attention on mangrove ecosystems, since they dominated the tropical Australian coast. The work began at study sites within an approximately 50 km^2 area of undisturbed and floristically diverse mangroves in Missionary Bay on Hinchinbrook Island, an essentially uninhabited and little visited National Park close by the mainland some 100 km north of Townsville. A 400 m boardwalk was constructed early in the program to facilitate regular access to the dominant *Rhizophora* mangrove forest communities. Adoption of a common study area for a considerable variety of research projects early proved advantageous in developing a comprehensive understanding of system processes within a particular and well-documented environmental setting.

However, the Hinchinbrook Island site could not be taken to characterize the mangroves throughout their northern Australian distribution. Surveys undertaken within a large number of previously unstudied estuaries along the tropical coast of Queensland and the Northern Territory revealed great variability in floristics, geomorphology and hydrology, at both small and large spatial scales (e.g. Boto and Bunt, 1981; Bunt and Williams, 1981; Bunt, Williams and Duke, 1982). The floristic patterns are dependent to some degree on both biogeographic history and direct responses to environmental and biological controls. Early observations also indicated that it was likely that there may be major differences in mangrove-dependent biota, trophic relationships and nutrient status on different parts of the tropical Australian coast. Were that so, it would be unreasonable to expect any single model to reflect the behaviour of all tropical Australian mangrove ecosystems.

With the objective of reviewing research directions and alternatives, five years after the mangrove program was established at AIMS, the Institute hosted a workshop of active mangrove researchers throughout Australia. The resultant papers were published in Clough (1982). One of the major findings of this workshop was the need for in- depth studies in a number of field sites, to test hypotheses arising from work at the Missionary Bay study site. In particular, greater emphasis needed to be given to studies of food chains and nutrient cycles linking mangrove and adjacent habitats. In the decade since this last review of research, the AIMS mangrove program has expanded its exploratory surveys and process-oriented hydrological and ecological research within estuaries in Australia and into Papua New Guinea, continued its work at the permanent study site at Missionary Bay and strengthened its laboratory-based programs. External specialist collaborations have been actively encouraged and have been extremely worthwhile.

In addition, in the 1980's, associations were developed with researchers elsewhere in the Indo-Pacific region, firstly through UNESCO, with its Research and Training Program on Mangrove Ecosystems in Asia and the Pacific and later through an ASEAN-Australia cooperative program in Marine Sciences, with studies of the regional mangroves a priority. This program is now well advanced and has been invaluable in developing the perspective of both Australian and Asian scientists (for a review of progress, see Alcala *et al.*, 1991).

As a separate initiative through the Australian Committee for Mangrove Research and with funding from the Australian Development Assistance Bureau, early in 1985 a Research for Development Seminar was hosted by AIMS. Discussions on the status, exploitation and management of the mangrove ecosystems of Asia and the Pacific brought together experts from a number of centres in Australia as well as from countries throughout the Indo-Pacific. Papers from this meeting were reported in proceedings edited by Field and Dartnall (1987) and recommendations called for intensified working interaction between Asian and Pacific scientists with the objective of matching research developments with the pressing needs of management in the region.

It now seems appropriate to evaluate the information on mangrove ecosystems presently available and to consider the extent to which its synthesis might be possible. It is important to bear in mind that the global mangrove ecosystem is highly variable throughout its distribution. Regional differences in floristic diversity are especially striking. Tree species richness is highest in south-east Asia (Tomlinson, 1986), and as a consequence forest communities in that region are often far more complex than those in the New World. Just as importantly, mangrove environments themselves, are also diverse in terms of local and regional climate, salinity regimes, tidal range and pattern, physiography, hydrology and other physical and physico-chemical variables.

At the same time, few if any mangrove research workers have close experience of the full diversity of expression within the ecosystem worldwide. Most are committed to quite limited numbers of study sites. Yet all would acknowledge and welcome the benefits of a wider perspective. Accordingly, attempts at global review and synthesis of accumulated regional and more local knowledge, both essential, present considerable challenges to intending interpreters. The problem is exacerbated by a literature-in-common which tends to be uneven in coverage and focus and, as often as not, is based on methods which hinder comparative analysis. There seems no entirely satisfactory solution to such difficulties. Those attempting reviews of mangrove research can do no more than exercise reasonable judgement in seeking generalizations.

Having said that, it is obvious that substantial progress in understanding tropical mangrove ecosystems is documented in the contributions making up this volume. Only those who have taken up research in the mangroves can fully appreciate the logistic difficulties and physical challenges such work entails. During the pioneering days of the AIMS mangrove program, for example, it took months of labour and vessel support backed by the skills of a group of Australian Army engineers to set up a timber walkway at a single site in Missionary Bay. Staff from the Australian Survey Office also spent considerably longer in the field to provide essential topographic maps of the densely forested intertidal terrain to serve as a base for planned long-term ecosystem studies (see Chapters 3, 9 and 10, this volume). The problems and the costs of accessing and conducting research within the mangroves along even more remote coastlines are daunting. Few, if any research groups, even today, have the resources necessary to cope with such tasks in a comprehensive fashion even at a limited number of sites.

Over the 16 or 17 years since its original construction, the boardwalk in Missionary Bay has been used as a working platform, not only by AIMS researches in a range of disciplines, but also by a steady stream of their collaborators from around the world. That combined experience is now yielding rewards at sites elsewhere in Australia as well as overseas.

This book has its foundations in the work done at Hinchinbrook Island and elsewhere in tropical northeast Australia. Its scope though, I believe, is far broader because those who have contributed its chapters, through direct overseas experience and collaboration with scientists outside Australia, are well placed to maintain a wide perspective. I am delighted to have been invited to introduce the volume and would like to do so on behalf of all those who worked out of Townsville during the early years of the AIMS mangrove program.

References

Alcala, A.C., 1991. (Ed.) *Proceedings of the regional symposium on Living Resources in Coastal Areas.* University of the Philippines, Quezon City, 597pp.

Bunt, J.S., 1982. Studies of mangrove litter fall in Australia. In: Clough, B.F., (Ed.), *Mangrove ecosystems in Australia: Structure, Function and Management,* pp.223-238, Australian Institute of Marine Science and Australian National University Press, Canberra.

Bunt, J.S., and Williams, W.T., 1981. Vegetational relationships in the mangroves of tropical Australia. *Marine Ecology Progress Series* **4**:349-359.

Bunt, J.S., Williams, W.T., and Duke, N.C., 1982. Mangrove distributions in north-east Australia. *Journal of Biogeography* 9.111-120.

Chapman, V.J., 1976. Mangrove vegetation. J. Cramer, Vaduz, 447pp.

Clough, B.F., (Ed.) 1982. *Mangrove ecosystems in Australia: Structure, Function and Management,* Australian Institute of Marine Science and Australian National University Press, Canberra 302pp.

Field, C.D., and Dartnall, A.J., (Eds.) 1987. *Mangrove ecosystems of Asia and the Pacific: status, exploitation and management.* Australian Institute of Marine Science, Townsville, 320pp.

Galloway, R.W., 1982. Distribution and physiographic patterns of Australian mangroves. In: Clough, B.F., (Ed.), *Mangrove ecosystems in Australia: Structure, Function and Management,* pp.31-54, Australian Institute of Marine Science and Australian National University Press, Canberra.

Hatcher, B.G., Johannes, R.E., and Robertson, A.I., 1989. Review of research relevant to conservation of shallow tropical marine ecosystems. *Oceanography and Marine Biology: An Annual Review* **27**:337-414.

Heald, E.J., 1969. The production of organic detritus in a South Florida estuary. PhD Diss. University of Miami.

Lugo, A.E., and Snedaker, S.C., 1974. The ecology of mangroves. *Annual Review of Ecological and Systematics* **5**:39-64.

Macnae, W., 1966. Mangroves of eastern and southern Australia. *Australian Journal of Botany* **14**:67-104.

Meehan, B., 1982. *Shell bed to shell midden.* Australian Institute of Aboriginal Studies, Canberra, 189pp.

Odum, W.E., 1970. Pathways of energy flow in a South Florida estuary. PhD Diss. University of Miami.

Ong, J.E., 1982. Mangroves and mariculture. *Ambio* **11**:252-257.

Rollet, B., 1981. *Bibliography of Mangrove Research 1600-1975.* UNESCO, Rome, 479pp.

Saenger, P., Hegerl, E.J., and Davie, J.D.S., 1983. Global status of mangrove ecosystems. *The Environmentalist* **3**:1-88.

Saenger, P., Specht, M.M., Specht, R.L., and Chapman, V.J., 1977. Mangal and coastal saltmarsh communities in Australasia. In: Chapman, V.J., (Ed.), *Wet Coastal Ecosystems: Ecosystems of the World*, pp.293-345, Elsevier, Amsterdam.

Soepadmo, E., Rao, A.N., and MacIntosh, D.J., (Eds.), 1984. *Proceedings of the Asian symposium on mangrove environment: research and management.* University of Malaysia and UNESCO, Kuala Lumpur, 828pp.

Tomlinson, P.B., 1986. *The Botany of Mangroves.* Cambridge University Press, 413pp.

Walsh, G.E., Snedaker, S.C., and Teas, H.J., (Eds.) 1975. *Proceedings of the International Symposium on the Biology and Management of Mangroves.* University of Florida, Gainesville, 846pp.

2

Mangrove Sediments and Geomorphology

Colin Woodroffe

2.1 Introduction

Mangrove ecosystems demonstrate close links between vegetation assemblages and geomorphologically-defined habitats. Mangrove species distribution is influenced by several environmental gradients which respond either directly or indirectly to particular landform patterns and physical processes. In addition, vegetation can change through time as landforms accrete or erode.

It is important to understand these physically-evolving habitats because significant changes can occur on timescales shorter than the life-history of the mangroves themselves, and thus past changes may explain present forest structure. On the other hand, the often rapid accumulation of sediments serves to preserve some of the record of past habitat change, such that stratigraphic and palaeoecological study can yield insights into the longer-term dynamics of the ecosystem.

Despite the close links between the ecology of mangrove communities and their sedimentary setting, the prevailing biological and geological views of mangrove ecosystems are somewhat contrary. Biologists now view mangrove forests as highly productive **sources** of organic matter, from which there is a net outwelling of energy, supporting complex estuarine and nearshore food webs (e.g. see Chapters 7 and 10, this volume). Geologists, on the other hand view mangrove shorelines as sediment **sinks**, characterised by long-term import of sediment, as indicated by the substantial accumulation of recent sediments which underlie mangrove forests and adjacent coastal plains. The physical processes involved in the flux of both organic matter and sediment are essentially the same, principally the freshwater and tidal flows, and the apparent contradiction of the role of mangrove forests as either a source or a sink, only serves to indicate the hydrodynamic complexity of determining a net budget for tidal wetlands in which large quantities of organic and inorganic material are constantly exchanged in both directions (see Chapter 3 and 10, this volume).

In this review the role of mangroves in accelerating the rate of mud accretion, and the relationship between species zonation and shoreline progradation are examined. There is an outline of the different environmental settings, within which mangrove forests occupy distinct geomorphologically-defined habitats, often following recurring patterns of species distribution. These habitats change over time, and the Quaternary palaeoecology of mangroves is examined. It is demonstrated that sea-level variations over the late Pleistocene and during the Holocene have totally disrupted mangrove forests. These longer-term studies are also shown to give a perspective on major depositional and erosional processes, which it is generally not possible to determine from shorter-term process studies.

2.2 Mangroves and Sedimentation

Mangrove forests are best developed on tropical shorelines where there is an extensive suitable intertidal zone (as found on low gradient or macrotidal coasts), with an abundant supply of fine-grained sediment, and are most luxuriant in areas of high rainfall or abundant freshwater supply through run-off or river discharge (Walsh, 1974; and see Chapter 8, this volume). While mangroves are generally associated with low- energy, muddy shorelines, particularly tropical deltas, they can grow on a wide variety of substrates, including sand, volcanic lava or carbonate sediments (Chapman, 1975). Where sediments, such as terrigenous muds, are brought in from outside the ecosystem, they are termed allochthonous. In carbonate areas where there is often not an abundant supply of terrestrial sediments, mangroves may be underlain by calcareous skeletal or reefal substrates, or calcareous mud, but are also often underlain by an organic peat derived largely from the roots of mangroves themselves. These sediments, produced within the ecosystem, are termed *in situ* or autochthonous.

It has proved difficult to determine the rate of mud sedimentation beneath mangroves. The identification of marker (brick-dust) horizons, used to measure salt marsh accretion rates, revealed sediment accumulation rates of around 1 mm yr^{-1} in New Zealand (Chapman and Ronaldson, 1958), but the technique has not been successfully applied elsewhere. Bird (1971) measured sedimentation rates of up to 8 mm yr^{-1} against stakes in a dwarf *Avicennia* forest in southern Australia. Stakes may alter the scour and sedimentary processes around them, and the pattern of sedimentation or erosion has been shown to vary over a range of -11 to + 4.6 mm yr^{-1} in experiments with grids of stakes simulating the pneumatophores of *Avicennia* in north-eastern Australia (Spenceley, 1977, 1982). More recently ^{210}Pb and ^{137}Cs isotopes have been used to show sedimentation rates of up to 3 mm yr^{-1} in mangroves from Mexico (Lynch *et al.*, 1989). Radiocarbon dating of mangrove sediments allows longer-term assessment of sedimentation rates, implying rates of up to 6 mm yr^{-1} under mangroves in the early Holocene in northern Australia (Woodroffe, 1990), but sedimentation rates over these timescales are closely constrained by rates of sea-level change, as is discussed below.

Accelerated sedimentation beneath mangrove forests may be inferred where the morphology of the intertidal zone under mangroves is steeper than where there are no mangroves (Chappell and Grindrod, 1984; Bird, 1986). The view is increasingly accepted that mangroves follow areas of mud accumulation, but that their establishment leads to more rapid accretion (Thom, 1967; Carlton, 1974; Zimmermann and Thom, 1982). Not only are mangrove roots and pneumatophores efficient sediment trappers effectively slowing water movement (Wolanski *et al.*, Chapter 3, this volume), but the fine roots also play an important role as sediment binders (Scoffin, 1970). However, mangrove shorelines can, and often do, undergo erosion (Carter, 1959). Extensive erosion by cliffing, sheet wash and tidal creek extension has been proposed in the mangrove-fringed estuary of King Sound, north-western Australia (Semeniuk, 1981).

2.3 Shoreline progradation and mangrove succession

Mangrove shorelines often demonstrate a zonation of species parallel to the shore (Chapman, 1970, 1975; Smith, Chapter 5, this volume). This has lead to a view that mangrove shorelines are successional, undergoing gradual change towards some non-halophytic climax community. Particularly influential in this respect was the study of mangroves in Florida by Davis (1938, 1940).

Davis viewed the seawardmost zone of *Rhizophora mangle* as pioneer, attributing it a role in the progradation of the shoreline. He considered the successive zones of mature *Rhizophora, Avicennia* with salt marsh species and *Conocarpus* to be seral stages in the sequence of replacement through time culminating in a tropical forest association climax. This approach, with its origin in the classical ecology of F.E. Clements, was challenged by Egler (1952), who realised that there had been changes of sea level. However, Chapman extended Davis' interpretation through his studies in similarly zoned mangrove forests in Jamaica (Chapman, 1944, 1975), and the view that zonation of mangrove forests reflects succession has become widely accepted (Richards, 1952; Kuenzler, 1974). The viviparous nature of mangrove propagules, which are particularly well-adapted to dispersal and establishment in shallow-water has been taken as further support for the notion that mangroves 'claim land from the sea' (Vaughan, 1909; Stephens, 1962).

Zonation need not indicate succession. Zonation of mangrove species reflects ecophysiological response of the plants to one or a series of environmental gradients. It is the combination of factors such as frequency and duration of inundation, waterlogging of substrate, pore water salinity and pore water potential, that determines which plants grow where. Thus in many cases mangrove zonation may be static, or steady-state, reflecting the pattern of distribution of those environmental factors (Lugo, 1980; Snedaker, 1982), and it has even been proposed that central zones of the mangrove shoreline may represent a climax community themselves (Johnstone, 1980).

Chappell and Grindrod (1984) demonstrate that zoned mangrove communities on the open shoreline of Princess Charlotte Bay, may be undergoing one or two different patterns of change: a rapid prograding mode and a cut and recover mode (Figure 1). The rapid prograding mode occurs when there is an abundant supply of mud, a gentle gradient (1:1000), and wide mangrove fringes, and preserves little if any evidence of periodic storm impacts (Figure 1a). The cut and recover mode occurs where there is little mud supply, a steeper gradient (1:200), and the mangrove fringe is narrower (100-150 m), and is devastated by severe storms which then concentrate sand and shell into chenier ridges landward of the mangroves (Figure 1a). Alternation between the progradational modes is considered to be a response to variations in mud supply and shell production (Chappell and Grindrod, 1984). Sedimentary, radiocarbon and palynological evidence supports the overall progradation of this shoreline and successive replacement of mangrove zones (Grindrod, 1985).

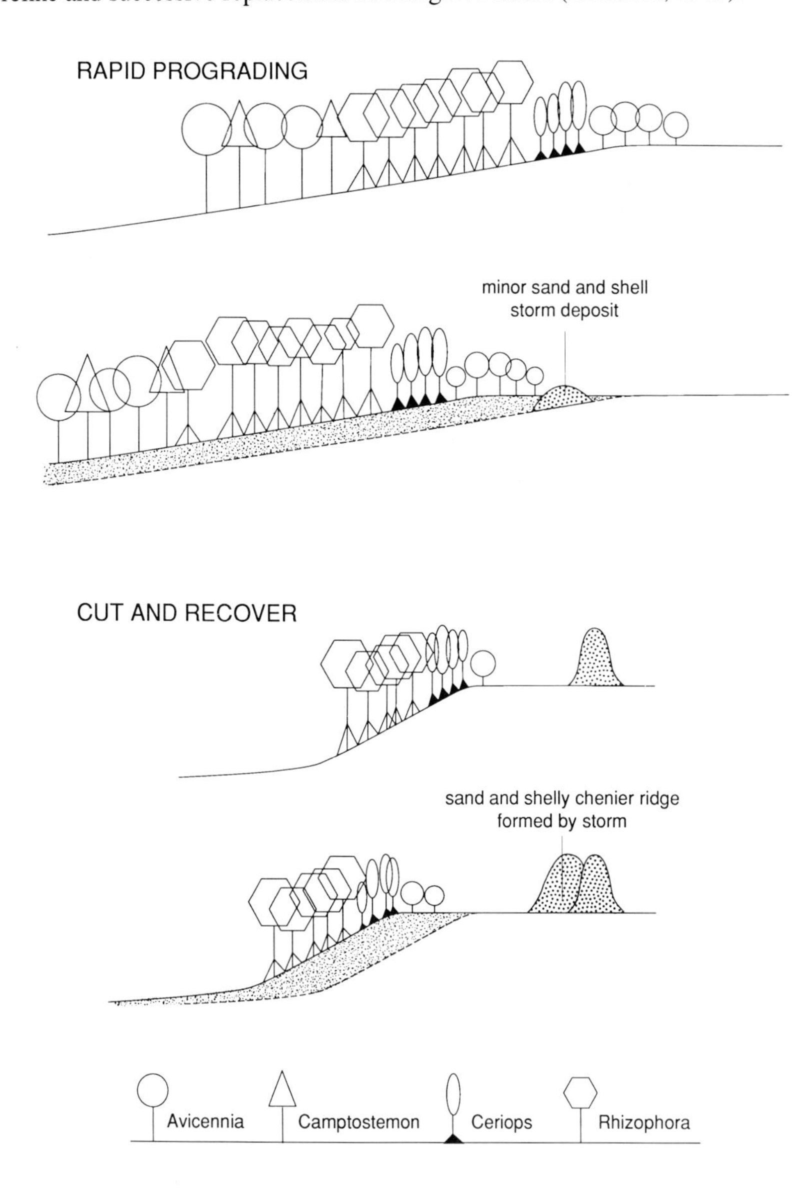

Figure 1. The characteristic zonation of mangrove species and the dominant geomorphological mode on open shorelines in northern Australia. (a) The rapid prograding mode and the cut and cover mode appear to have alternated during the Holocene progradation of a chenier plain shoreline, Princess Charlotte Bay (after Chappell and Grindrod, 1984).

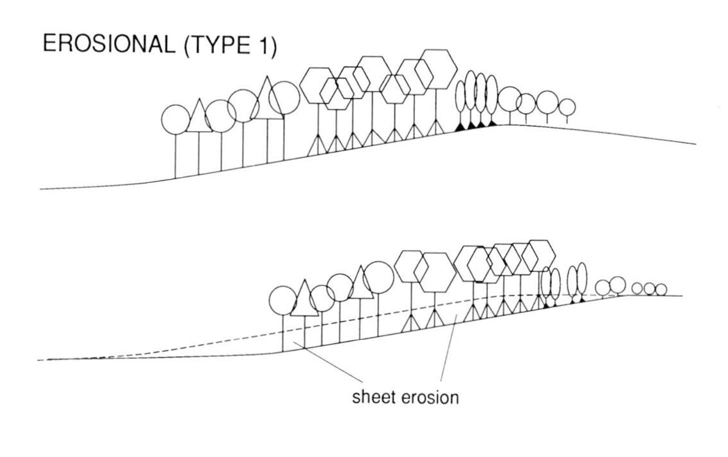

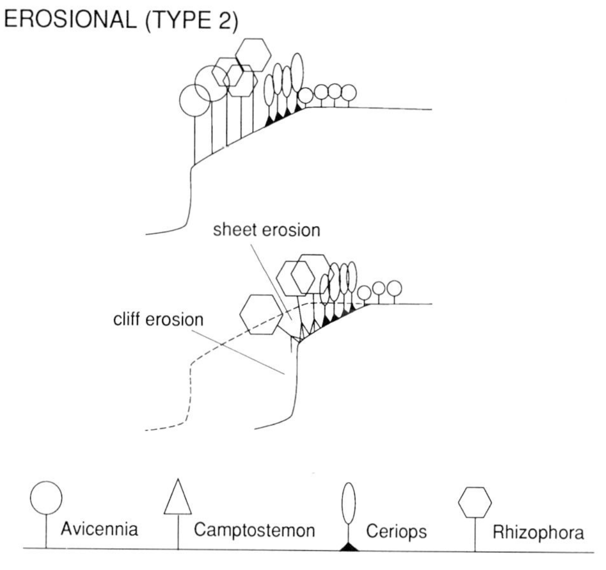

Figure 1(b) The erosional Type 1 and erosional Type 2 shorelines represent similar coastal plains which are undergoing retreat predominantly through sheet erosion and cliff erosion, King Sound, Western Australia (after Semeniuk, 1980).

Semeniuk (1980) has demonstrated that similar zoned mangrove shorelines in King Sound, Western Australia, may actually be undergoing erosion and retreating. Despite broad Holocene progradational plains with chenier ridges (Jennings and Coventry, 1973: Jennings, 1975), the dominant trend at present is erosional with retreat through sheet erosion, cliff erosion and tidal creek extension (Semeniuk, 1981). In terms of mangrove zonation Semeniuk has identified several types of coastline (Semeniuk, 1980). Figure 1b illustrates schematically Type 1, a gentle gradient shoreline with a broad mangrove fringe prominently zoned which appears to be undergoing gradual retreat through sheet erosion, and Type 2, a steeper gradient shoreline on which the mangrove fringe is narrower and less clearly zoned, which is often retreating through wave-induced cliff erosion of the muddy substrate on which the mangroves are rooted. The vegetation patterning, illustrated schematically in Figure 1b, is very similar to the two modes identified by Chappell and Grindrod, and yet it appears that in these cases the shoreline is eroding, and that mangrove zones are propagating landwards. This serves to illustrate very effectively that the zonation of mangroves reflects the physiological control by environmental gradients across the intertidal zone, and that it may be a poor indicator of the dynamics of mangrove habitats.

While zones dominated by individual species or associations of species may be identifiable on open shorelines where there is a steep topographic gradient, they are not characteristic or easy to discriminate in many areas of coastal wetland (Lugo and Snedaker, 1974). In a detailed study of deltaic mangroves in Tabasco, Mexico, Thom (1967) has advanced the view that in such a geomorphologically dynamic environment mangroves are opportunistic in colonising available substrate. Geomorphologically- distinct mangrove habitats in which mangrove species patterns recur can be recognised within a series of broad environmental settings.

2.4 Environmental settings

Mangrove shorelines occur in a number of different environmental settings, comprising particular suites of recurring landforms and differing in the physical processes responsible for sediment transport and deposition. Thom (1982) initially defined 5 terrestrial settings, based upon the classification of deltas proposed by Wright *et al.*, (1974) and that of bedrock embayments proposed by Roy *et al.*, (1980). These consist of Setting I, river-dominated (allochthonous); Setting II, tide- dominated (allochthonous); Setting III, wave-dominated barrier lagoon (autochthonous); Setting IV, composite river and wave-dominated; and Setting V, drowned bedrock valley.

Subsequently Thom has identified a further 3 carbonate settings (Thom, 1984); consisting of Setting VI, carbonate platform; Setting VII, sand/shingle barrier; and Setting VIII, Quaternary reef top. To these may be added a further carbonate setting, inland mangroves and depressions, which form a distinct category within the mangroves of Pacific Islands (Woodroffe, 1987).

Each setting comprises 3 components; its geophysical characteristics (climate, tides and sea-level), its geomorphological characteristics (the dynamic history of the land surface and contemporary geomorphological processes), and biological characteristics (microtopographic, elevation and sediments, drainage and nutrient status), which combine to define the setting (Thom, 1982, 1984). The supply of sediment in terrestrial settings, or its production in the case of carbonate settings, and the subsequent transport and deposition are important in relation to mangrove ecology. There may also be a relationship to the trophodynamic structure of the mangrove and associated systems.

The settings are not discrete categories, but intergrade such that many particular coastlines might fall between two or more individual settings. This is particularly true of the carbonate settings which all share the characteristics that they are systems where calcareous sediment is produced *in situ,* either as reef growth, as skeletal biogenic sediments, or as precipitated carbonate, and mangrove forests are generally underlain by these calcareous sediments or by mangrove-derived peat.

In this account the five distinct terrestrial settings identified by Thom (1982) will be described, and carbonate settings will be outlined, but without subdividing the latter into the distinct or exclusive classes adopted by Thom (1984). The relationship of these settings (Figure 2) to the functional classification of mangrove ecosystems of Lugo and Snedaker (1974) will then be considered.

2.4.1 Setting I: River-dominated

The most extensive mangrove forests are developed in the deltas of large tropical rivers (i.e. in the Bay of Bengal in the deltas of the Ganges and Brahmaputra Rivers; in the Gulf of Papua associated with the deltas of the Fly and Purari Rivers). Such deltas receive enormous sediment loads from geologically young and tectonically active headwaters, and are extremely dynamic. They can be divided into an active deltaic plain, dominated by those distributaries which carry most of the fluvial discharge, and an abandoned deltaic plain, associated with distributaries which are no longer active (Figure 2a). The subaerial delta can also be subdivided into an upper deltaic plain which is generally remote from marine or tidal influence and a lower deltaic plain and their distribution is often taken to mark the limit of the lower deltaic plain. Mangrove forests are frequently restricted in the active deltaic plain because the strong freshwater flows, often over a well-developed salt wedge in the deeper parts of the channel, do not favour extension of mangroves upstream. By contrast mangroves can be extensive along former distributaries in the abandoned deltaic plain. This pattern is shown in Figure 2a, which is based largely on the Purari Delta (after Thom and Wright, 1983), but can be seen in many other deltas (i.e. Brahmaputra).

Within the delta, mangroves are distributed in response to microtopographic characteristics related to elevation and frequency of inundation (Baltzer, 1970). Their patterning and dynamics are largely a function of the geomorphological development of such features as distributary channels, point bars, natural river levees and interdistributary basins (Thom, 1967). Such deltaic coasts can change very rapidly (Vann, 1980); rates of seaward extension of up to 125 m yr^{-1} have been recorded in Java (Macnae, 1968) and lateral migration of distributaries has been recorded at rates of up to 800 m yr^{-1} in the Brahmaputra River (Coleman, 1969).

Interdistributary basins are strongly influenced by climate. They are dominated by saline flats bare of vegetation in the more arid areas (i.e. Senegal and Gambia, Giglioli and King, 1966; northern Australia, Spenceley, 1976), by mangroves in wetter areas (i.e. Tabasco, Mexico, Thom, 1967) and by freshwater wetlands or peat swamp forest in the most perhumid equatorial areas (i.e. Fly and Purari Rivers, Thom and Wright, 1983).

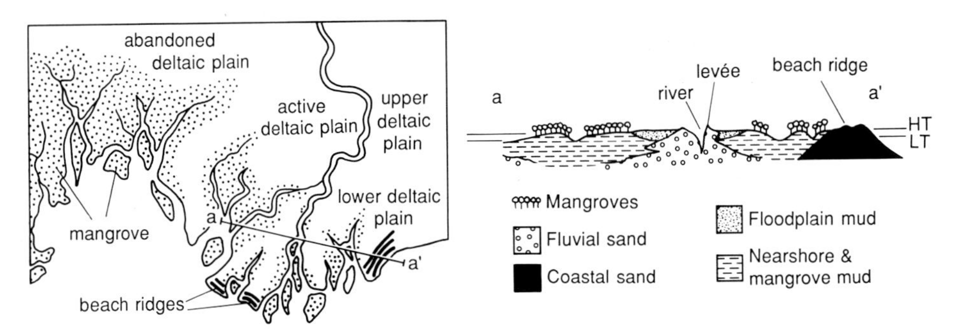

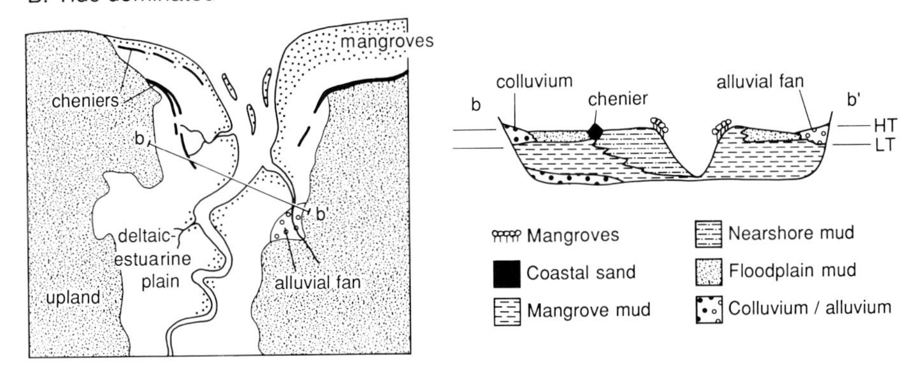

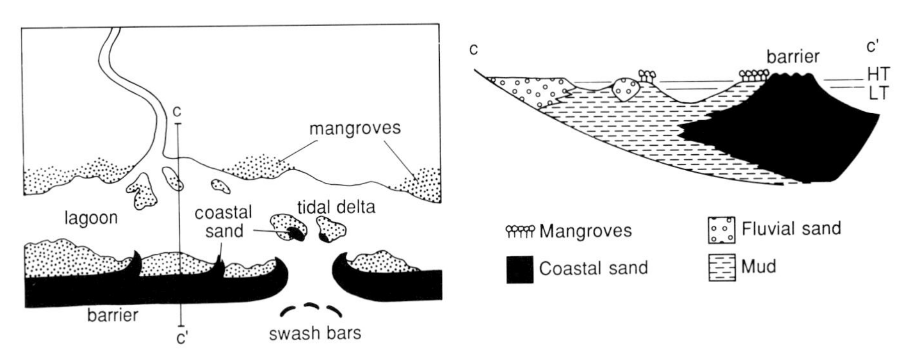

Figure 2. Environmental settings: typical planform, and stratigraphic cross-section. A) River-dominated, based on the Purari delta; B) Tide-dominated, based on macrotidal estuaries in northern Australia; C) Wave-dominated.

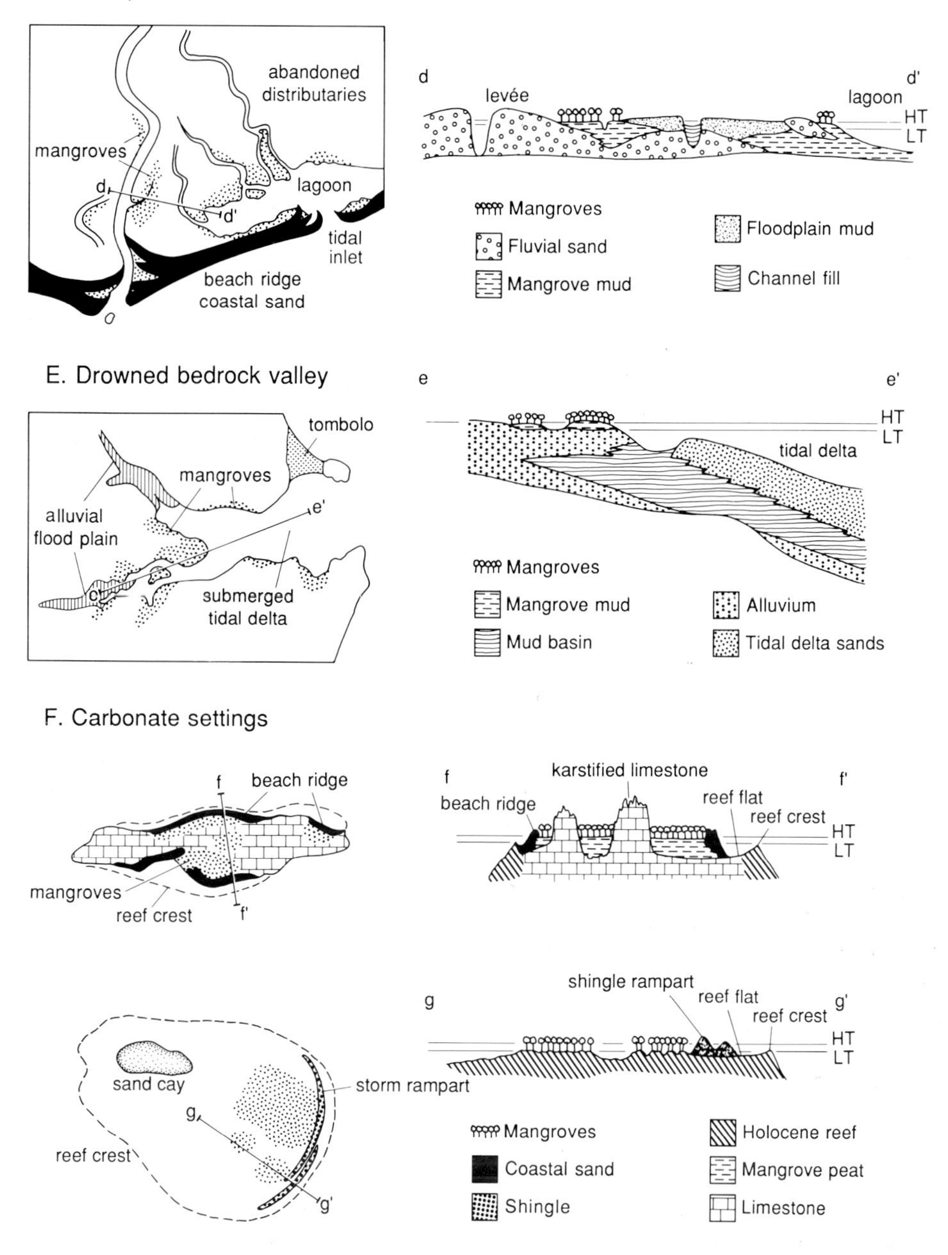

Figure 2. Environmental settings: typical planform, and stratigraphic cross-section. D) Composite river and wave-dominated, based on the Grijalva; E) Drowned bedrock valley, based on estuaries in south-eastern Australia; F) Carbonate settings, based on West Indian limestone islands and low wooded islands of the Great Barrier Reef of Australia.

2.4.2 Setting II: Tide-dominated

Tide-dominated mangrove settings are characteristic of areas of high tidal range where there is an extensive, low-gradient intertidal zone available for mangrove colonisation. Strong bi-directional tidal velocities characterise macrotidal (tidal range > 4 m) estuaries, which have broad funnel-shaped mouths and often linear islands or shoals paralleling flow directions (Wright *et al.*, 1973). While such estuaries do not have the complex channel-margin habitats associated with active delta distributaries, there are nevertheless recurrent patterns of mangrove species related to channel-margin habitats (Thom *et al.*, 1975), and a series of geomorphologically-defined habitats on the saline flats (Semeniuk, 1983). Figure 2b illustrates a typical patterning of such habitats in mangrove forests of north and north-western Australia, where colluvial and alluvial deposits, and chenier ridges marking former shorelines, provide habitats for mangroves (Semeniuk, 1983, 1985a, 1985b). Abandoned portions of some deltaic plains also contain former distributaries that are now dominated by tidal flows. These show many of the morphological characteristics, such as the rapidly tapering funnel-shaped form, of tide-dominated estuaries.

2.4.3 Setting III: Wave-dominated

On those coasts which are dominated by wave energy and on which there is an abundant supply of sand, the coast will be formed of shore-parallel sandy ridges, often as barrier islands enclosing a series of elongate lagoons (Figure 2c). High wave energy and sandy substrates are not generally favourable conditions for mangrove establishment, but mangrove forests may occur in the sheltered lagoons. The most extensive mangroves on wave-dominated coasts occur in either of two settings. Firstly the extensive fluvially-dominated environments developed where a large river reaches a wave-dominated shoreline, as for instance with the Shoalhaven River in south-eastern Australia, which is a large river that has now largely filled the coastal embayment into which it empties and is discharging fluvial sediment directly onto a sandy plain of otherwise shelf-derived sand (Wright 1970). Secondly, mangroves can be extensive in drowned river valleys, which are coastal embayments which have not filled in entirely with fluvial sediments; such embayments are common along the coast of south-eastern Australia (Roy *et al.*, 1980; Roy, 1984). These are described under Setting IV and Setting V respectively.

Extensive lagoonal systems, formed as a result of wave processes, represent an important mangrove setting in the New World in particular (i.e. Mexico, Lot-Helgueras *et al.*, 1975; Brazil, Schaeffer-Novelli *et al.*, 1990). Mangrove habitats reflect gradients of salinity which are controlled by climatic factors and the volumes of freshwater river discharge that reaches different parts of the lagoonal system. These systems change slowly, largely as a result of diversion of river flow or the formation or infill of breaches through the seaward ridge.

2.4.4 Setting IV: Composite river and wave-dominated

Where rivers bring large volumes of sediment to a wave-dominated coast a complex of landforms develops within which there can be extensive mangrove forests. Lagoons are well-developed in many deltas where fluvial sands are redistributed along the coast from a large delta (i.e. Niger Delta, Allen, 1965; John and Lawson, 1990). In the Grijalva Delta, in Tabasco, Mexico, Thom (1967) has demonstrated a well-marked segregation of mangrove species according to landform type: point bar, levee, lagoon, distributary channel and interdistributary basin, while the dynamics of landforms reflect the shift of active sedimentation and the subsidence of former deltaic plains (Figure 2d).

2.4.5 Setting V: Drowned bedrock valley

Many large coastal embayments in the tropics and subtropics have been downed by the post-glacial rise in sea level, and are known are rias. These provide sheltered environments within which mangrove forests develop on muddy substrates (Figure 2e). Many of the estuarine systems of south-eastern Australia result from drowning of bedrock valleys, and a detailed classification has been proposed by Roy *et al.,* (1980). Embayments go through a series of evolutionary stages as a result of infill with mud from the hinterland. Mangroves are initially associated with embayment-head alluvial and deltaic environments; but are most extensive during intermediate stages when complex shoals and intertidal habitats develop; and tend to be eradicated in the later stages when rivers traverse an almost entirely infilled plain and discharge directly into the sea (Roy, 1984). Similar ria shorelines can be recognised in the Kimberley region of north-western Australia (Semeniuk, 1985a).

2.4.6 Carbonate settings

Carbonate settings are those in which terrestrial sediment supply is low or absent but calcareous sediment production dominates. Included are mangroves of oceanic islands, coral reefs and carbonate banks. In many cases the tidal range is low, and the mangrove substrate consists of mangrove-derived peat. Thom (1984) has identified 3 settings, carbonate platform (such as the extensive mangrove forests of the Bahamas and Florida Bay), sand-shingle barrier, and Quaternary reef top, to which may be added inland mangroves and depressions (Lugo, 1981). However these settings often intergrade and are not exclusive; they will be treated as one broad setting in this account, two examples of which are illustrated in Figure 2f.

The most detailed descriptions of mangrove forests in carbonate settings come from south-western Florida, where Davis (1940) initially proposed his successional model of mangrove zone replacement. It is now clear that mangroves in Florida Bay form a part of a broader dynamically-changing environment and that their role varies according to distinct habitats (Wanless, 1974). In particular they appear to have played a significant role, both in a transgressive and regressive mode in the formation of islands (Wanless, 1974; Enos and Perkins, 1979).

On the Great Barrier Reef of Australia mangrove forests occupy a significant proportion of the surface area of many of the islands, particularly the low wooded islands. The habitats occupied by mangroves reflect the distribution of other landforms, such as the storm ramparts which provide protection. Mangrove forest area and dynamics appear to be a function of the evolutionary stage of the reefal substrate and its suitability for colonisation (Stoddart, 1980).

2.5 Mangrove Habitats and Functional Ecology

The broad environmental settings described above represent regional entities within which first order differentiation of mangrove forests is possible. Each of the settings contains suites of landforms, some of which may be common to several settings, within which particular environmental gradients control the distribution of mangrove species through ecophysiological factors. Areas of the greatest geomorphic diversity tend to contain the highest degree of habitat variation.

The dominant processes in each setting tend to differ, but settings are not entirely process-related. Thus riverine and tidal processes represent important, and contrasting influences on mangrove forests, the former characterised by unidirectional flows and net export of material from mangroves while the latter are bi-directional with less clear flux (Wolanski *et al.*, Chapter 3 this, volume). Few settings, however, demonstrate a total dominance by either; the abandoned distributaries of a delta being dominated by tidal rather than fluvial processes, and tidal estuaries often receiving seasonal freshwater riverine floods.

More research needs to be done to discriminate landform-related mangrove habitat within each environmental setting, following the pioneering work of Thom in the Grijalva Delta (1967), and of Thom *et al.*, (1975), and Semeniuk (1983) in northern Australia.

Mangrove ecologists have made extensive use of a functional classification of mangrove forests derived by Lugo and Snedaker (1974). Lugo and Snedaker recognised six categories (Figure 3): i) Overwash mangroves, small islands, generally composed of *Rhizophora* (in the New World), completely overwashed, and often underlain by mangrove peat but not characterised by litter accumulation. ii) Fringe mangrove, a *Rhizophora* - dominated littoral fringe inundated by daily tides, but with litter accumulation. iii) Riverine mangrove, tall, productive *Rhizophora* - dominated mangrove stands flanking a river channel receiving nutrient- rich freshwater flushing. iv) Basin mangrove, typically mixed, or *Avicennia* - dominated characteristic of interior areas of mangrove forests. v) Scrub mangrove, a dwarfed stand especially of *Rhizophora* <1.5 m tall, often in nutrient-poor areas. vi) Hammock mangrove, a special form of basin mangrove found in the Everglades, forming small islands of mangrove over a mangrove-derived peat which infills a depression in the underlying limestone substrate. Lugo and Snedaker's scheme was initially developed for, and has had its greatest application in, carbonate settings of the New World, where mangrove patterning in extensive low-lying coastal wetlands does not demonstrate clear zonation of species. While primarily based upon mangrove physiognomy, the classification has also been shown to relate to structural complexity, soil salinity, productivity measured by litter fall and physiological properties of the trees (Pool *et al.*, 1975, Twilley *et al.*, 1986).

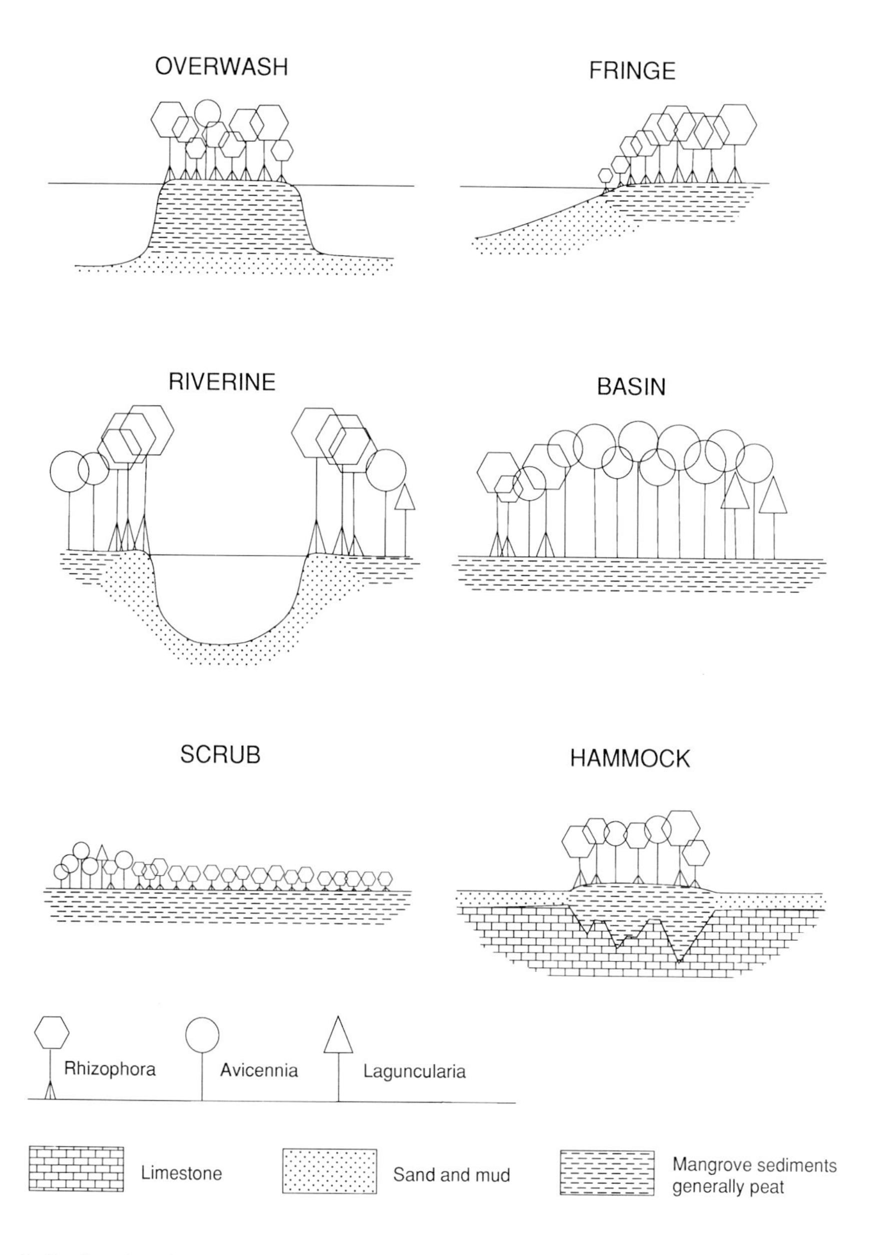

Figure 3. Six functional types of mangrove forest (after Lugo and Snedaker, 1974).

These functional types of mangrove need not only be applied to carbonate settings, but could be recognised in the various habitats identified within other environmental settings; thus riverine mangroves dominate major distributaries in the river-dominated deltas, hammock mangroves are similar to inland mangroves and depressions found on islands in the Pacific, and overwash mangroves are typical of linear shoals found within the macrotidal mouths of tide-dominated estuaries or incipient islands within river-dominated deltas.

However, being based primarily on mangrove physiognomy, the classes are not always directly indicative of the mangrove habitat or the physical processes operating therein. Scrub mangrove, for instance, is defined largely on the basis of mangrove height, whereas the habitat occupied by those mangroves might be just as effectively described as basin when they are interior stands of mangrove, and fringe or overwash mangrove when such stands are open to frequent inundation. While these functional classes are often easily recognisable in the species-poor mangrove forests of the New World, they can be less clearly recognised within the diverse mangrove forests of parts of the Old World. In northern Australia patterns of soil salinity and litter fall have been more effectively related to geomorphologically-defined habitats (Semeniuk, 1985b; Woodroffe *et al.*, 1988).

Recognition of the different functional role of mangroves in different habitats, is an important step in understanding the ecology of mangrove forests. Different physical processes control such fundamental factors as the transport and deposition of sediment, the supply of nutrients, and the significance of organic outwelling from a mangrove forest. The two most important physical processes are riverine unidirectional flows and tidal bi-directional flows (Wolanski *et al.*, Chapter 3, this volume). River-dominated habitats are characterised by high nutrient influx and strong outwelling from mangrove forests. Tide-dominated habitats are characterised by bi-directional fluxes of water and suspended material often with little net export, and even with overall import of material (eg. see Chapters 9 and 10, this volume).

In contrast to those mangrove forests which are dominated by river or tidal processes, many mangrove forests are insulated from either of these processes because they occur in the interior of wetlands. The most extreme examples are the inland mangroves and depressions, which like the hammock mangroves of the Florida Everglades, are entirely isolated from any tidal action, and occur in areas where there are no rivers. Interior mangroves often show a much clearer response to climatic factors, such as the effects of rainfall, run-off and evaporation. Within interior mangroves, flooding may be infrequent but water often has a long residence time (Wolanski *et al.*, 1990).

Mangrove habitats may be thought of as occupying a broad continuum between those that are river-dominated, those that are tide-dominated and those that are interior. Figure 4 illustrates this continuum and its relationship to Lugo and Snedaker's functional classification. Basin mangroves tend towards an interior classification, though may be variously influenced by either river or tidal processes. Scrub mangroves generally occupy a broad range between tide-dominated and interior habitats. Within a large delta various extremes of each of these hydrodynamic cases is likely; active distributaries may be river-dominated; abandoned distributaries tide-dominated; while interdistributary basins represent interior mangroves. Even within tidally-dominated mangrove settings in northern Australia, where for much of the year the development of an evaporation-driven hypersaline water mass at the mouth, and a coastal boundary layer, inhibits exchange between estuarine systems and nearshore waters (Wolanski and Ridd, 1990), river discharges in the monsoon season render these systems seasonally leaky (Wolanski *et al.*, Chapter 3, this volume).

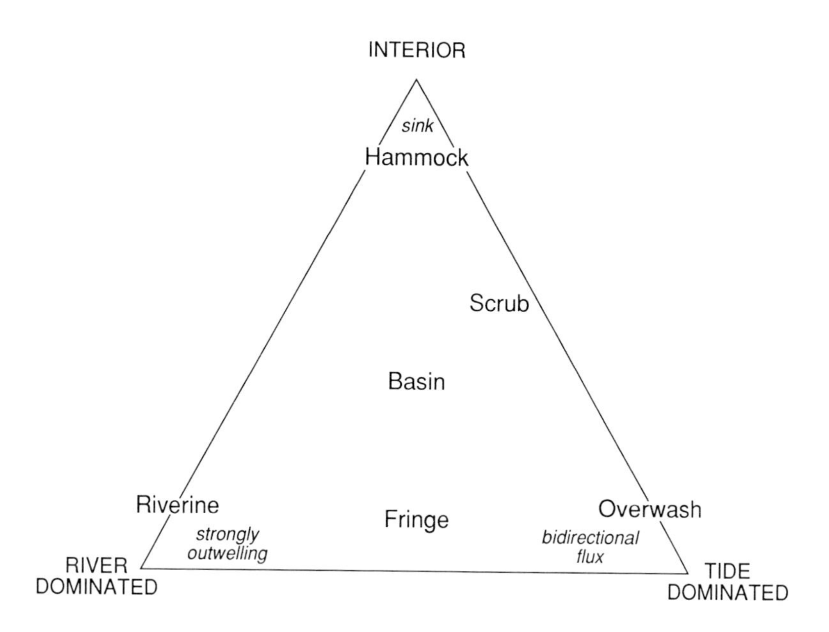

Figure 4. The relationship between functional types of mangrove forests and the dominant physical processes. River-dominated habitats are characterised by strongly outwelling mangrove forests; tide-dominated habitats are characterised by bi-directional fluxes; and interior habitats are typically sinks for sediment, organic matter and nutrients.

2.6 Stratigraphy and paleoecology

Mangrove ecosystems are very dynamic and undergo change at several different timescales. Climatic and sea-level changes, which generally operate at 10^2-10^4 years, have been important controls on the longer-term development of mangrove forests; while tidal processes operate over much shorter timescales. The timescales at which geomorphological processes operate overlap with those at which ecological processes function (Figure 5). Many of these processes occur too slowly to observe directly; some may be detected by the comparison of time-series aerial photography (and increasingly satellite imagery); and others must be determined either by stratigraphic or paleoecological reconstruction.

2.6.1 Late Quaternary sea-level change

Although there are isolated records of mangrove from Tertiary sediments (i.e. Chandler, 1957; Churchill, 1973), mangrove deposits are not widespread in Pleistocene sediments. The Quaternary Period (approximately the last two million years, comprising the Pleistocene and the Holocene) was characterised by the extension and retreat of polar ice sheets (the Ice Ages), and fluctuations of sea level over a range of 100-150 m. During the warm periods, the interglacials, the sea-level was high, generally at a level close to present sea level. However, mangrove sediments from previous interglacials are far less well-known, and generally less well preserved, than coral reef deposits of the same age (with some significant exceptions;

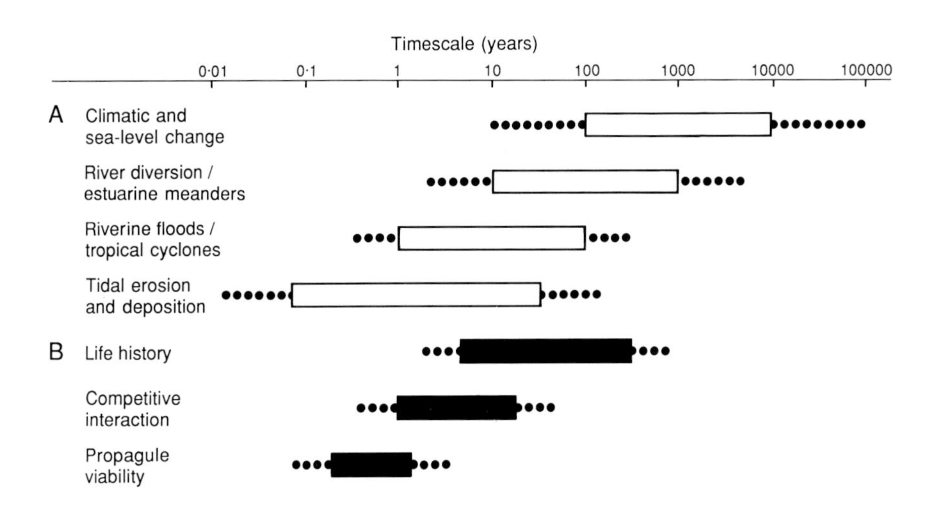

Figure 5. Timescales over which a) geomorphological processes, and b) ecological processes operate.

i.e. van der Hammen, 1974; Caratini and Tissot, 1988). By contrast, sediments deposited during the Holocene (the last ten thousand years) are well-preserved, and contain macrofossils, microfossils and sedimentary structures which can yield valuable information about mangrove ecosystem dynamics over recent millenia.

Over the last few thousand years mangrove ecosystems have undergone almost continual disruption, as a result of fluctuations of sea level. The pattern of ocean volume changes has been reconstructed using oxygen isotope records from foraminifera in deep-sea cores. This pattern has been calibrated against palaeosea level determined from reef terraces dated by U-series dating on rapidly uplifting coasts, and a detailed sea-level curve for the past 250,000 years has been derived (Figure 6, after Chappell and Shackleton, 1986). The general pattern for the last interglacial-glacial cycle is one of overall sea-level fall through a series of oscillations, associated with ice-sheet accumulation, reaching the last glacial maximum at around 18,000 years B.P. (Before Present). Since then the ice has melted and sea level has risen rapidly, at average rates of 10-15 mm y^{-1}.

These late Quaternary sea-level fluctuations must have had dramatic consequences for intertidal mangrove communities. The effect is likely to have been different on different coasts. Figure 6 illustrates the topography typical of a mid-ocean coral atoll and that of a broad continental shelf. Many oceanic islands rise up steeply from an ocean floor 4000 m or more deep. At lowest sea levels these must have existed as islands surrounded by near-vertical limestone cliffs up to 100 m high as the present drop-off of these coral atolls was exposed subaerially. Subsequently on atolls this surface must have been entirely drowned in early Holocene as the sea rose above the level of Last Interglacial limestones. Such a pattern must have resulted in the total eradication of terrestrial biota (including mangroves) from most atolls approximately 8000 years B.P., and its re-establishment only after reef islands formed 3000-4000 years B.P. (Woodroffe and Grindrod, 1991).

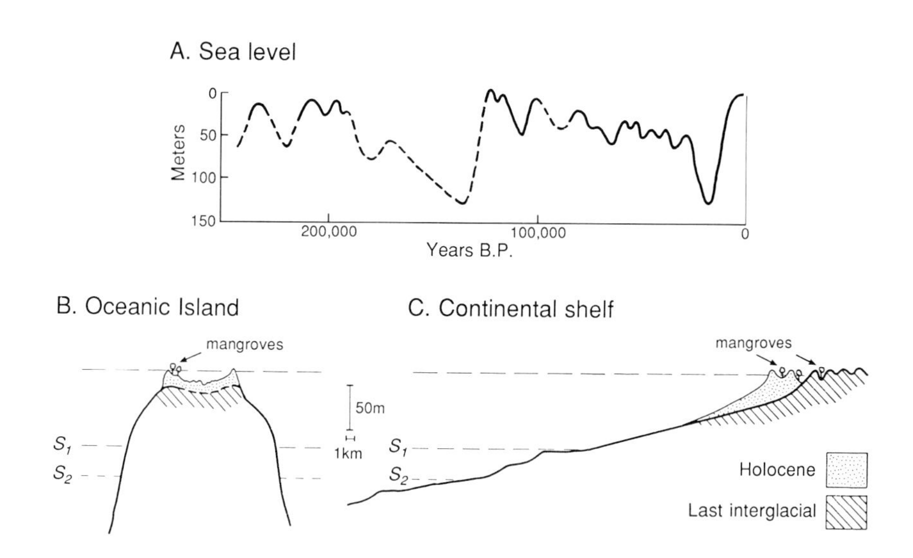

Figure 6. The effect of Quaternary sea-level fluctuations and nearshore topography on the persistence of mangrove forests (after Woodroffe and Grindrod, 1991); A) the pattern of sea-level fluctuation over the last 240,000 years (after Chappell and Shackleton, 1986); B) schematic topography of an oceanic island (based upon a coral atoll), C) schematic topography of a continental shelf.

Local eradication (and recolonisation) of mangroves is less likely to have occurred on broad low gradient continental shelves where rapid postglacial sea-level rise has been expressed as horizontal translation of the shoreline in some instances over 100s of kilometres, as on the Sunda Shelf of Southeast Asia or the Sahul Shelf of northern Australia. The extent of mangrove forests may have fluctuated considerably during this period of sea-level rise (van Campo, 1986).

The sea-level curve for the last 240,000 years shown in Figure 6, is essentially a global 'eustatic' curve reflecting the volume of water in the oceans. It demonstrates that, during the Quaternary, sea level has been changing rapidly for a larger proportion of the time than it has been stable. The Holocene comprises two phases of change, the early phase (approximately 10,000-6,000 years B.P.) during which sea level was rising rapidly worldwide (the postglacial marine transgression), and a later phase (since approximately 6,000 years B.P.), during which the sea was relatively stable. In fact the details of relative sea-level change over the last 6,000 years vary for different parts of the world. This is because the most significant phase of polar ice-melt is considered to have finished by around 6,000 years B.P., so that there ceased to be any significant overall increase in ocean volume (eustatic change), which previously masked local isostatic adjustments to the distribution of ice and water on the earth's surface.

Within the tropics there was negligible ice accumulation during glaciations so that glacio-isostatic adjustments are relatively minor, but hydro-isostatic adjustments in response to redistributed water loads, assume a relatively more important role. The response of a visco-elastic earth to redistributed ice/water loads was modelled by Clark *et al.*, (1978), and a series of zones of different relative sea-level change were identified. More sophisticated geophysical models (Peltier, 1988; Nakada and Lambeck, 1988: Tushingham and Peltier, 1991) indicate considerably more intrazonal variation than predicted by the Clark *et al.*, (1978) model. Nevertheless, sea-level reconstructions from many sites indicate a broad concurrence with the curves proposed by Clark *et al.*, (Hopley, 1982). Some of the evidence used to reconstruct the curves consists of mangrove material, generally sampled from *in situ*; much of these data and the problems of compaction and contamination associated with such sea-level reconstructions are discussed by Woodroffe (1988, 1990).

In particular a different pattern of sea-level change has been consistently proposed for 'Atlantic' sites as opposed to 'Pacific' sites (Adey, 1978); thus in the West Indies sea level appears to have approached present level at a decelerating rate, whereas in Australia sea level appears to have reached present level 6,000 years ago, since when it has remained stable, or even fallen 1-2 m (Thom and Chappell, 1975; Thom and Roy, 1985). Broad envelopes of sea-level change, allowing for some intrazonal variability, are shown in Figure 7.

2.6.2 Shoreline response to sea-level change

During phases of sea-level rise the shoreline moves landwards, known as a transgression; whereas when sea level falls, or when it is stable, the shoreline moves seaward, known as a regression. While sea-level change is likely to be a major cause of transgressive or regressive sedimentary sequences, sediment deposition also serves to build out the coast (regression) and sediment erosion causes retrogradation (transgression). Thus it is possible to envisage sea-level rising but when sedimentation is rapid, a regressive sedimentary sequence may result. Furthermore, with sea level falling erosion could occur causing shoreline retreat (Curray, 1964). These patterns are illustrated schematically in Figure 8. During the initial phase of the Holocene, which was characterised by rapid sea-level rise the shoreline has generally been transgressive (Grindrod and Rhodes, 1984). However, during the mid and late-Holocene many shorelines in zones V and VI of Figure 7 have been regressive. In Florida and the West Indies under conditions of decelerating sea-level rise the switch from transgressive to regressive appears to have occurred only in the last 3000-3500 years, after the rate of rise slowed to less than the average rates of sediment accretion (Parkinson, 1989). Mid and late-Holocene sedimentary sequences are particularly well-developed for two reasons, firstly because this relatively long period of comparative sea-level stabilisation allowed large regressive sediment bodies to build up, and secondly because the broad coincidence of present sea level with previous interglacial sea-level maxima has ensured that there is an extensive horizontal or low gradient surface upon which Holocene coastal landforms may be superimposed (Figure 6).

The pattern of geomorphological development has been different for different settings and three settings will be examined in detail: firstly tropical deltas which tend to be river-dominated; secondly estuarine embayments, often tide-dominated; and thirdly carbonate banks or reefs.

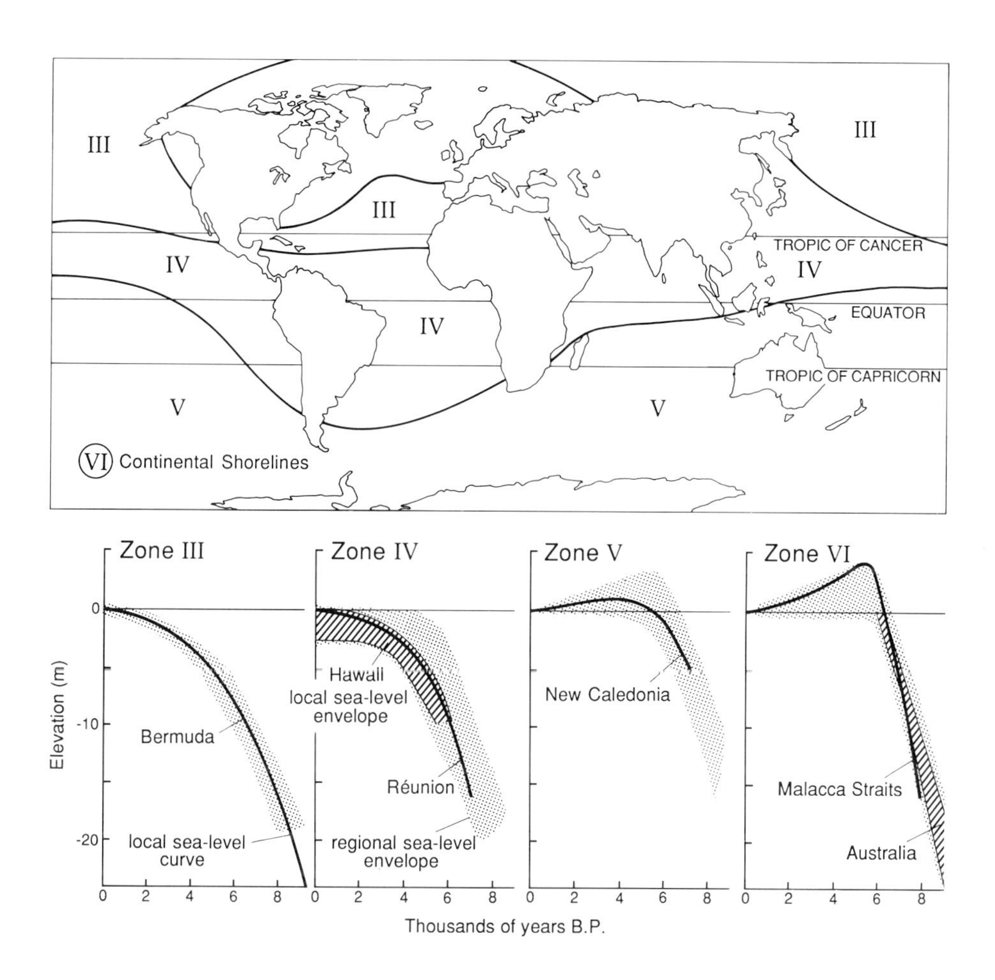

Figure 7. Patterns of Holocene sea-level history for 4 tropical zones (zone VI represents continental shorelines) based upon modelling on a visco-elastic earth (after Clark *et al.*, 1978; adapted from Hopley, 1982). Specific sea-level curves are shown, together with broad envelopes representing intrazonal variation.

Deltas build out as a result of the deposition of river-borne sediment, often with frequent distributary switching (Coleman and Wright, 1975; Wright, 1985). The Mississippi Delta has been examined in great detail (Frazier, 1967; Coleman, 1988), having formed a series of delta lobes during the Holocene. However, mangroves are restricted to the most peripheral habitats in the Mississippi Delta, which is close to the latitudinal limit for mangroves. The distribution of landforms in a typical mangrove-dominated delta has been described by Thom (1967) and is outlined above. However, it is becoming clear that there are many different types of delta in the tropics (Wright, 1989), and the role of mangrove forests in the Holocene evolution of these has not been studied in the detail that it has for estuaries.

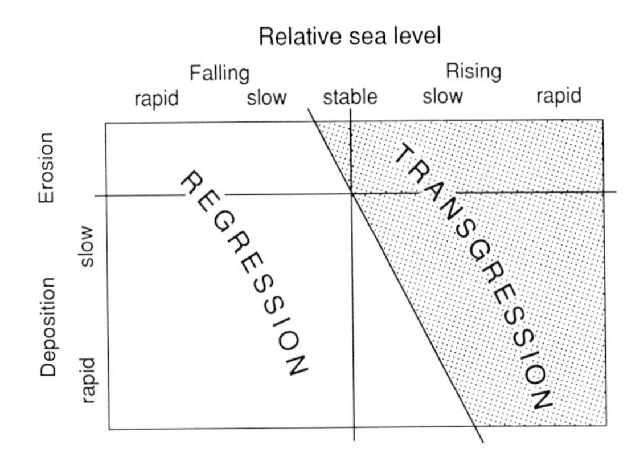

Figure 8. Transgression and regression, and their dependence upon the pattern of relative sea level change and erosion/deposition (Curray, 1964).

2.6.3 Holocene evolution of estuaries

Mangrove-fringed estuaries, such as those found in northern and eastern Australia, generally consist of extensive Holocene estuarine and coastal plains, and at least three models of development have been proposed (Figure 9). In each case in Figure 9 it is assumed that initial inundation of the prior river valley occurred as sea level rose (transgression) and that mangrove forests during this phase migrated landwards with marine incursion. Since sea-level stabilisation (around 6,000 years B.P. in zones V and VI of Figure 7, and sometime later in other zones), embayment infill can be described either by the progradational model, the big swamp model, or the barrier estuary/mud basin model.

Progradational model

The progradational model is perhaps the easiest to understand. The shoreline of the estuary gradually builds out into the estuary, which is generally fringed by littoral mangrove forests. Estuarine plains form to landward of the mangroves as the muddy sediments accrete to an elevation at which they only rarely, or never, get inundated by tides (Figure 9). Within the Australian region the detailed radiocarbon chronology of bayside chenier plains demonstrate such a progradational infill within Broad Sound and Princess Charlotte Bay, Queensland (Cook and Mayo, 1977; Chappell and Grindrod, 1984). In this model, the mangrove sediments beneath the estuarine plains, get progressively younger towards the present shoreline.

Big swamp model

The big swamp model is one in which sediments accumulated rapidly as sea level stabilised. Mangrove forests were extensive throughout much of the plains and were

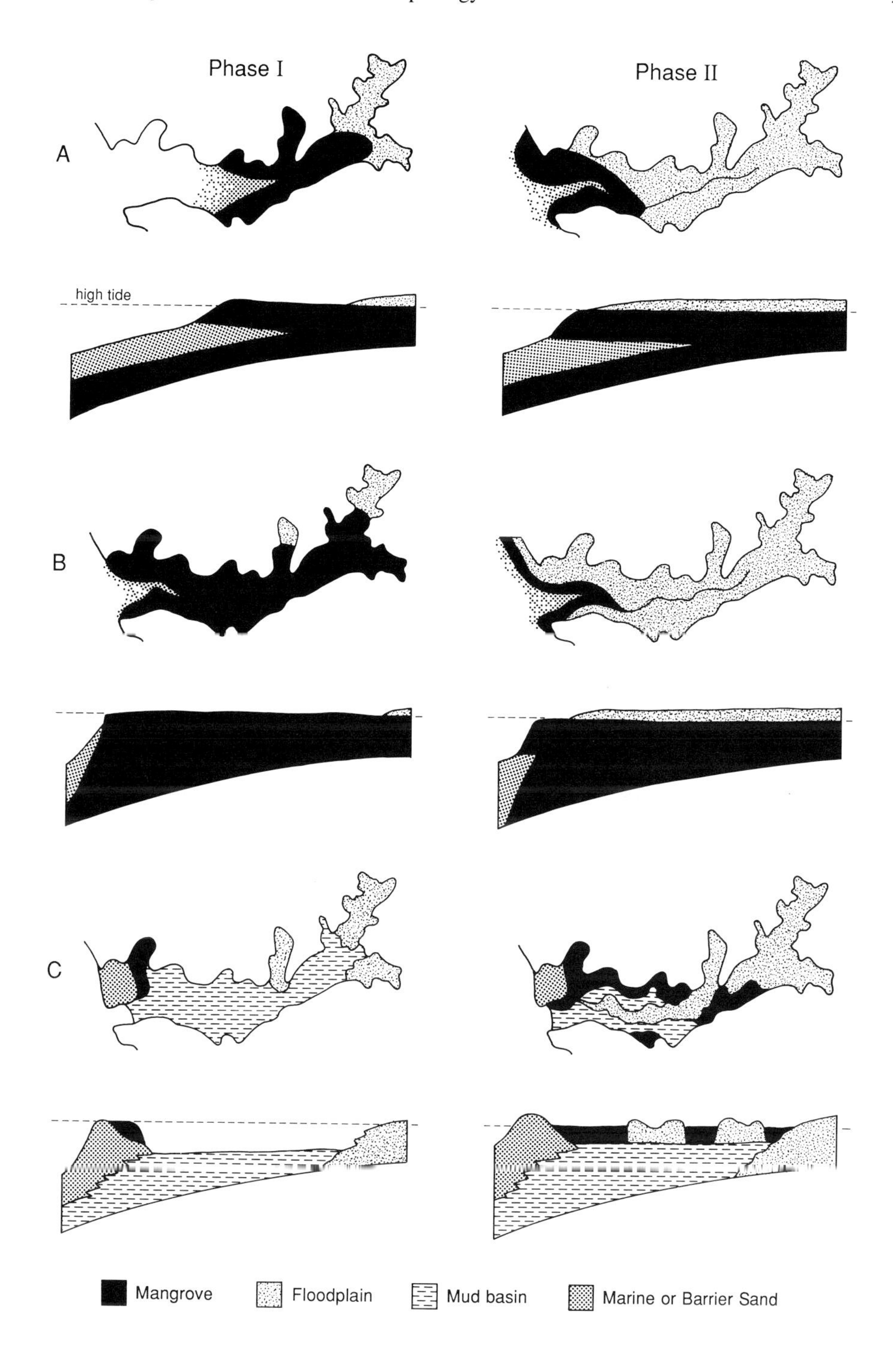

Figure 9. Schematic representation of 3 models of estuarine infill: A) progradational model; B) big swamp model, and C) barrier estuary/mud basin.

subsequently replaced by more landward ecosystems through vertical accretion of sediment. This model of infill differs from the progradational model in that there is little or no age differentiation of the mangrove sediments (Figure 9). It appears typical of much of northern Australia; there are mangrove muds dated at around 6,000 years B.P. infilling many estuarine embayments in the Northern Territory and Western Australia (Woodroffe *et al.*, 1985). The pattern of infill has been examined in detail for the South Alligator River, a macrotidal seasonal river draining lateritic lowlands and a sandstone plateau in northern Australia.

Three phases can be recognised in the evolution of the South Alligator estuarine plains, and these are shown in Figure 10 (based upon detailed stratigraphy and palynology described by Woodroffe *et al.*, 1986, 1989). During an initial transgressive phase (8000-6000 years B.P.) there was marine incursion and mangrove establishment as the prior valley was inundated through sea-level rise. This was followed by the big swamp phase (6800-5300 years B.P.) with mangrove forest widespread throughout most of the area of the present plains (Figure 10). Since that time mangrove forests have disappeared from most of the plains as a result of vertical sediment accretion with final replacement by a terrestrial wetland (Grindrod, 1988; Clark and Guppy, 1988), and the tidal river has adopted a sinuous channel pattern (since 5300 years B.P.), with the subsequent development of cuspate channel morphology in part of the estuary since about 2500 years B.P. During this last sinuous/cuspate phase the tidal channel has migrated within a narrow meander tract within the axis of the valley, reworking big swamp mangrove muds and depositing younger channel-margin lateral-accretion deposits.

The channel morphologies and stratigraphic pattern found in the South Alligator River and estuarine plains have also been identified in other estuarine systems in northern Australia and Southeast Asia (Woodroffe, 1988). Big swamp mangrove muds appear to be widespread, but the extent to which estuarine floodplains have been reworked and tidal channels have meandered within the valley axis depends upon a subtle balance of fluviotidal parameters, such that large sandy rivers such as the Daly River migrate rapidly and have extensive scroll plains, while smaller muddy rivers such as the Adelaide River appear to have migrated little if at all (Chappell, 1988; Chappell *et al.*, in press).

The composition of the terrestrial community that occupies the plains after the demise of the mangrove forest depends upon the regional climate. In the Alligator Rivers region seasonally-flooded herbaceous wetlands and paperbark (*Melaleuca* spp.) swamps dominate the estuarine floodplains. These seem to have first developed around 4000 years B.P., but became particularly extensive in the last 1500 years. Freshwater clays settle out from suspension in wet season floods and the plains vertically accrete so that in these monsoonal areas (average annual rainfall around 1500 mm) a convex floodplain forms with greater elevations close to the river channel, and lower elevations in backwater swamps (Figure 11). In the semiarid estuaries of Western Australia there appears to have been a big swamp phase synchronous with that in the Alligator Rivers region, but in the much drier Ord and Fitzroy Rivers (average annual rainfall about 700 mm) the estuarine plains which have replaced those mangrove forests are saline and support samphire or are unvegetated (Jennings, 1975; Thom *et al.*, 1975; Semeniuk, 1982; Kendrick and Morse, 1990). These plains are near-horizontal and have not built up a convex form (Figure 11).

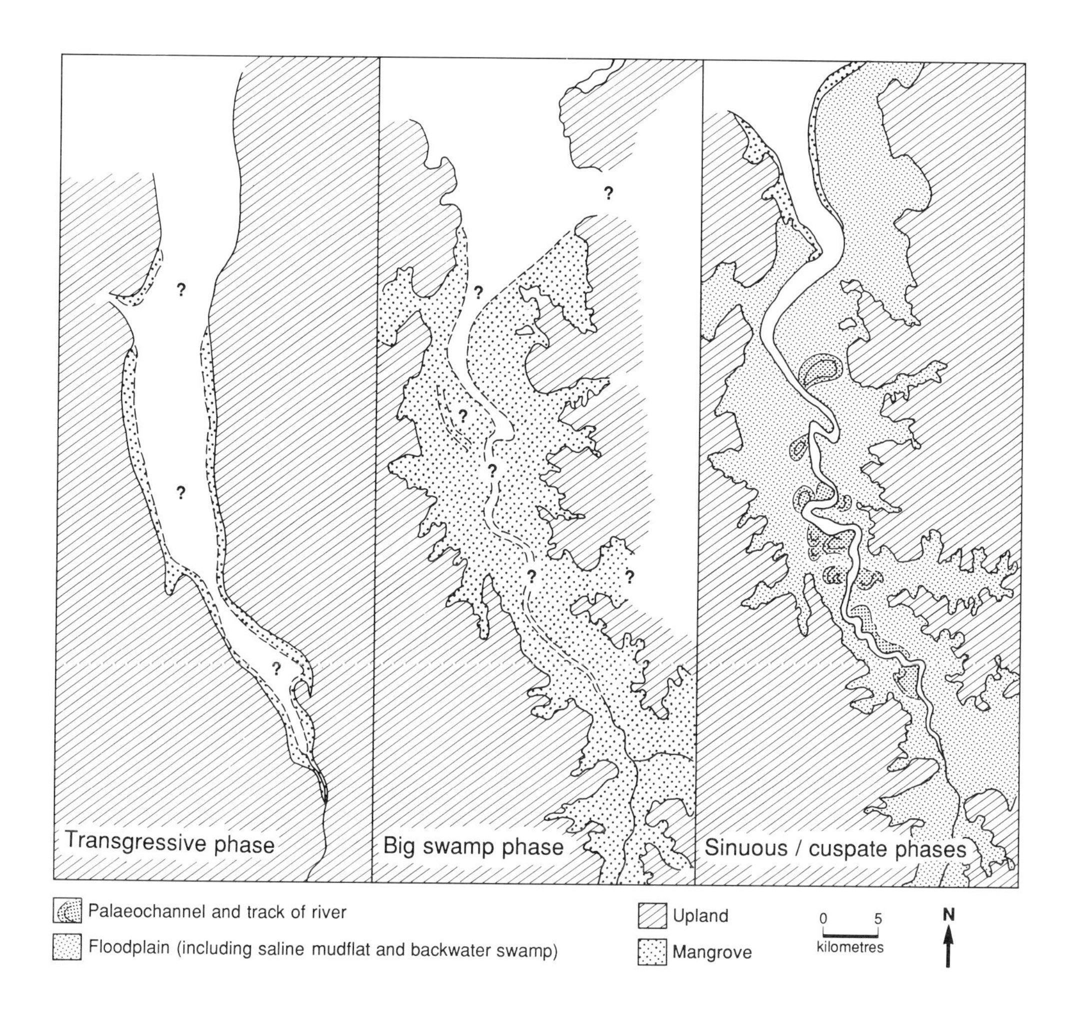

Figure 10. Holocene evolution of the South Alligator tidal river showing a transgressive phase, a big swamp phase, and sinuous/cuspate phase (after Woodroffe, 1988).

In the more equatorial parts of Southeast Asia (with average annual rainfalls in excess of 3000 mm), extensive species-poor tropical forests, termed peat swamp forests, develop on coastal and estuarine plains with a substrate of woody peat (Anderson, 1964). These ombrogenous peats are particularly well-developed in Sarawak where they can be more than 10 m thick, domed and overlie marine clays deposited in a mangrove environment (Figure 11). Pollen analysis has demonstrated a long-term succession from mangrove through to peat swamp (Anderson and Muller, 1975; Bosch, 1988). Radiocarbon dating of the transition suggests that mangrove forests persisted until 4500-4000 years B.P. before being replaced by freshwater peat swamp (Coleman *et al.*, 1970; Anderson and Muller, 1975). The stratigraphy and the pattern of change appears broadly similar to that in northern Australia, where transition to monsoonal forest is also possible (Crowley *et al.*, 1990), although from the few

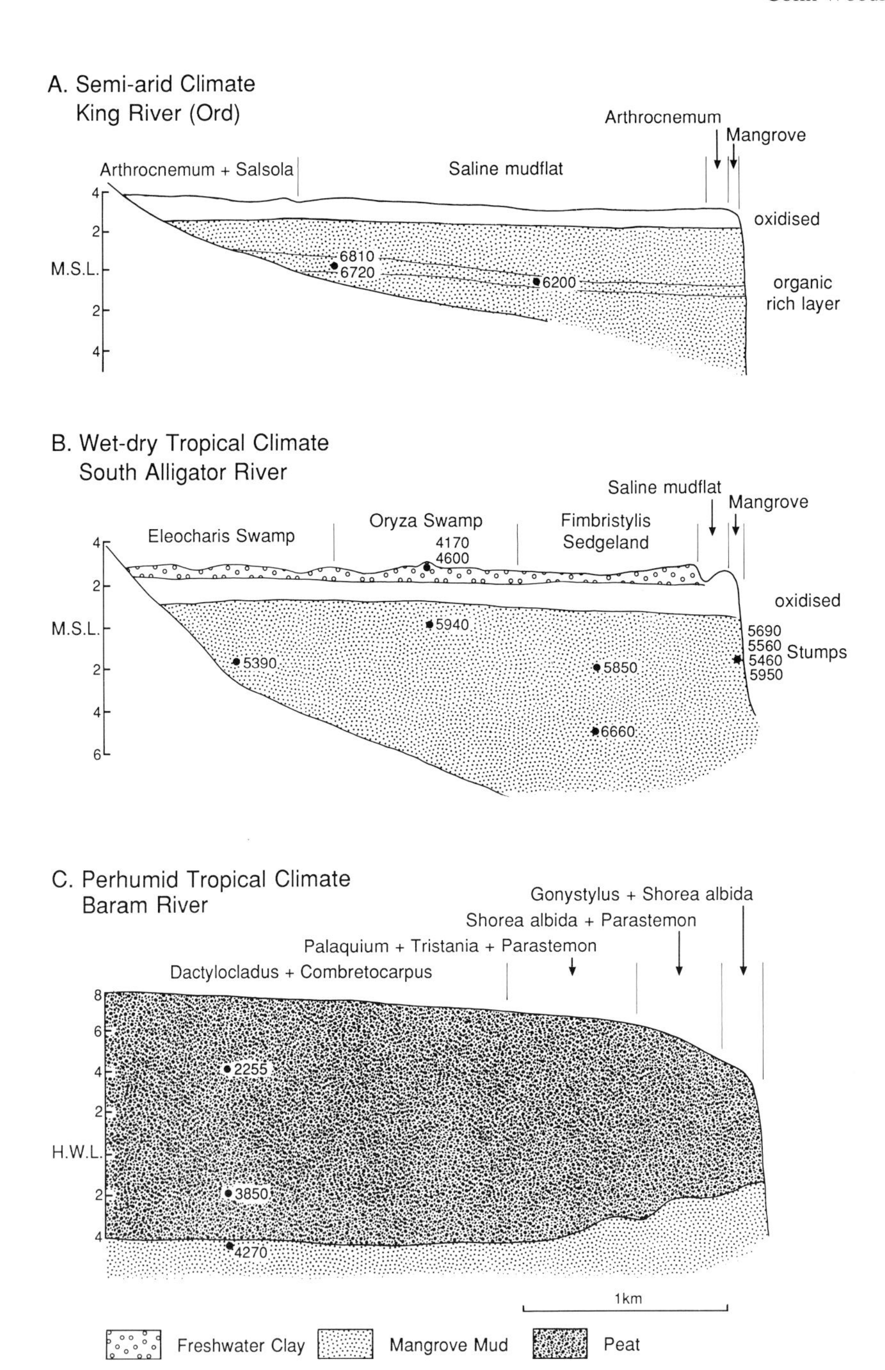

Figure 11. Characteristic cross-section of estuarine sediments showing radiocarbon dates in A) semi-arid, B) wet-dry tropical and C) perhumid tropical climates (after Woodroffe, 1988): King (Ord) River in north-western Australia based on Thom *et al.*, (1975); South Alligator River in northern Australia based on Woodroffe *et al.*, (1989), and Baram River in northern Borneo, based on Anderson and Muller (1975).

dates available the mangrove forests appear to have persisted longer in Malaysian coastal environments. This may reflect a slightly different pattern of Holocene sea-level change with sea level continuing to rise after 6000 years B.P. and reaching a maximum a few metres above present about 4500-4000 years B.P. (Geyh *et al.*, 1979; Woodroffe, 1988).

Barrier estuary/mud basin

The barrier estuary/mud basin model of Holocene infill is characteristic of wave-dominated coasts, such as the New South Wales coast of south-eastern Australia. An evolutionary model of estuaries on this embayed coast has been developed by Roy *et al.*, (1980). Embayments may be classified into drowned river valleys (i.e. Sydney Harbour, Hawkesbury River system), barrier estuaries (i.e. Lake Macquarie, Lake Illawarra), and saline coastal lakes (i.e. Dee Why and Manly Lagoons). Each receives sand from seaward, reworked off the shelf during the postglacial marine transgression and forming either a sandy barrier, or a flood tidal delta. Each lagoon is a sink for fluvial sand and mud brought from terrestrial sources by rivers (Figure 12). Infilling occurs at different rates depending upon estuary type, sediment supply and the size of the estuarine basin (Roy, 1984). Fine-grained sediments are deposited in a mud basin landward of the barrier or tidal delta. This mud basin gradually contracts as it is infilled by the expansion of deltaic floodplains, and estuarine waters are increasingly dominated by shoals often with an extension of the area covered by mangrove forests. The final stage of infill is characterised by channelisation, ultimately carrying river discharge directly to the coast, as with the Shoalhaven River in flood (Wright, 1970). This evolutionary model also has broad applications beyond the area for which it has been developed; similar patterns of development can be inferred for the estuarine embayments of northern New Zealand and the Kimberleys.

2.6.4 Carbonate settings

Carbonate settings are generally not areas receiving large volumes of terrestrial sediment, but instead sediment is produced *in situ*. Studies of the geomorphology and development of shorelines in Florida have had a particularly significant impact on our interpretation of the dynamics of mangrove forests in carbonate settings. It was here that Davis (1940) initially proposed his successional model of mangrove zone replacement. Davis identified a lower calcareous sediment which he considered was marine in origin, overlain by mangrove peat, in turn overlain by more calcareous sediments (Figure 13a). Egler (1952) contested this essentially regressive interpretation, recognising that the great thicknesses of mangrove peat (3-4 m)must have been deposited as the sea had risen, and that the sequence was predominantly transgressive (Figure 13b). This view was validated by Scholl (1964a, 1964b), and others (i.e. Gleason *et al.*, 1974), who demonstrated that the basalcalcitic muds were of freshwater (not marine) origin (Figure 13c). Subsequently, Spackman *et al.*, (1966, 1969) demonstrated that the peat profile actually contains the transgression within it, with a change from intertidal mangrove peat to freshwater peat (Figure 13d), rather than at its base. While the mid- Holocene record in southern Florida is transgressive, the late Holocene is largely

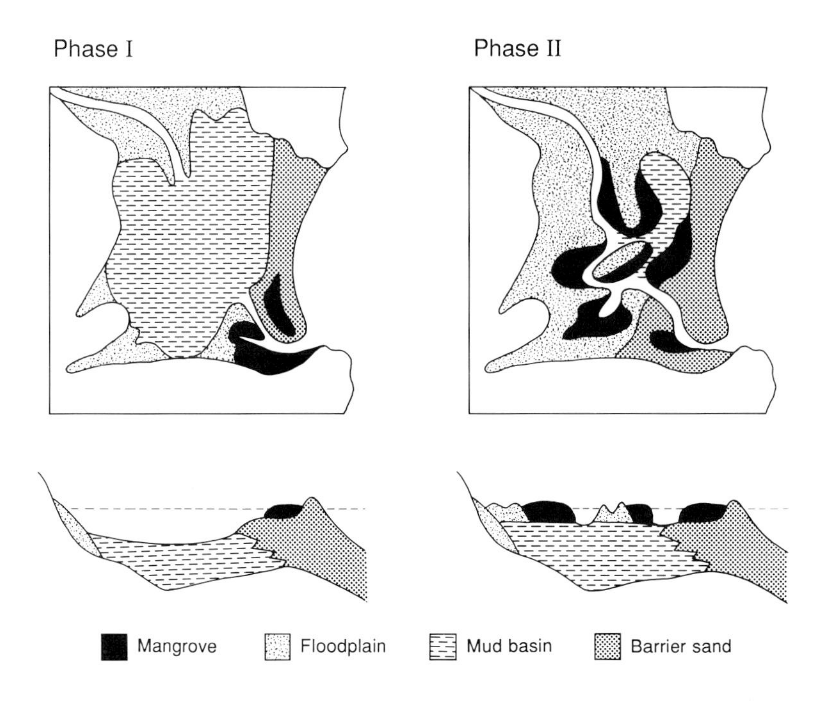

Figure 12. Schematic representation of stages of infill of barrier estuary (after Roy, 1984).

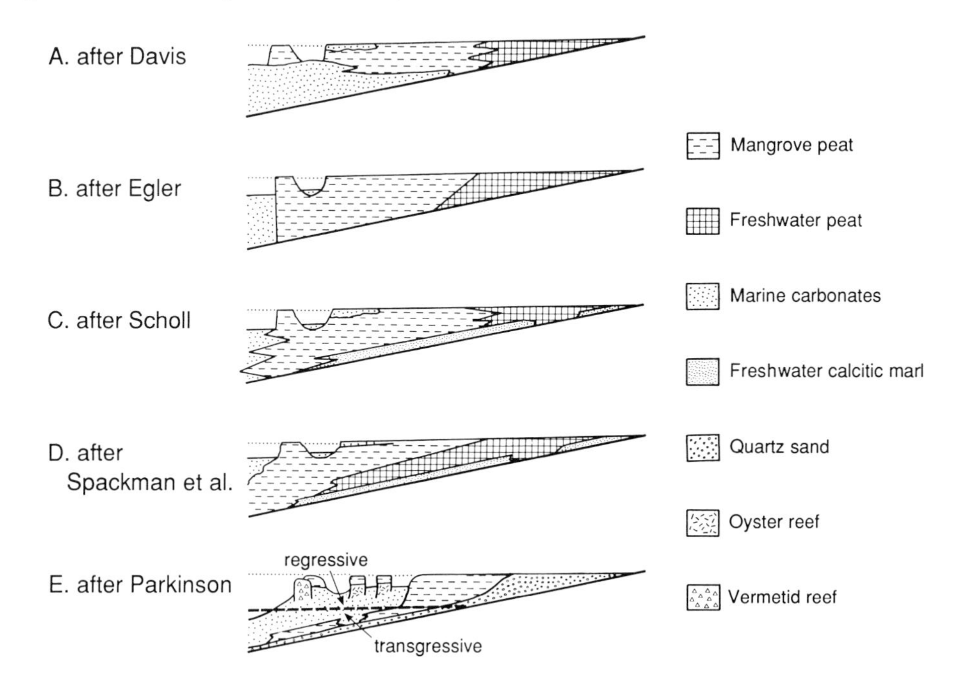

Figure 13. Interpretations of the stratigraphy of south-western Florida; a) after Davis (1940), b) after Egler (1952), c) after Scholl (1964a, 1964b), d) after Spackman *et al.* (1966, 1969), e) after Parkinson (1989).

regressive. This has been shown in the Ten Thousand Islands area by Parkinson (1989), who notes that oyster reefs to landward, and a vermetid reef to seaward have only formed since about 3500-3200 years B.P., after a deceleration in the rate of sea-level rise (Figure 13e). Here as in Florida Bay (Wanless, 1974; Turmel and Swanson, 1976; Enos and Perkins, 1979), the rate of sea-level rise appears to have been critical in determining whether the shoreline has moved landwards (transgressive) or seawards (regressive); mangrove sedimentation being able to keep pace with a rate of sea-level rise of little more than 2 mm yr^{-1} (Parkinson, 1989).

Mangrove distributions in carbonate settings in the Old World also reflect topography of the calcareous sediments, though in some instances rates of carbonate sediment production may not be as high as on the broad banks of the Florida-Bahama region. Steers (1937) postulated that the extent of mangrove on low wooded islands of the Great Barrier Reef of Australia was a function of infill of the reef top. Examination of the reefal substrate has shown that since 6000 years B.P. vertical reef growth has been superseded by horizontal reef flat consolidation and microatoll formation (Stoddart *et al.*, 1978). The relative fall of sea level during the mid- and late Holocene (Chappell, 1982) on the Great Barrier Reef has brought much of this emergent fossil reef flat into an elevational range which is suitable for mangrove colonisation. It appears that the spread of mangroves on the low wooded islands is dependent upon the evolutionary stage of the reefal substrate and its suitability for colonisation (Stoddart, 1980).

2.7 Conclusion

It is clear that mangrove forests have persisted through the Quaternary despite substantial fluctuations of sea level, although these must have caused repeated total disruption to mangrove shorelines. In each of the major environmental settings within which mangrove forests occur, substrate characteristics and patterns of sedimentation or substrate topography are a control on the morphology of mangrove forests. These have altered as the relationship between land and sea has adjusted during the Holocene.

River-dominated systems receive an allochthonous sediment supply. The volume of sediment deposited is broadly a function of the catchment size, though it is becoming apparent that there are many more deltaic types within the tropics than previously realised. Tide-dominated systems also contain abundant allochthonous sediment, but in these it is primarily the tides which account for sediment redistribution. Large tidal range and rapid tidal currents lead to bank erosion and tidal resuspension of sediment. Carbonate settings are often less geomorphologically dynamic, being dependent upon *in situ* production of either calcareous sediment, or mangrove peat.

Much research remains to be done on the interactions of environmental factors, sedimentation patterns and mangrove ecophysiology within mangrove forests. To a large extent mangroves appear to opportunistically colonise areas that are favourable. These complex ecosystems play a very valuable role in estuarine and nearshore food webs, and yet they have evidently been through a very changeable history throughout the Quaternary as sea level has fluctuated. As an ecosystem, mangrove forests must be well adapted to change; they have demonstrated a long-term resilience at the timescales over which geomorphological evolution of landforms occurs. Mangrove forests may be disrupted on shorter timescales by natural disturbances, such as catastrophic storms or lightning strikes. Whereas these are devastating, given time (decades), the forest may recover. Human influence on those ecosystems, particularly through reclamation, represents a catastrophe from which the ecosystem is often given little chance to recover. Such anthropogenic disturbance may represent the ultimate disruption to many large areas of mangrove forests.

2.8 References

Adey, W.H., 1978. Coral reef morphogenesis: a multidimensional model. *Science* **202**: 831-837.

Allen, J.R.L., 1965. Coastal geomorphology of Eastern Nigeria: beach- ridge barrier islands and vegetated tidal flats. *Geologie en Mijnbouw* **44**: 1-21.

Anderson, J.A.R., 1964. The structure and development of the peat swamps of Sarawak and Brunei. *Journal of Tropical Geography* **18**: 7-16.

Anderson, J.A.R., and Muller, J., 1975. Palynological study of a Holocene peat and a Miocene coal deposit from NW Borneo. *Review of Palaeobotany and Palynology* **19**: 291-351.

Baltzer, F., 1970. Datation absolue de la transgression Holocene sur la cote ouest de la Nouvelle Caledonie sur des echantillons de tourbes a paletuviers: Interpretation Neotectonique. *Compte Rendu Academie Sciences Paris* **271D**: 2251-2254.

Bird, E.C.F., 1971. Mangroves as land-builders. *Victorian Naturalist* **88**: 189-197.

Bird, E.C.F., 1986. Mangroves and intertidal morphology in Westernport Bay, Victoria, Australia. *Marine Geology* **69**: 251-271.

Bosch, J.H.A., 1988. The Quaternary deposits in the coastal plains of Peninsular Malaysia. Geological Survey Malaysia Quaternary Geology Report 1/86, 87pp.

Caratini, C., and Tissot, C., 1988. Paleogeographical evolution of the Mahakam Delta in Kalimantan, Indonesia during the Quaternary and late Pliocene. *Review of Palaeobotany and Palynology* **55**: 217-228.

Carlton, J.M., 1974. Land-building and stabilisation by mangroves. *Environmental Conservation* **1**: 285-294.

Carter, J., 1959. Mangrove succession and coastal change in Southwest Malaya. *Transactions of the Institute of British Geographers* **26**: 79-88.

Chandler, M.E.J., 1957. Note on the occurrence of mangrove in the London Clay. *Proceedings of the Geologists Association* **62**: 271-272.

Chapman, V.J., 1944. 1939 Cambridge University Expedition to Jamaica. *Journal of the Linnean Society Botany (London)* **52**: 407-533.

Chapman, V.J., 1970. Mangrove phytosociology. *Tropical Ecology* **11**: 1- 19.

Chapman, V.J., 1975. *Mangrove vegetation.* J. Cramer,Vaduz, 447pp.

Chapman, V.J., and Ronaldson, J.W., 1958. The mangrove and salt marsh flats of the Auckland Isthmus. New Zealand, Department of Scientific and Industrial Research, Bulletin 125, 79pp.

Chappell, J., 1982. Evidence for smoothly falling sea level relative to north Queensland, Australia, during the past 6000 years. *Nature* **302**: 406-408.

Chappell, J., 1988. Geomorphological dynamics and evolution of tidal river and floodplain systems in northern Australia. In: Wade- Marshall, D., and Loveday, P., (Eds.), *Northern Australia: Progress and Prospects, vol 2: Floodplains research,* pp. 34-57, Northern Australia Research Unit, Australian National University Press.

Chappell, J., and Grindrod, J., 1984. Chenier plain formation in northern Australia, In: Thom, B.G., (Ed.), *Coastal Geomorphology in Australia*, pp. 197-232, Academic Press, Sydney.

Chappell, J., and Shackleton, N.J., 1986. Oxygen isotopes and sea level. *Nature* **324**: 137-140.

Chappell, J., Woodroffe, C.D., and Thom, B.G., (in press). Tidal river morphodynamics and vertical and lateral accretion of deltaic-estuarine plains in Northern Territory, Australia. *Zeitschrift fur Geomorphologie.*

Churchill, D.M., 1973. The ecological significance of tropical mangrove in the early Tertiary floras of southern Australia. *Geological Society of Australia, Special Publication* **4**: 79-86.

Clark, J.A., Farrell, W.E., and Peltier, W.R., 1978. Global change in post glacial sea level: a numerical calculation. *Quaternary Research* **9**: 265-287.

Clark, R.L., and Guppy, J.C., 1988. A transition from mangrove forest to freshwater wetland in the monsoon tropics of Australia. *Journal of Biogeography* **15**: 665-684.

Coleman, J.M., 1969. Brahmaputra river: channel processes and sedimentation. *Sedimentary Geology* **3**: 129-239.

Coleman, J.M., 1988. Dynamic changes and processes in the Mississippi River Delta. *Geological Society of America Bulletin* **100**: 999-1015

Coleman, J.M., and Wright, L.D., 1975. Modern river deltas: variability of process and sand bodies. In: Broussard, M.L., (Ed.), *Deltas: models for exploration,* pp. 99-149, Houston Geological Society.

Coleman, J.M., Gagliano, S.M., and Smith, W.G., 1970. Sedimentation in a Malaysian high tide tropical delta. In: Morgan, J.P., (Ed.), *Deltaic sedimentation: modern and ancient,* pp. 185-197, Special Publication of the Society of conomic Paleontologists and Mineralogists, Tulsa.

Cook, P.J., and Mayo, W., 1977. Sedimentology and Holocene history of a tropical estuary (Broad Sound, Queensland). *BMR Bulletin* **170**, 260pp.

Crowley, G.M., Anderson, P., Kershaw, A.P., and Grindrod, J., 1990. Palynology of a Holocene marine transgressive sequence, lower Musgrave River valley, North-east Queensland. *Australian Journal of Ecology* **15**: 231-240.

Curray, J.R., 1964. Transgressions and regressions. In: Miller, P.L., (Ed.), *Papers in Marine Geology*, pp. 175-203, Shepard Commemorative Volume.

Davis, J.H.Jr., 1938. Mangroves, makers of land. *Nature Magazine* **31**: 551-553.

Davis, J.H.Jr., 1940. The ecology and geologic role of mangroves in Florida. *Papers of the Tortugas Laboratory* **32**: 303-412.

Egler, F.E., 1952. Southeast saline Everglades vegetation and its management. *Vegetatio* **3**: 213-265.

Enos, P., and Perkins, R.D., 1979. Evolution of Florida Bay from island stratigraphy. *Geological Society of America Bulletin* **90**: 59-83.

Frazier, D.E., 1967. Recent deltaic deposits of the Mississippi River: their development and chronology. *Transactions of the Gulf Coast Association of Geological Societies* **17**: 287-315.

Geyh, M.A., Kudrass, H.R., and Streif, H., 179. Sea level changes during the late Pleistocene and Holocene in the Strait of Malacca. *Nature* **278**: 441-443.

Giglioli, M.E.C., and King, D.F., 1966. The Mangrove Swamps of Keneba, Lower Gambia River Basin. III Seasonal variations in the chloride and water content of swamp soils with observations on the water level and chloride concentration of free soil water under a barren mud flat during the dry season. *Journal of Applied Ecology* **3**: 1-19.

Gleason, P.J., Cohen, A.D., Smith, W.G., Brooks, H.K., Stone, P.A., Goodrick, R.O., and Spackman, W., 1974. The environmental significance of Holocene sediments from the Everglades and saline tidal plain. In: Gleason, P.J., (Ed.), *Environments of South Florida: present and past*, pp. 287-341, Memoir 2, Miami Geological Society.

Grindrod, J., 1985. The palynology of mangroves on a prograded shore, Princess Charlotte Bay, North Queensland, Australia. *Journal of Biogeography* **12**: 323-348.

Grindrod, J., 1988. The palynology of Holocene mangrove and saltmarsh sediments, particularly in northern Australia. *Review of Palaeobotany and Palynology* **55**: 229-245.

Grindrod, J., and Rhodes, E.G., 1984. Holocene sea-level history of a tropical estuary, Missionary Bay, north Queensland. In: Thom, B.G., (Ed.), *Coastal Geomorphology in Australia*, pp. 151-178, Academic Press, Sydney.

Hopley, D., 1982. *Geomorphology of the Great Barrier Reef*. Wiley Interscience.

Jennings, J.N., 1975. Desert dunes and estuarine fill in the Fitzroy estuary, north-western Australia. *Catena* **2**: 215-262.

Jennings, J.N., and Coventry, R.J., 1973. Structure and texture of a gravelly barrier island in the Fitzroy Estuary, Western Australia, and the role of mangroves in the shore dynamics. *Marine Geology* **15**: 145-167.

John, D.M., and Lawson, G.W., 1990. A review of mangrove and coastal ecosystems in West Africa and their possible relationships. *Estuarine, Coastal and Shelf Science* **31**: 505-529.

Johnstone, I.M., 1980. Succession in zoned mangrove communities: where is the climax? In: Teas, H.J., (Ed.), *Biology and ecology of mangroves*, pp. 131-139, Junk, The Hague.

Kendrick, G.W., and Morse, K., 1990. Evidence of recent mangrove decline from an archaeological site in Western Australia. *Australian Journal of Ecology* **15**: 349-353.

Kuenzler, E.J., 1974. Mangrove swamp systems. In: Odum, H.T., Copeland, B.J., and McMahan, E.A., (Eds), *Coastal Ecological Systems of the U.S.*, pp. 346-371, The Conservation Foundation, Washington.

Lot-Helgueras, A., Vazquez-Yanes, C., and Menendez, F., 1975. Physiognomy and floristic changes near the northern limit of mangroves in the Gulf coast of Mexico. In: Walsh, G.E., Snedaker, S.C., and Teas, H.J., (Eds.), *Proceedings of the International Symposium on the Biology and Management of Mangroves*, pp. 52-61, University of Florida, Gainesville.

Lugo, A.E., 1980. Mangrove ecosystems: successional or steady state? *Biotropica* (Suppl.) **1980**: 65-72.

Lugo, A.E., 1981. The inland mangroves of Inagua. *Journal of Natural History* **15**: 845-852.

Lugo, A.E., and Snedaker, S.C., 1974. The ecology of mangroves. *Annual Review of Ecology and Systematics* **5**: 39-64.

Lynch, J.C., Meriwether, J.R., McKee, B.A., Vera-Herrera, F., and Twilley, R.R., 1989. Recent accretion in mangrove ecosystems based on ^{137}Cs and ^{210}Pb. *Estuaries* **12**: 264-299.

Macnae, W., 1968. A general account of the fauna and flora of mangrove swamps and forests in the Indo-Pacific region. *Advances in Marine Biology* **6**: 73-270.

Nakada, M., and Lambeck, K., 1988. The melting history of the late Pleistocene Antarctic ice sheet. *Nature* **333**: 36-40.

Parkinson, R.W., 1989. Decelerating Holocene sea-level rise and its influence on south-western Florida coastal evolution. *Journal of Sedimentary Petrology* **59**: 960-972.

Peltier, W.R., 1988. Lithospheric thickness, Antarctic deglaciation history, and ocean basin discretization effects in a global model of postglacial sea level change: a summary of some sources of nonuniqueness. *Quaternary Research* **29**: 93-112.

Pool, D.J., Lugo, A.E. and Snedaker, S.C., 1975. Litter production in mangrove forests of southern Florida and Puerto Rico. In: Walsh, G.E., Snedaker, S.C., and Teas, H.J., (Eds.), *Proceedings of the International Symposium on the Biology and Management of Mangroves*, pp. 213-237, University of Florida, Gainesville.

Richards, P.W., 1952. *The Tropical Rain Forest: an ecological study*. Cambridge University Press.

Roy, P.W., 1984. New South Wales estuaries: their origin and evolution. In: Thom, B.G., (Ed.), *Coastal Geomorphology in Australia*, pp. 99-122, Academic Press.

Roy, P.S., Thom, B.G., and Wright, L.D., 1980. Holocene sequences on an embayed high-energy coast: an evolutionary model. *Sedimentary Geology* **26**: 1-19.

Schaeffer-Novelli, Y., Mesquita, H.deS.L., and Cintron-Molero, G., 1990. The Cananeia Lagoon estuarine system, Sao Paulo, Brazil. *Estuaries* **13**: 193-203.

Scholl, D.W., 1964a. Recent sedimentary record in mangrove swamps and rise in sea level over the south-western coast of Florida. *Marine Geology* **1**: 344-366.

Scholl, D.W., 1964b. Recent sedimentary record in mangrove swamps and rise in sea level over the south-western coast of Florida. *Marine Geology* **2**: 343-364.

Scoffin, T.P., 1970. The trapping and binding of subtidal carbonate sediments by marine vegetation in Bimini Lagoon, Bahamas. *Journal of Sedimentary Petrology* **40**: 249-273.

Semeniuk, V., 1980. Mangrove zonation along an eroding coastline in King Sound, North-western Australia. *Journal of Ecology* **68**: 789-812.

Semeniuk, V., 1981. Long-term erosion of the tidal flats, King Sound, north-western Australia. *Marine Geology* **43**: 21-48.

Semeniuk, V., 1982. Geomorphology and Holocene history of the tidal flats, King Sound, north-western Australia. *Journal of the Royal Society of Western Australia* **65**: 47-68.

Semeniuk, V., 1983. Mangrove distribution in North-western Australia in relationship to regional and local freshwater seepage. *Vegetatio* **53**: 11- 31.

Semeniuk, V., 1985a. Development of mangrove habitats along ria shorelines in north and north-western tropical Australia. *Vegetatio* **60**: 3-23.

Semeniuk, V., 1985b. Mangrove environments of Port Darwin, Northern Territory: the physical framework and habitats. *Journal of the Royal Society of Western Australia* **67**: 81-97.

Snedaker, S.C., 1982. Mangrove species zonation: Why? In: Sen, D.N., and Rajpurohit, K.S., (Eds.), *Tasks for vegetation science*, pp. 111-125, Junk, The Hague.

Spackman, W., Dolsen, C.P., and Riegel, W., 1966. Phytogenic organic sediments and sedimentary environments in the Everglades-mangrove complex. Part I. Evidence of a transgressing sea and its effect on environments of the Shark River area of southwest Florida. *Palaeontographica* **B117**: 135-152.

Spackman, W., Riegel, W., and Dolsen, C.P., 1969. Geological and biological interactions in the swamp-marsh complex of southern Florida. In: Dapples, E.C., and Hopkins, M.B., (Eds.), *Environments of Coal Deposition*, **114**: 1-35, Geological Society of America Special Paper.

Spenceley, A.P., 1976. Unvegetated saline tidal flats in north Queensland. *Journal of Tropical Geography* **42**: 78-85.

Spenceley, A.P., 1977. The role of pneumatophores in sedimentary processes. *Marine Geology* **24**: M31-M37.

Spenceley, A.P., 1982. Sedimentation patterns in a mangal on Magnetic Island near Townsville, North Queensland, Australia. *Singapore Journal of Tropical Geography* **3**: 100-107.

Steers, J.A., 1937. The coral islands and associated features of the Great Barrier Reef. *Geographical Journal* **89**: 1-28, 119-139.

Stephens, W.M., 1962. Trees that make land. *Sea Frontiers* **8**: 219-230.

Stoddart, D.R., 1980. Mangroves as successional stages, inner reefs of the Great Barrier Reef. *Journal of Biogeography* **7**: 269-284.

Stoddart. D.R., McLean, R.F., Scoffin, T.P., Thom, B.G., and Hopley, D., 1978. Evolution of reefs and islands, northern Great Barrier Reef: synthesis and interpretation. *Philosophical Transactions of the Royal Society of London* **B284**: 149-160.

Thom, B.G., 1967. Mangrove ecology and deltaic geomorphology: Tabasco, Mexico. *Journal of Ecology* **55**: 301-343.

Thom, B.G., 1982. Mangrove ecology: a geomorphological perspective. In: Clough, B.F., (Ed.), *Mangrove ecosystems in Australia, structure, function and management,* pp. 3-17, Australian National University Press, Canberra.

Thom, B.G., 1984. Coastal landforms and geomorphic processes. In: Snedaker, S.C., and Snedaker, J.G., (Eds.), *The mangrove ecosystem: research methods,* pp. 3-17, UNESCO, Paris.

Thom, B.G., and Chappell, J., 1975. Holocene sea levels relative to Australia. *Search* **6**: 90-93.

Thom, B.G., and Roy, P.S., 1984. Relative sea levels and coastal sedimentation in southeast Australia in the Holocene. *Journal of Sedimentary Petrology* **55**: 257-264.

Thom, B.G., and Wright, L.D., 1983. Geomorphology of the Purari Delta. In: Petr, T., (Ed.), *The Purari - Tropical environment of a high rainfall river basin,* pp. 47-65, Junk, The Hague.

Thom, B.G., Wright, L.D., and Coleman, J.M., 1975. Mangrove ecology and deltaic-estuarine geomorphology, Cambridge Gulf-Ord River, Western Australia. *Journal of Ecology* **63**: 203-222.

Turmel, R.J., and Swanson, R.G. 1976. The development of Rodriguez Bank, a Holocene mudbank in the Florida reef tract. *Journal of Sedimentary Petrology* **46**: 497-518.

Tushingham, A.M., and Peltier, W.R., 1991. Ice-3G: a new global model of late Pleistocene deglaciation based upon geophysical prediction of post- glacial relative sea level change. *Journal of Geophysical Research* **96**: 4497-4523.

Twilley, R.R., Lugo, A.E., and Patterson-Zucca, C., 1986. Litter production and turnover in basin mangrove forests in southwest Florida. *Ecology* **67**: 670-683.

van Campo, E., 1986. Monsoon fluctuations in two 20,000 yr B.P. oxygen- isotope/pollen records off southwest India. *Quaternary Research* **26**: 376- 388.

van der Hammen, T., 1974. The Pleistocene changes of vegetation and climate in tropical South America. *Journal of Biogeography* **1**: 3-26.

Vann, J.H., 1980. Shoreline changes in mangrove areas. *Zeitschrift für Geomorphologie*, N.F. *Suppleband* **34**: 255-261.

Vaughan, T.W., 1909. The geologic work of mangroves in southern Florida. *Smithsonian Miscellaneous Collections* **52**: 461-464.

Walsh, G.E., 1974. Mangroves: a review. In: Reimold, R.J., and Queen, W.H., (Eds.), *Ecology of Halophytes*, pp. 51-174, Academic Press, New York.

Wanless, H.R., 1974. Mangrove sedimentation in geological perspective. In: Gleason, P.J., (Ed.), *Environments of South Florida: present and past*, pp. 190-200, Memoir 2, Miami Geological Society.

Wolanski, E., and Ridd, P., 1990. Mixing and trapping in Australian tropical coastal waters. In: Cheng, R.T., (Ed.), *Residual currents and long-term transport*, pp. 165-183, Springer Verlag, New York.

Wolanski, E., Mazda, Y., King, B., and Gay, S., 1990. Dynamics, flushing and trapping in Hinchinbrook Channel, a giant mangrove swamp, Australia. *Estuarine, Coastal and Shelf Science* **31**: 555-579.

Woodroffe, C.D., 1987. Pacific Island mangroves: distribution and environmental settings. *Pacific Science* **41**: 166-185.

Woodroffe, C.D., 1988. Changing mangrove and wetland habitats over the past 8000 years, northern Australia and Southeast Asia. In: Wade- Marshall, D., and Loveday, P., (Eds.), *Northern Australia: Progress and Prospects, Vol. 2: Floodplains research*, pp. 1-33, North Australia Research Unit, Australian National University Press, Canberra.

Woodroffe, C.D., 1990. The impact of sea-level rise on mangrove shorelines. *Progress in Physical Geography* **14**: 483-520.

Woodroffe, C.D., Thom, B.G., and Chappell, J., 1985. Development of widespread mangrove swamps in mid-Holocene times in northern Australia. *Nature* **317**: 711-713.

Woodroffe, C.D., Chappell, J., Thom, B.G., and Wallensky, E., 1986. Geomorphological dynamics and evolution of the South Alligator tidal river and plains. *North Australia Research Unit Monograph* **3**: 1-190.

Woodroffe, C.D., Chappell, J., Thom, B.G., and Wallensky, E., 1989. Depositional model of a macrotidal estuary and floodplain, South Alligator River, Northern Australia. *Sedimentology* **36**: 737-756.

Woodroffe, C.D., Bardsley, K.N., Ward, P.J., and Hanley, J.R., 1988. Production of mangrove litter in a macrotidal embayment, Darwin Harbour, N.T., Australia. *Estuarine, Coastal and Shelf Science* **26**: 581-598.

Woodroffe, C.D., and Grindrod, J., 1991. Mangrove biogeography: the role of Quaternary environmental and sea-level change. *Journal of Biogeography*.**18**: 479-492.

Wright, L.D., 1970. The influence of sediment availability on patterns of beach ridge development in the vicinity of the Shoalhaven River delta, NSW. *Australian Geographer* **11**: 336-348.

Wright, L.D., 1985. River deltas. In: Davis, R.A., (Ed.), *Coastal Sedimentary Environments*, pp. 1-76, Springer-Verlag, New York.

Wright, L.D., 1989. Dispersal and deposition of river sediments in coastal sea: Models from Asia and the tropics. *Netherlands Journal of Sea Research* **23**: 493-500.

Wright, L.D., Coleman, J.M., and Thom, B.G., 1973. Processes of channel development in a high-tide-range environment: Cambridge Gulf - Ord River delta. *Journal of Geology* **81**: 15-41.

Wright, L.D., Coleman, J.M., and Erickson, M.W., 1974. Analysis of major river systems and their deltas: morphologic and process comparisons. Coastal Studies Institute, Louisiana State University, Technical Report 156, 114pp.

Zimmerman, R.C., and Thom, B.G., 1982. Physiographic plant geography. *Progress in Physical Geography* **6**: 45-59.

3

Mangrove Hydrodynamics

Eric Wolanski, Yoshiro Mazda and Peter Ridd

3.1 Introduction

Studies of the physical processes in mangrove swamps and mangrove-fringed estuaries are few and far between compared to those of temperate estuaries. The main reason is that these swamps are mostly located in tropical areas in less-developed countries, where research funding is inadequate and field conditions are harsh. Mangrove research has been better funded in some of these countries under the umbrella of international agencies; however as a rule these are 'applied' agencies and the studies are site-specific, short-term, and addressed to answer specific resource assessment and social, environmental and pollution problems. The estuarine physicists are usually seen only as a necessary evil, a peripheral discipline, that life scientists need to deal with in order to obtain information for management on the pathways for water, nutrients and sediment. This information is usually inadequate or poorly understood. In most of the less-developed countries, management of the mangrove resources is hampered by a large human population or prawn farms encroaching on mangroves, which are seen as wasteland. Management, where it exists, is usually carried out with inadequate scientific understanding and basically represents mismanagement that may cost mankind dearly in the future. A sound understanding of the physical processes in mangrove swamps is needed for proper management. Physical processes control several key biological and chemical processes in the mangroves.

Lugo and Snedaker (1974) have classified mangrove forests into 5 types:

a. RIVERINE FOREST: Floodplains along river drainages, which are inundated by most high tides and flooded during the wet season,

b. BASIN FOREST: Partially impounded depressions, which are inundated by few high tides during dry season, and most high tides during wet season,

c. FRINGE FOREST: Shorelines with steep elevation gradient, which are inundated and flushed by all high tides,

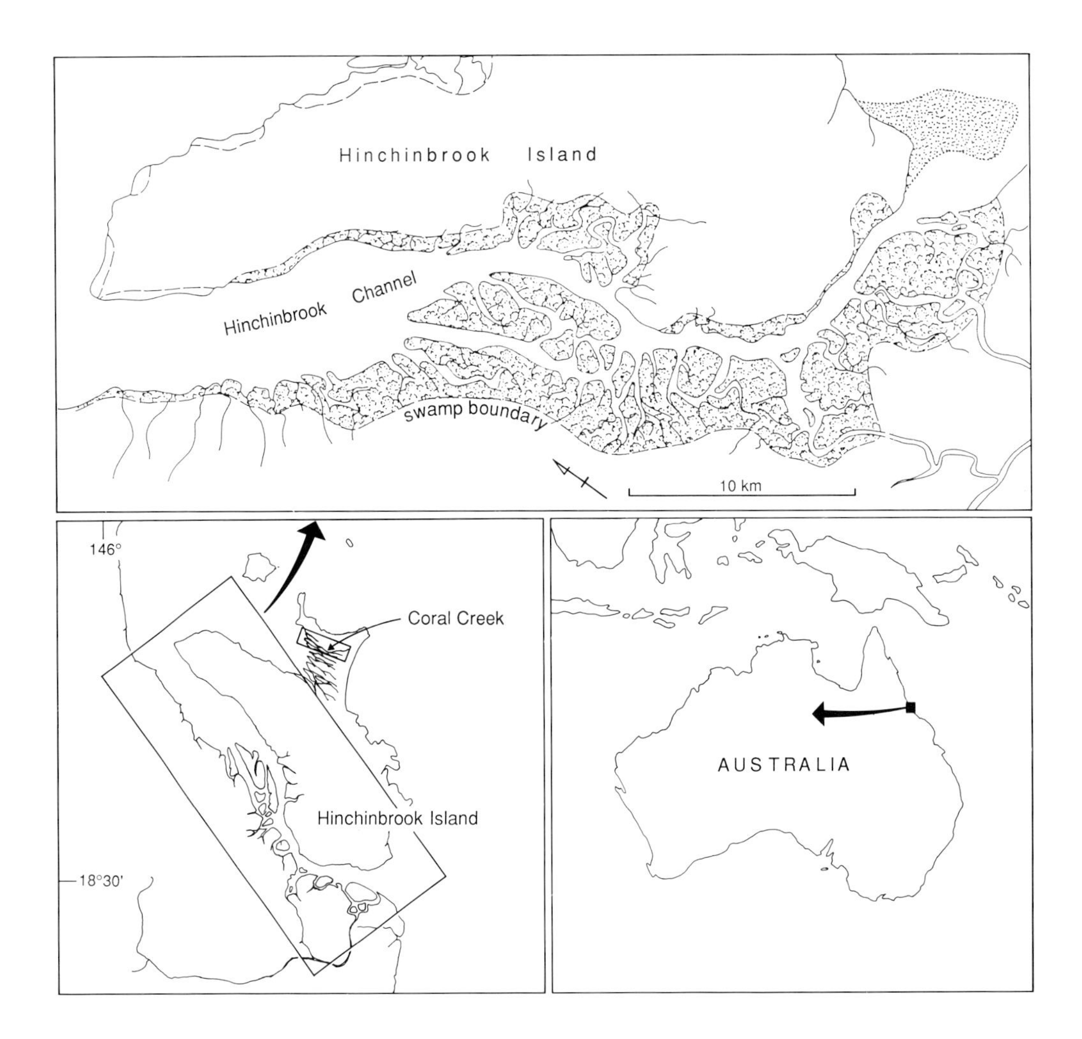

Figure 1. Map of Hinchinbrook Channel and its mangrove swamps in northern Australia.

d. OVERWASH FOREST: Low islands and small peninsulas, which are completely overwashed on all high tides,

e. DWARF FOREST: Topographic flats above mean high water, which are tidally inundated only during wet season, and are dry for most of year.

The water circulation in the riverine forest type has been studied somewhat more than the other types. This type of forest has also suffered the most devastation following the construction of prawn farms and land reclamation. This type of forest is exemplified by Hinchinbrook Channel, northern Australia (Figure 1). The mangrove area is a complex system comprising intricate, long tidal creeks (mangrove creeks) and surrounding mangrove swamps. A typical mangrove creek is Coral Creek in Hinchinbrook Island (see Figure 1).

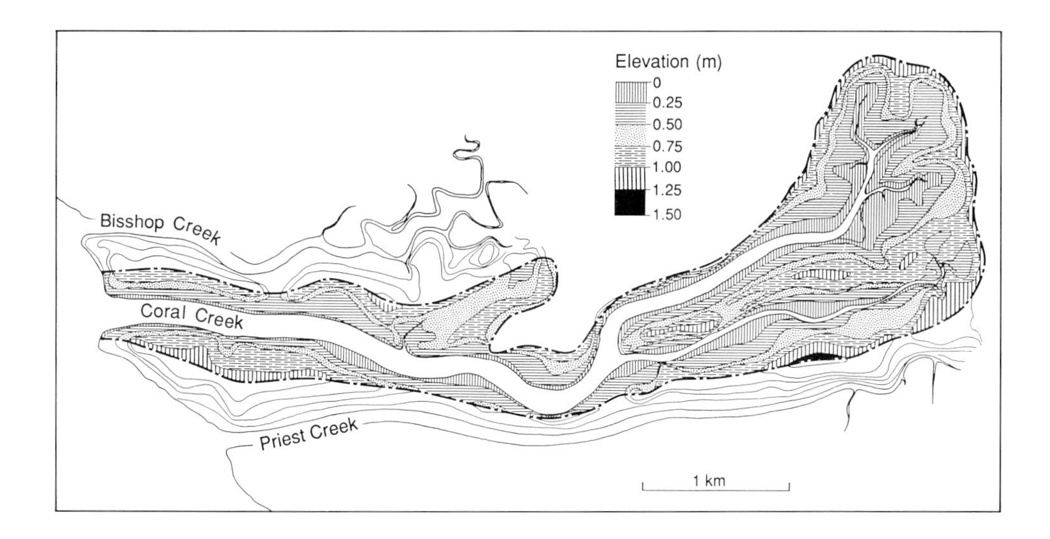

Figure 2. Map of Coral Creek, a mangrove creek in Missionary Bay, northern Australia, and elevation of the substrate in the surrounding mangrove swamp.

Figure 2 shows Coral Creek, together with the topography of the surrounding mangrove swamp. Near the mouth, the bottom of the deeply scoured channel is up to 6 m below the swamp level immediately adjacent to the channel. Upstream the channel nearly dries up at spring low tide. The swamps substrate rise from approximately 0.3m above mean sea level to 1.3m above at the boundary of the swamp, typically less than 300m from the creek. The slope of the substrate is usually very small (1×10^{-3} - $\sim 4 \times 10^{-3}$); see Table 1. Based on the few mangrove creeks that have been modelled, the ratio of the swamp area to the creek area appears to be of order 2-10 (Table 1). The swamp thus increases considerably the tidal prism of the estuary.

Table 1. Physical Characteristics of Mangrove Swamps

Location	**Bottom Slope of the Swamp**	**Ratio of Swamp Area (Ss) to Creek Area (Sc)**	**Ss+Sc (km^2)**
Hinchinbrook Channel (Australia)	2/1000	2.1	163.5
Coral Creek (Australia)	3/1000	5.5	5.8
Dickson Inlet (Australia)	******	6.2	2.8
Tuff Crater (New Zealand)[1]	1/1000	44.0	0.2
Klong Ngao (Thailand)[2]	4/1000	2.7	13.3
Nakama-Gawa (Japan)	3/1000	3.9	1.3
Tudor Creek (Kenya)[3]	3/1000	7.2	31.4

1. Calculated from Woodroffe (1985). 2. Calculated from Wattayakorn *et al.* (1990).
3. Calculated from Wolanski (1989), and personal communication from Vanden Berghe.

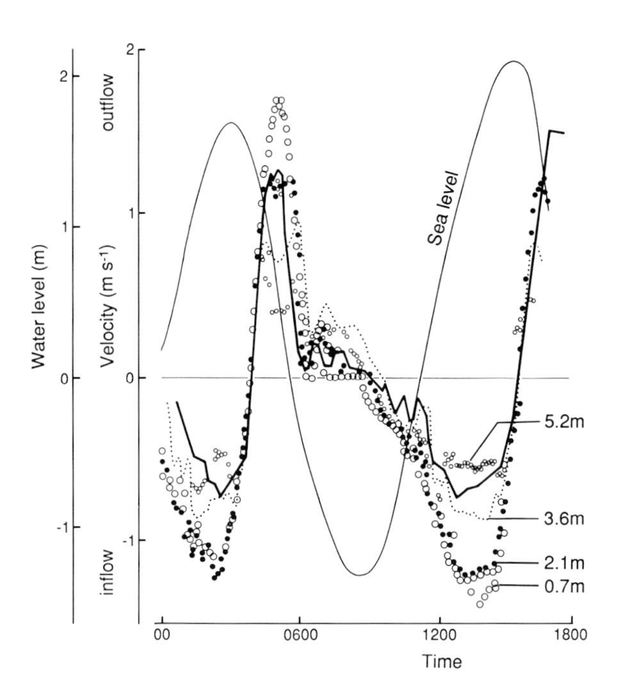

Figure 3. Time series plot of the sea level and of the computed and observed currents (at four different depths, 0.7 m, 2.1 m, 3.6 m and 5.2 m) at the mouth of Coral Creek at spring tide.

3.2 Tidal Circulation

The tidal circulation in such systems is the dominant cause of water movement. The circulation within the creeks is characterised by a pronounced asymmetry between the ebb and flood tides, with the ebb tide being slightly shorter in duration but with stronger currents than the flood tide. The water velocities within the creek often exceed 1 ms^{-1}. The velocities within the swamp rarely reach 0.1 ms^{-1}. The peak tidal velocities in creeks which have a very large ratio of swamp volume to creek volume, are much higher than if there were no fringing swamps. This is simply because a considerably larger volume of water must pass through the creek in order to fill the swamp.

Wolanski *et al.*, (1980) were the first to explain quantitatively aspects of tidal flow in mangroves using the typical example of Coral Creek on Hinchinbrook Island (Figure 2). Because of the very complex nature of the swamp/creek system, Wolanski *et al.,* (1980) used a numerical model to describe the flow. The system was divided into two basic components, firstly the tidal creek where acceleration, sea level gradients, frictional and inertial effects are important and secondly the mangrove swamp where inertial effects can be neglected because the water velocities never exceeded 0.1 ms^{-1}.

Figure 3 shows the time series of observed currents in Coral Creek together with the computed velocities for a spring tide. The model successfully reproduces the main features of the depth averaged flow, characterized by a tidal asymmetry with peak flood currents

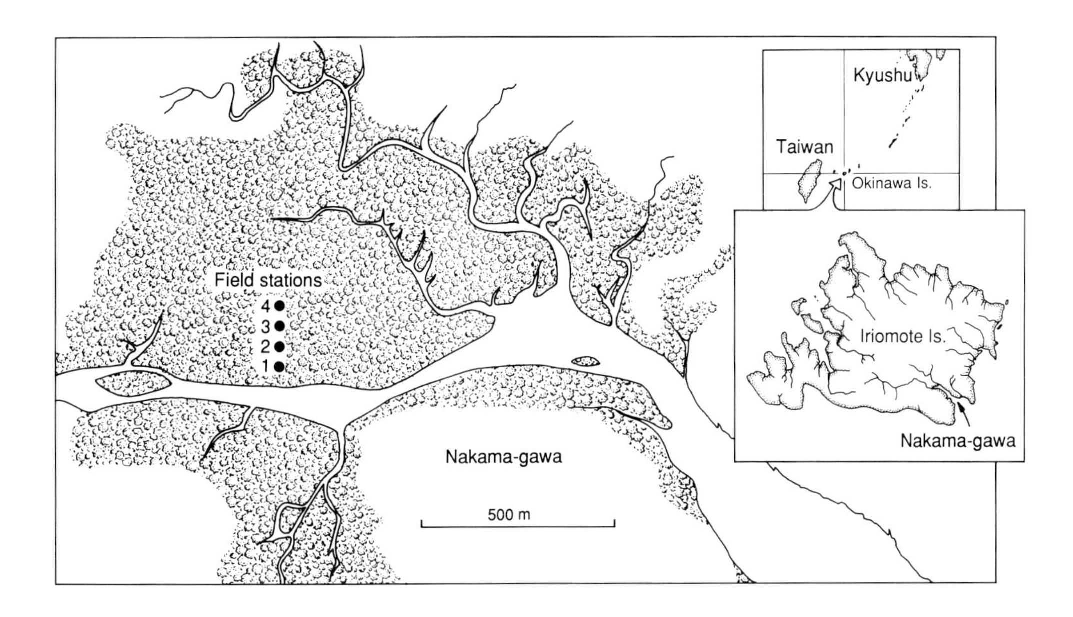

Figure 4. Map of Nakama Gawa, a mangrove swamp in Iriomote Island, Japan, and location of field stations.

being only about two thirds the peak ebb tidal currents.

Wolanski *et al.*, (1980) also used their computer model to show that the currents within the swamp of Coral Creek are small, of the order of a few cms^{-1}, in good agreement with observed values. These small currents are due to the high friction to flow in the mangrove because of the high vegetation density. They parameterised friction using a Manning friction coefficient, n. This coefficient was first introduced to deal with friction in a fully turbulent, open channel flow bed, but has been found to be also applicable to flow in densely vegetated flood plains (Petryk and Bosmajan, 1975). The Manning coefficient can be expressed in MKS units, as

$$n = \frac{1}{u} I^{1/2} H^{2/3}$$

where u is the depth mean current velocity, I the gradient of the water surface and H the water depth. Wolanski *et al.,* (1980) suggested that the Manning friction coefficient, n, is of order 0.2-0.4 in the swamp, an order of magnitude larger than in the tidal creeks where, in MKS units, n = 0.02-0.04, but the data was insufficient for a more detailed assessment. Burke and Stolzenbach (1983) analysed the surface flow through salt marsh grass, and also found a high value of the Manning friction coefficient ($n \cong 0.2$).

More detailed measurements of the Manning friction coefficient were undertaken recently in the mangrove swamp of Nakama Gawa, Iriomote Island, Japan (Figure 4). Figure 5 shows the time series plot of the water levels at stations 3 and 4 located respectively at 90 and 135m from the bank (see location map in Figure 4), the water velocity, the gradient of the water surface, and the resulting values of the Manning friction coefficient, n. This coefficient is found to be, in MKS units, of order 0.4, in a range of 0.2 to 0.7. The flow is thus highly frictional. Note in

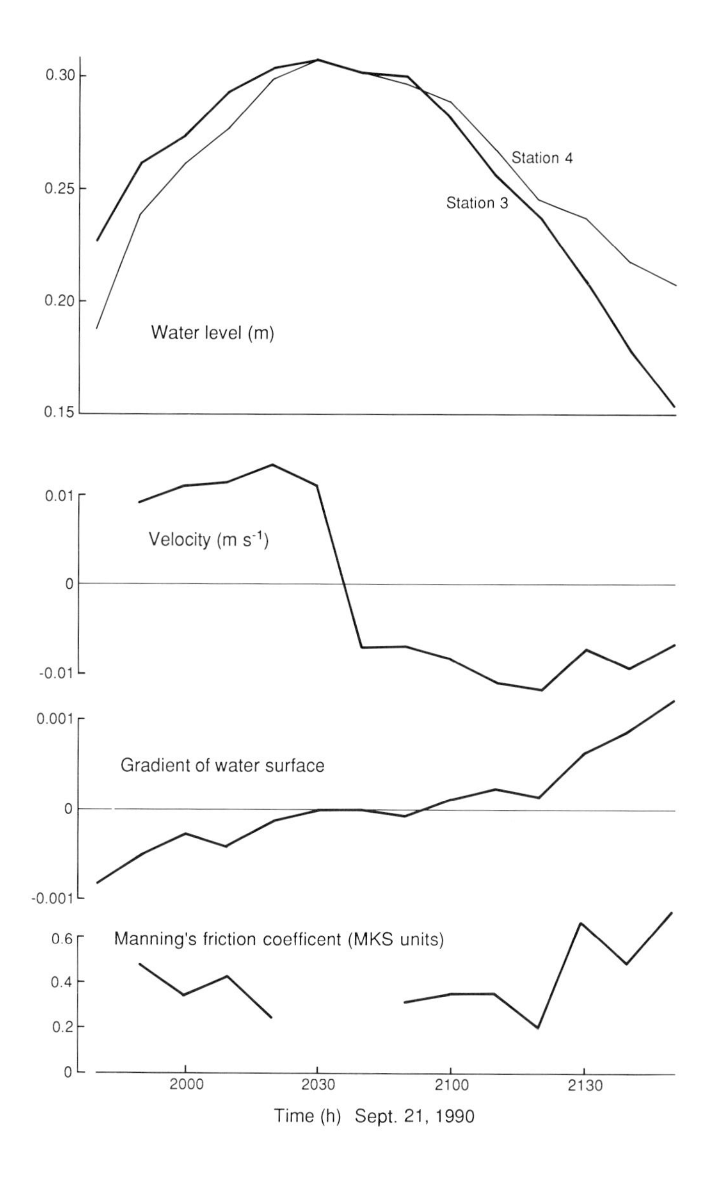

Figure 5. Time series plot of the water velocity, the gradient of the water surface and the Manning roughness coefficient in the mangrove of Nakawa Gawa.

Figure 5 the large water surface gradients, up to 10^{-3}, i.e. 1 m per 1000 m, yet the velocities are very small of order a few cm per sec. The large gradients imply that differences of several tens of cm can exist in the water level in the swamp. Thus the full tidal water level range observed at the creek does not reach the edges of the swamp which may be hundreds of m from the creek. The large water surface gradients are the result of the large frictional effects caused by the vegetation.

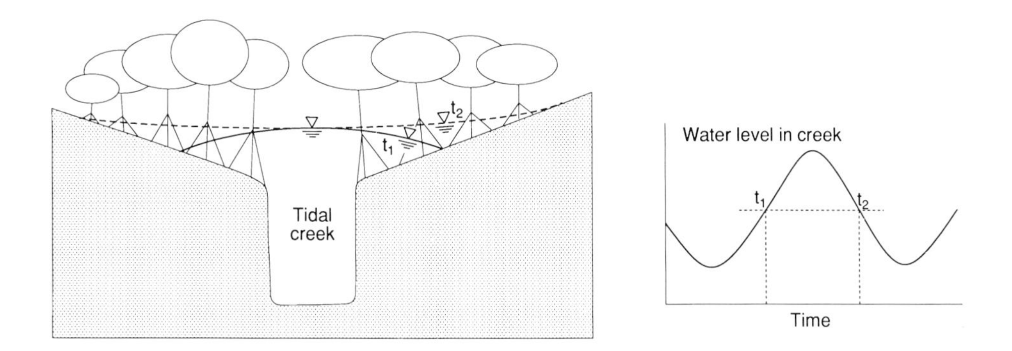

Figure 6. Sketch of the water level in the creek-mangrove swamp system at ebb tide (time t_2) and flood tide (time t_1).

3.3 Tidal Asymmetry

The peak ebb tidal currents in tidal creeks with a large area of fringing mangroves are often 20 to 50% higher than the peak flood tidal currents (Wolanski *et al.*, 1980; Woodroffe, 1985(a) and (b)). This result is opposite to the situation in many estuaries without mangroves (Aubrey, 1986; Friedrichs *et al.*, 1990). The key to understanding the ebb-flood asymmetry lies in the phase changes that occur in the tidal signals from the mouth of the creek to the head of the creek. There will always be a small but significant time lag (phase lag) between the time of high tide at the head and the mouth of the creek, due to friction from the bed and mangrove roots. When the tide reaches high tide at the head of the creek, it is already falling at the mouth, and this provides the necessary water slope to accelerate the water back toward the mouth as the ebb phase commences.

This situation is shown schematically in Figure 6. where the heights of water in the system are shown at ebb (time t_1) and flood tides (time t_2) and where the difference in the water level is greatly exaggerated. Though in Figure 6 the water levels in the creek at the mouth are the same, the water levels and the water gradients in the swamp are different. The processes can be modelled mathematically, taking into account the time lags between the time of high tide at the mouth and at the head of the creek. The phase angle when the water is not flooding the mangroves is less than the phase angle when the water is flooding the mangroves. The swamp fills in the latter half of the rising tide with a considerable portion on either side of the time when the tidal forcing at the mouth has reached its peak. The falling tide in the swamp occurs when the water level at the mouth is already falling and thus there is less time for the water to leave the system on the ebb. The result is that ebb tidal currents are stronger than the flood tidal currents.

The ebb-flood tide asymmetry is crucial in the maintenance of the geometry of the mangrove/creek system. The larger ebb currents tend to export sediments from the creek and maintain a deep, navigable, tidal channel. Reduction in the size of the swamp, for instance by prawn farming or land reclamation, reduces the tidal asymmetry and the peak ebb tidal currents. As a result the creek silts rapidly (in a few years) and becomes shallower until a suitable ratio between the remaining swamp and creek dimensions is re-established in order for the self scouring action to resume. This increased siltation and reduced navigability (to the point where the creek can dry up totally at low tide) is a common occurrence in the South-East Asia region for areas where prawn farms have been constructed in mangrove forests (E. Wolanski and K. Boto, personal observations).

3.4 Exchange with the Nearshore Zone

The exchange of water, and thus nutrients or other properties of interest, between the nearshore zone and the mangrove swamps can occur by either direct tidal exchange, or through diffusive processes. For very short creeks that drain directly into the sea, direct tidal exchange is the primary mechanism as at low tide the quantity of water left in the creek is often minimal. In the presence of a small longshore current, only a small fraction of the water leaving the creek on the ebb will return on the flood tide, and thus complete exchange of the water may occur in just one tidal cycle.

More commonly, coastal waters near mangroves are very shallow, often with vast inter-tidal mud flats. Friction effects dominate the tidal dynamics in these shallow waters. This prevents the formation of ebb tidal jets at the mouth of mangrove-fringed tidal creek. Instead the ebb tide flow is fan-like (Wolanski and Ridd, 1990). The flood tide flow is also friction-dominated (Wolanski and Imberger, 1987) and the flow is also fan-like. This implies that, in the absence of longshore currents, the return coefficient (i.e., the fraction of the water that leaves the estuary at ebb tide that returns at the following flood tide) is large. Even if the tidal currents are oriented longshore, the absence of tidal jets implies that water leaving a mangrove creek at ebb tide stays in nearshore waters. If the coastline is mangrove-fringed, all creeks along a length of a coastline are linked. It may well be that the water leaving one creek at ebb tide subsequently reenters another creek at flood tide. Water becomes coastally trapped and mixes only slowly with offshore water. This coastal trapping' phenomenon was demonstrated by Wolanski and Ridd (1990) for Bowling Green Bay, a shallow, mangrove-fringed, bay in tropical Australia, and by Wolanski *et al.*, (1990) for Hinchinbrook Channel. They suggested that coastal waters may be trapped for typically two to six weeks in calm weather in these two mangrove-fringed, shallow bays.

Direct exchange of mangrove creek water and coastal waters can also occur under the influence of freshwater inflow, as in temperate estuaries. Classical salt wedge estuaries are likely to occur in deep creeks so that the outflow of brackish creek water on the surface is compensated for by an inflow of seawater close to the bottom. If the freshwater inflow is large all salt water can be flushed out of the estuary. This classical secondary circulation effect enhances the importation of offshore water into the estuary and thus the linkage of nearshore and creek water.

After all saltwater is flushed out of an estuary in a short-lived river flood, saltwater returns in the system through the deep channels. However, as the seawater intrudes back in these deep channels, freshwater is trapped at high tide in the mangrove and at low tide as a

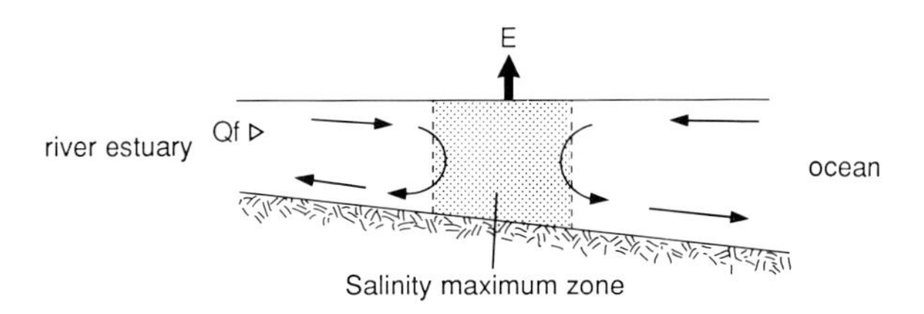

Figure 7. Sketch of the salinity maximum zone in a tropical mangrove-fringed estuary in the dry season, adapted from WOLANSKI (1986).

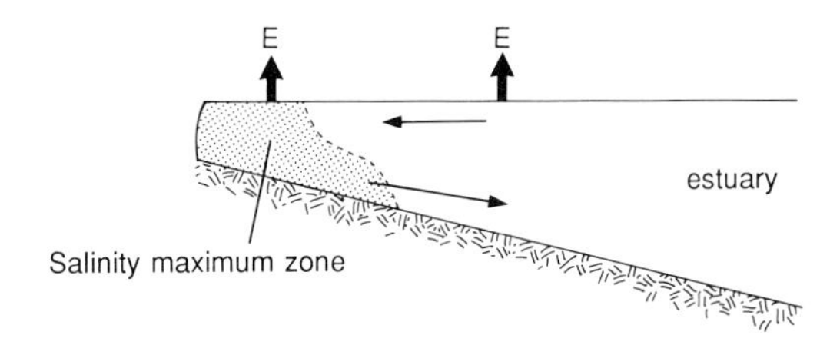

Figure 8. Sketch of the inverse estuarine circulations in a mangrove creek in the dry season, adapted from WOLANSKI (1989).

freshwater plume along the banks of the river (Wolanski and Ridd, 1986). It may take several weeks for freshwater to be flushed out of a mangrove-fringed river after the flood has receded and seawater intrudes again in the system.

In the dry season, under conditions of low freshwater inflow into the creek, and with high evaporation rates, the upper reaches of the mangrove creek can become completely isolated from the nearshore zone. This can occur if the water acquires a salinity greater than that of seawater because of evaporation over the creek and swamp. A salinity maximum zone thus develops (Figure 7). An inverse estuary circulation exists on the lower side of this zone effectively isolating the upper reaches of the estuary from the coastal waters. In tropical Australia, this unusual circulation can last several months (Wolanski, 1989; Ridd *et al.*, 1988)

Mangrove creeks that receive no freshwater runoff in the dry season can have poor water quality naturally, even without man-made influences. This is because the evapotranspiration drives an inverse estuarine circulation pattern sketched in Figure 8 (Wolanski, 1989). This inverse circulation is weakest when the tidal currents are strong enough to maintain vertical homogeneity in salinity. When vertical gradients in salinity exist, the bottom waters are not aerated and continuously receive mangrove detritus. By decomposition this detritus extracts dissolved oxygen from the water. As a result low (<1 mg l^{-1}) dissolved oxygen (DO) concentration are commonly found near the bottom in such tidal creeks, and anoxic conditions occasionally result (Wolanski, 1989; Mazda *et al.*, 1990b). When anoxic conditions occur, nutrients are released from the mangrove mud (Sato *et al.*, 1990) and water quality further degrades.

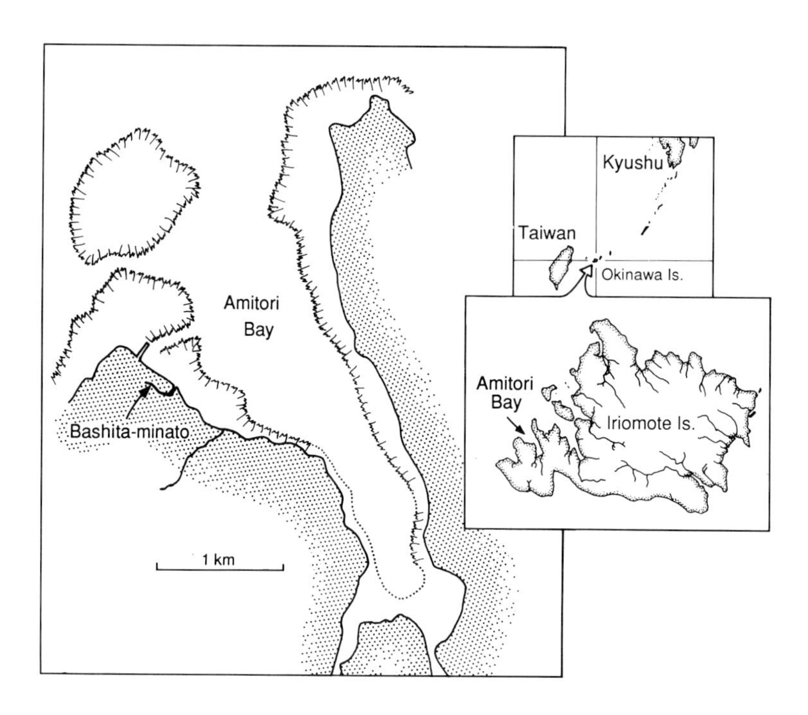

Figure 9. Map of the Bashito-Minato mangrove swamp and nearby coral reef in Japan.

The importance of exchange processes between the mangroves and the nearshore zone has been illustrated by Mazda *et al.*, (1990b) in the small Bashito-Minato swamp on Iriomote Island, Japan (Figure 9). The creek drains into a coral reef area which has DO levels completely different from those existing in the swamp. Figure 10 shows time series plot of the solar radiation, the swamp water level and the dissolved oxygen concentration in the swamp and in the reef area. The DO in the coral reef area rose and fell with the diurnal cycle, reflecting primary productivity. The DO in the swamp however, rose sharply at commencement of the flood tide and fell slowly thereafter, irrespective of the solar radiation. It is thus apparent that for this swamp, the main input of DO was from the incoming reef water rather than production within the swamp. The DO levels in the reef strongly influenced those in the swamp, and presumably also vice versa. Thus when two different systems such as these are in close proximity, exchange processes between the water in the swamp and in the coastal zone are very important.

3.5 Tidal Diffusion

Large undisturbed mangrove creeks such as Coral Creek and others of similar or larger size have a considerable quantity of water remaining at low tide and this water will return to the head of the creek or swamp when the flood tide resumes. Assuming that there is no net freshwater discharge the only method by which this water can exchange with the sea is by diffusive processes, i.e., turbulence mechanisms in the creek or swamp. These transport suspended or dissolved material from zones of high concentrations to low concentrations.

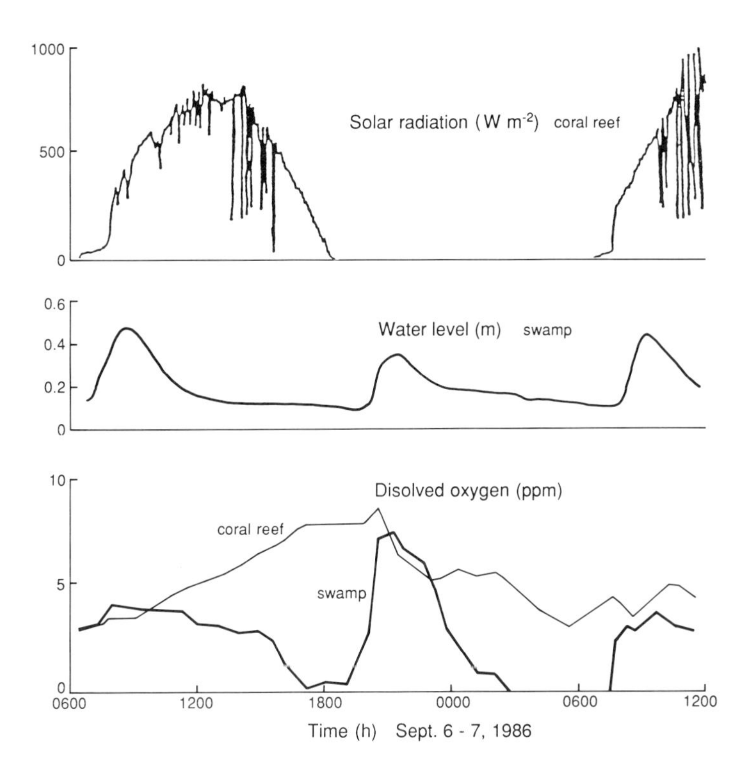

Figure 10. Time series plot of the solar radiation, the sea level in the swamp and the dissolved oxygen concentration in both the swamp and the reef area, in the Bashito-Minato mangrove swamp when it was open to the ocean.

Although turbulent motion is the cause of diffusion, by itself it cannot account for the magnitude of the diffusion effect that is observed in estuaries, but must be enhanced by another process. In classical estuaries with no mangrove swamp, trapping of water in small lateral embayment and the shear in the cross-stream water velocity (with the velocity close to the bank being smallest) are very effective in enhancing the diffusion. This can be understood most easily by considering the release at a point in the middle of a creek of a quantity of nutrients or pollutant. Cross-stream turbulent diffusion drives some of the pollutant toward the edge of the creek where it enters the slower boundary flow along the creek bank, or is temporarily trapped in embayments or irregularities. In either case, the pollutant slows down while the material remaining in midstream continues to move up the creek. The result is that the material is spread out longitudinally very effectively.

In mangrove fringed creeks this shear dispersion process is magnified many times by the presence of the mangrove swamps (Wolanski and Ridd, 1986, and Ridd *et al.*, 1990), because of the shearing effect caused by the trapping of water in the swamps. Again consider a point source of a pollutant that is dumped in the creek at rising tide. As the pollutant moves up the creek a proportion of it continuously moves into the swamp and thus by high tide the initial point source can be spread out over the swamp and part may still remain in the creek. When the tide turns, the pollutant reenters the creek and is spread out over a considerable length of the creek. This diffusion process is conceptually similar to the conventional shear diffusion that occurs in all rivers or estuaries, in which the cross-stream diffusion coefficient drives the transport of contaminants to the edge of the stream where the velocity is lower than the central stream. The swamps enhance this process. The trapping effect in mangroves is analogous to the boundary flow along the channel except the process driving the lateral transport of contaminant into the swamp is water slope and associated advective flow rather than cross-stream diffusion coefficient.

Wolanski and Ridd (1986) and Ridd *et al.*, (1990) used the pioneering work of Okubo (1973) to develop formulae for the longitudinal mixing coefficient, B, in a tidal creek due to the enhanced shear diffusion effect caused by the trapping of water in the mangroves. They predicted

$$B = \frac{A}{(1+\varepsilon)} + \frac{\varepsilon\, u^2\, a^2\, T}{48\,(1+\varepsilon)} \qquad (1)$$

where A is the ambient turbulent mixing coefficient, e is the ratio of water in the swamps to that in the creek at high tide, u is the characteristic water velocity and the elevation of the swamp substrate was taken to be higher than the mean sea level so that the swamp is not inundated for a fraction (1-a) of the tidal cycle of period 2T.

The residence time t_o for water in the swamp/creek system can be estimated from the above equation using

$$t_o = l^2/B \qquad (2)$$

where l is the length of the mangrove fringed creek.

Wattayakorn *et al.*, (1990) applied Equation 1 to the 4 km long Klong Ngao mangrove creek in Thailand giving a mixing coefficient of about 10-40 m^2s^{-1}, a very high value for a non-mangrove-fringed tidal creek, but a value that agrees well with observations. Equations 1 and 2 have been applied to the giant mangrove swamp of Hinchinbrook Channel (Fig. 1) where it was found that B ~ 100 m^2s^{-1} and t_o = 50 days, implying long term trapping of water (Wolanski *et al.*, 1990). Experimental evidence using the flushing of brackish water from the channel as a tracer, indicated an exponential decay in the salinity deficit with an e-fold time of 54 days, in surprisingly good agreement with the prediction.

3.6 Transverse Circulation

All estuaries including those fringed by mangroves, exhibit secondary circulation phenomena superimposed on the primary tidal circulation and net downstream transport from freshwater inflow. The most common form is the longitudinal circulation caused by the net saltwater inflow at the bottom of the creek and freshwater flow at the surface due to density effects. As mentioned previously, this influences the direct exchange between offshore water and creek water.

Perhaps less important but certainly less well documented in mangrove creeks are transverse circulations. This may enhance the formation of long lines of floating mangrove leaves and seeds. These floating leaves and seeds are commonly found aggregated in long lines, often provide shelter for small fish, and constitute the bulk of the outwelling of particulate carbon from mangrove swamps (Boto and Bunt, 1981; Wattayakorn *et al.*, 1990). The processes forming these long lines of floating mangrove detritus have not been studied in detail but they are presumably similar to the linear oceanographic features observed near coral reefs (Wolanski and Hamner, 1988; Kingsford, 1990; Kingsford *et al.*, 1991).

Many mangrove swamps are highly sinuous. As water passes around a meander, a secondary transverse circulation cell develops with water at the surface moving to the outside of the bend and water at the creek bed moving to the inside. This is caused by the small but significant water level rise on the outside of the bend that is required to balance the centrifugal force for flow in curves. Because the longitudinal water velocity at the creek bed is reduced by friction, the cross river water slope is greater than the centrifugal force required and the bottom water moves to the inside of the bend. This secondary flow in meanders has not been studied in detail in the context of mangrove creeks but is likely to increase vertical mixing and to transport buoyant material such as floating mangrove leaves to the outside of a bend.

Another secondary circulation regime which requires investigation in mangrove creeks are the axial convergence cells described by Nunes and Simpson (1985) for temperate estuaries. These cells can form a very long convergence zone along the centre of the river whereby floating material can accumulate, and are caused by an interaction of longitudinal salinity gradients with transverse velocity shear.

Buoyancy effects also create small-scale oceanographic fronts when water leaves the swamps at ebb tide to return to the tidal creek. A foam line and an aggregation of mangrove detritus are sometimes observed along this front.

Because such secondary transverse circulations produce the fronts aggregating mangrove leaves and seeds, they are likely to be important in the ultimate export of these material. More research needs to be done on this aspect of mangrove creek hydrology.

3.7 Tidal Currents and Sediment Fluxes

Tidal currents are of prime importance in determining sediment transport rates both in the creek and swamp, and also the sediment grain sizes of both the suspended and bed load material. An understanding of sediment movement is obviously very important to mangrove geomorphology, but is also essential in water chemistry studies as the silt and clay-size particles, that usually dominate the sediment composition in mangrove swamps, can carry heavy metals and other important chemicals.

There is obviously a large difference in the sediment transport regimes in the swamps and in the creek due to the large difference in currents. The sediment budget of mangrove creeks has not been studied in sufficient detail to assess the net budget. This net budget is believed to reflect primarily an import of coastal sediments during storms and the self-scouring of the creeks because of the tidal asymmetry. Sediment suspended by the strong currents in the creek can settle out when it enters the swamp. The rate of trapping of fine sediment in the swamp does not appear to have been studied at all.

In the creek, the strong tidal currents make bedloads possible. Larcombe and Ridd, (1991) have studied bed load transport in Ross Creek, a sandy bottom mangrove-fringed tidal creek in Northern Australia. The bed had well developed sandripples (up to 20 cm high) and the onset of sediment movement was determined by electromagnetic sediment level sensors (Ridd, 1992) as the ripples migrated. The sediment movement threshold velocities were found for this sediment (modal size of 0.2 mm) to be about 0.40 ms^{-1}. The peak ebb and flood tidal currents, at 0.4 m above the bed, were 0.8 ms^{-1} and 0.4 ms^{-1} respectively and thus very little bedload transport occurred on the flood tides. This confirms the role of the ebb-flood asymmetry in keeping undisturbed mangrove creeks deep and well scoured.

3.8 Groundwater Flow

Although movement of water through the swamp substrate is generally much smaller in magnitude than the tidal currents, it is crucial in determining the chemistry surrounding the mangrove roots and the organisms such as crabs which inhabit the swamp. It also has an important affect on the chemistry of the creek water.

We have observed that at the initial stages of the inundation of the Coral Creek mangrove swamp at rising tide, the first few millimetres of flooding water at a given point are not due to surface water advancing across the swamp but come through numerous crab holes in the ground. This is an highly effective mechanism which prevents accumulations of salt due to evapotranspiration (Wolanski and Gardner, 1981). The importance of crab burrows is also apparent in the surface and porewater chemistry of the Itacuruca Swamp in Brazil (Ovalle *et al.*, 1990) where rainfall infiltrates readily in the soil producing no surface runoff, in contrast with the Klong Ngao swamp in Thailand where the density of crab holes is low, and direct runoff is observed (Wattayakorn *et al.*, 1990).

Wada and Takagi (1988) have suggested that groundwater flow is indispensable for plant respiration in Thailand mangroves. Ovalle *et al.*, (1990) showed that the mixing of surface water with groundwater in the mud flats in front of a mangrove swamp is an important buffer mechanism for nutrient exchange between coastal and mangrove waters.

Mazda *et al.*, (1990a) have shown that considerable exchange of nutrients occur through groundwater flow between the ocean and the Bashito-Minato mangrove swamp in Japan (Figure 9). This swamp occasionally becomes ponded by the formation of a sill. Water level differences between the ocean and the mangrove creek can then reach 0.7 m. This water slope drives a groundwater flow, with periods of inflow and outflow. It has a high impact on the water quality of the mangrove creek. As shown in Figure 11 for the Bashito-Minato

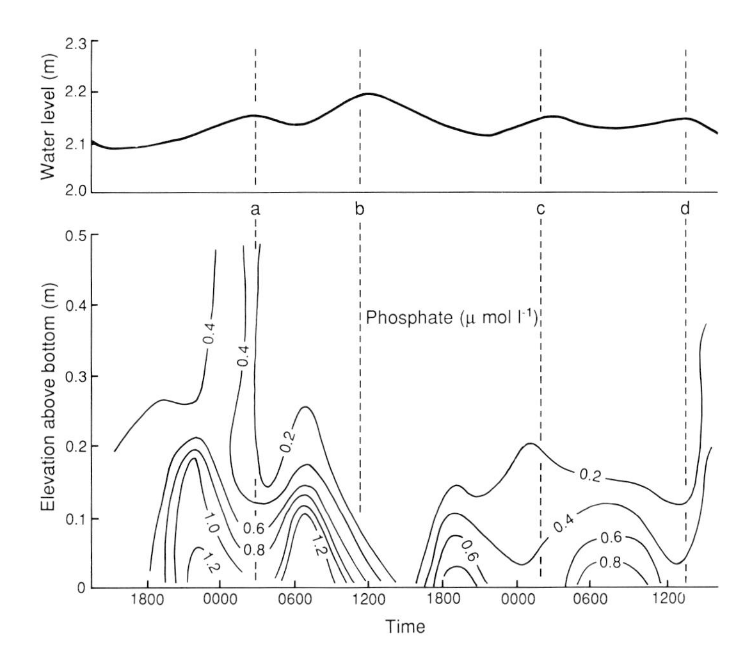

Figure 11. Time series plot of water level and the distribution of phosphate in the Bashito-Minato swamp when it was ponded by a sill at the mouth. The water level fluctuations are due to groundwater flow. The events marked 'a', 'b', 'c' and 'd' denote the seepage of groundwater out of the bottom into the swamp. At these times the concentration of phosphate near the bottom decreases considerably (adapted from Mazda *et al.*, 1990).

mangrove swamp when it is ponded, the momentum of the inflowing groundwater mixes the bottom water with the overlying water, resulting in removing the anoxic conditions and the high phosphate concentration near the bottom. When an outflow occurs, i.e. when water leaves the swamp by groundwater flow, the benthic algae photosynthesizing on the bottom sediment are entrained in the sediment.

Extensive areas of tropical salt flats adjacent to many mangrove areas exist along the estuaries of the Gulf of Carpentaria in tropical Australia. Cracks in the ground form after prolonged periods (months) of dry weather and no tidal flooding. These cracks are many tens of centimetres deep and provide a passage for localised ground water flow. There are no crab holes. However, as soon as the ground becomes fully water logged either after rain or tidal inundation, the cracks close up and groundwater flow presumably becomes negligible. Certainly we have noted pools of water standing on the salt flats for a few weeks indicating minimal groundwater flow. The high content of clay material in the mangrove swamp or salt flat substrate is probably the primary cause for the minimal groundwater flow, in the absence of crab holes or cracks. These observations suggest that crab holes are a key feature in the hydrology of mangroves, preventing hypersaline conditions.

3.9 Climatic and Tidal Influences

Seasonal climatic and tidal variations are very important in determining the moisture and thus salinity content of the porewater within the sediment of the mangrove swamp, thus influencing the conditions under which the mangrove trees must live. Tidal variations in water height occur on time scales from 12 hours for the semi-diurnal tides, to one month for the spring-neap cycle and up to six months for climatic seasonal fluctuations such as the monsoon cycle. Clearly the variations determine the times at which the mangrove swamp is covered by water, with the areas at higher elevation being flooded only at the highest tides with often longer periods when no flooding occurs. In tropical areas with little rainfall and high evaporation, very high water losses and consequently high concentration of salt in the porewater can occur.

The balance between rainfall and evaporation, in conjunction with the tidal variations, is the key factor determining if the upper levels of the swamp are mangrove swamp or tidal salt flat (Oliver, 1982). Because of their higher elevations, the salt flats are often only flooded during the spring tide. This is the case for the estuaries of Tanzania (A. Semesi, pers. comm). In the absence of any rainfall in the dry season, hypersaline conditions result.

In the Gulf of Carpentaria in tropical Australia, the mangrove swamps are fringed by large areas of salt flats (Rhodes, 1980; Ridd *et al.*, 1988). In this region alone, the area of salt flats is approximately 4000 km^2, compared with the total area of mangroves in Australia of 11500 km^2 (Galloway, 1982). This part of the coast is characterised by very low rainfall from June to November. Further, in the southern area of this Gulf, the wind stress causes the mean sea level to vary by up to 1 m from summer to winter upon which the normal tidal fluctuations are superimposed (Forbes and Church, 1983). As a result, the salt flats are not flooded during these months, even at the spring tides. Hypersaline conditions thus prevail for six months of the year. Detritus accumulate in these salt flats for six months. When the mean sea level increases following a change in seasonal winds, spring tides inundates these salt flat region. The initial tidal inundation results in a readily measurable export of salt, silicate and orthophosphate from the salt flats to the estuaries and coastal waters (Ridd *et al.*, 1988). This observation suggests that salt flats may be ecologically important.

It must be stressed that the absolute magnitudes of the tides are not the important factor in determining if hypersaline conditions exist, but rather variations in total tide (astronomical and meteorological) height which leave the swamp dry for long periods of time. Also, although direct input of rainfall onto the swamp does not usually influence the hydrodynamics to any great extent, the influence of the tidal height variations, in conjunction with rainfall and evaporation are crucial in determining the salt balance.

Given that rainfall significantly affects porewater salinity level, it is likely that it also affects nutrient levels within the swamp substrate, particularly in areas where regular flooding by the tides does not occur.

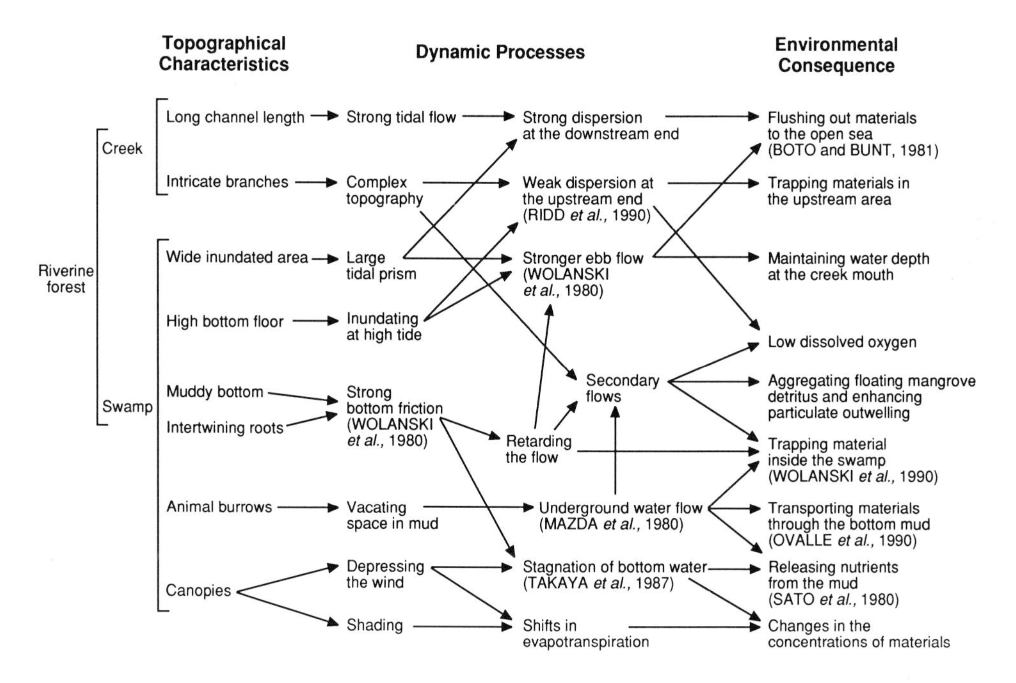

Figure 12. Links between physical, chemical and biological processes in mangrove swamps.

3.10 Links between Physical, Biological and Chemical Processes

Figure 12 illustrates a number of such links. The water circulation is quite different in mangrove creeks and in the swamps. The creeks have long, often branching channels. The swamps are wide and heavily vegetated with a complex substrate. Strong tidal flows exist in the creeks as a result of the tidal prism due to the surrounding swamp. This results in a strong dispersion in the downstream direction, which help flush out material to the open sea. The dispersion in the upstream end of a tidal creek traps material there where water quality is least.

Frictional forces and the presence of swamps lead to an asymmetry of the tidal currents. This asymmetry maintains a deep, self-scoured, tidal channel. Mangrove land reclamation for prawn farms or other uses, result in siltation of the channel.

The high vegetation density in the swamp leads to high friction which retards the flow, resulting in trapping in the swamp. Because of trapping phenomena, anoxic conditions can occur in waters near the sediment surface, and nutrient efflux from the sediment may increase.

The complex topography leads to secondary three-dimensional currents and small scale topographically-controlled fronts, aggregating floating mangrove detritus in long lines, and enhancing particulate export.

Groundwater flow is enhanced by the presence of decaying roots, crab burrows etc. and is

Groundwater flow is enhanced by the presence of decaying roots, crab burrows etc. and is thus an important pathway for water, salt and nutrients. It prevents excessive accumulation of salt from evapotranspiration, and can help transport nutrients in and out of the swamp.

The extensive canopy prevents wind mixing and also solar radiation reaching the mud surface. The inhibition of wind mixing enhances stagnation, and the inhibition of solar radiation decreases surface evaporation and prevents excessive water temperature which are otherwise felt in exposed mud banks where the temperature can reach 50°C.

These examples illustrate some intricate feedbacks between physical, biological and chemical processes in mangrove swamps. This is a rich area of research in the future.

3.11 Conclusion

The mangrove swamp/creek system is a masterpiece of engineering which has been produced by nature. While the water circulation is poorly understood compared with temperate estuaries, many advances have been achieved in the last decade. The usual circulation processes found in temperate estuaries are complicated in mangrove fringed estuaries by the massive areas of swamps, often with an intricate dendritic drainage system. This, combined with a large evapotranspiration, creates novel estuarine circulation patterns. The swamps affect the tidal currents, the longitudinal and transverse mixing rates and the secondary circulations as well as the sediment transport regimes in the estuary. The hydrodynamics of the system controls the geomorphology, in particular the creek and swamp dimensions, and in turn the geomorphology controls the hydrodynamics to produce a natural system which is, if not in equilibrium, at least highly stable.

The water circulation in the mangrove swamp controls the chemistry and biology of the swamps and the estuary. Also the swamps and the nearshore waters often share a common body of water called a coastal boundary layer which only slowly mixes with offshore waters.

3.12 References

Aubrey, D.G., 1986. Hydrodynamic controls on sediment transport in well-mixed bays and estuaries. In: J. van de Kreeke (Ed.), *Physics of shallow estuaries and bays*, pp. 245-258, Springer-Verlag, New York.

Boto, K.G., and Bunt, J.S., 1981. Tidal export of particulate organic matter from a northern Australian mangrove system. *Estuarine, Coastal and Shelf Science* **13**: 247-255.

Burke, R.W., and Stolzenbach, K.H., 1983. Free surface flow through salt marsh grass. Massachusetts Institute of Technology, Sea Grant College Program, Publ. Nv. MITSG 83-16, Cambridge, Ma., 252 pp.

Forbes, A.M., and Church, J.A., 1983. Circulation in the Gulf of Carpentaria. II. Residual currents and mean sea level. *Australian Journal of Marine and Freshwater Research*, **34**: 11-22.

Friedrichs, C.T., Aubrey, D.G., and Peer, P.F., 1990. Importance of relative sea level rise on evolution of shallow estuaries. In: Cheng, R.T., (Ed.), *Residual currents and longterm transport,* pp. 105-122 Springer-Verlag, New York.

Galloway, R.W., 1982. Distribution and physiographic communities of Australian Mangroves. In: Clough B.F. (Ed.), *Mangrove ecosystems in Australia,* pp 31-54, Australian National University Press, Canberra.

Kingsford, M.J., 1990. Linear oceanographic features: a focus for research on recruitment processes. *Australian Journal of Ecology* **15**, 10-20.

Kingsford, M.J., Wolanski, E., and Choat, J.H., 1991. Influence of tidally induced fronts and Langmuir circulations on distribution and movements of presettlement fish around a coral reef. *Marine Biology* **109**: 167-180.

Larcombe, P.A., and Ridd, P.V., 1991. Measurement of bedform migration using sediment level sensors. Unpublished manuscript.

Lugo, A.E., and Snedaker, S.C., 1974. The ecology of mangroves. *Annual Review of Ecology and Systematics* **5**: 39-64.

Mazda, Y., Yokoch, H., and Sato, Y., 1990a. Groundwater flow in the Bashita-Minato mangrove area, and its influence on water and bottom mud properties. *Estuarine, Coastal and Shelf Science* **31**: 621-638.

Mazda, Y., Sato, Y., Sawamoto, S., Yokochi, H., and Wolanski, E., 1990b. Links between physical, chemical and biological processes in Bashita-Minato, a mangrove swamp in Japan. *Estuarine, Coastal and Shelf Science* **31**: 817-833.

Nunes, R.A., and Simpson, J.H., 1985. Axial convergence in a well-mixed estuary. *Estuarine, Coastal and Shelf Science* **20**: 637-649.

Okubo, A., 1973. Effects of shoreline irregularities on streamwise dispersion in estuaries and other embayments. *Netherlands Journal of Sea Research* **6**: 213-224.

Oliver, J., 1982. The geographic and environmental aspects of mangrove communities: Climate. In: Clough B.F. (Ed.), *Mangrove ecosystems in Australia*, pp. 19-30, Australian National University Press, Canberra.

Ovalle, A.R.C., Rezende, C.E., Lacerda, L.D., and Silva, C.A.R., 1990. Factors affecting the hydrochemistry of a mangrove tidal creek, Sepetiba Bay, Brazil. *Estuarine, Coastal and Shelf Science* **31**: 639-650.

Petryck, S., and Bosmajan, G., 1975. An analysis of flow through vegetation. *Journal of the Hydraulic Division of the American Society of Civil Engineers* **101**: 871-874.

Rhodes, E.G., 1980. Modes of Holocene coastal progredation; Gulf of Carpentaria. PhD. thesis, Dept. of Biogeography and Geomorphology, Australian National University, Canberra.

Ridd, P.V., 1992. A sediment level sensor for erosion and siltation. *Estuarine, Coastal and Shelf Science* (in press).

Ridd, P.V., Sandstrom, M.W., and Wolanski, E., 1988. Outwelling from tropical tidal salt flats. *Estuarine, Coastal and Shelf Science* **26**: 243-253.

Ridd, P.V., Wolanski, E., and Mazda, Y., 1990. Longitudinal diffusion in mangrove fringed tidal creeks. *Estuarine, Coastal and Shelf Science* **31**: 541-554.

Sato, Y., Mazda, Y., and Okabe, S., 1990. Changes in water properties of a mangrove area and their physical-chemical mechanisms. *Bulletin of Coastal Oceanography* **28**: 51-62.

Takaya, A., Mazda, Y., and Sato, Y., 1987. Environmental characteristics of mangrove basin: Hydrography and water quality in Bashita-Minato, Iriomote Island. *Bulletin of Coastal Oceanography* **25**: 52-60.

Wada, H., and Takagi, T., 1988. Soil-water-plant relationships of mangroves in Thailand. *Galaxea,* **7**, 257-270.

Wattayakorn, G., Wolanski, E., and Kjerfve, B., 1990. Mixing, trapping and outwelling in the Klong Ngao mangrove swamp, Thailand. *Estuarine, Coastal and Shelf Science* **31**: 667-688.

Wolanski, E., 1989. Measurements and modelling of the water circulation in mangrove swamps. COMARF Regional project for Research and Training on Coastal Marine Systems in Africa - RAF/87/038. Serie Documentaire No. 3:1-43.

Wolanski, E., Jones, M., and Bunt, J.S., 1980. Hydrodynamics of a tidal creek-mangrove swamp system. *Australian Journal of Marine and Freshwater Research* **31**: 431-450.

Wolanski, E., and Gardner, R., 1981. Flushing of salt from mangrove swamps. *Australian Journal of Marine and Freshwater Research* **32**: 681-683.

Wolanski, E., and Imberger, J., 1987. Friction-controlled selective withdrawal near inlets. *Estuarine, Coastal and Shelf Science* **24**: 327-333.

Wolanski, E., and Ridd, P.V., 1986. Tidal mixing and trapping in mangrove swamps. *Estuarine, Coastal and Shelf Science* **23**: 759-771.

Wolanski, E., and Hamner, W.M., 1988. Topographically controlled fronts in the ocean and their biological influence. *Science* **241**: 177-181.

Wolanski, E., and Ridd, P.V., 1990. Mixing and trapping in Australian tropical coastal water. In: Cheng R.T., (Ed.), *Residual currents and long-term transport in estuaries and bays.* pp 165-183, Springer-Verlag.

Wolanski, E., Mazda, Y., King, B., and Gay, S., 1990. Dynamics, flushing and trapping in Hinchinbrook Channel, a giant mangrove swamp, Australia. *Estuarine, Coastal and Shelf Science* **31**: 555-580.

Woodroffe, C.D., 1985a. Studies of a mangrove basin, Tuff Crater, New Zealand: III. The flux of organic and inorganic particulate matter. *Estuarine, Coastal and Shelf Science* **20**: 447-461.

Woodroffe, C.D. 1985b. Studies of a mangrove basin, Tuff Crater, New Zealand: II. Comparison of volumetric and velocity-area methods of estimating tidal flux. *Estuarine, Coastal and Shelf Science* **20**: 431-445.

4

Mangrove Floristics and Biogeography

Norman C. Duke

4.1 Introduction

Mangroves are a diverse group of predominantly tropical trees and shrubs growing in the marine intertidal zone where conditions are usually harsh, restrictive and dynamic. Here, they are subject to both shorter term rhythms of tides and seasons, as well as longer term changes of climate and sea level. As a group, they share several highly specialized and collectively well-known adaptations, notably exposed breathing roots, support roots and buttresses, salt-excreting leaves, and viviparous water-dispersed propagules. However, as individuals, we know less about them, exemplified by the mistaken belief that these characters might be shared equally by all species. Therefore, in this chapter, it was necessary first to clarify and enhance the concept and definition of what is a mangrove, prior to discussing their biogeography, and why they occur in certain localities and not others. It will also be seen that the genetic diversity of these plants belies their ecological uniformity, raising serious doubts for ideas of shared ancestry with their co-inhabitors. Finally, the concept of the mangrove habitat will be enhanced by knowing more about the individuals that provide the structural framework and trophodynamic coherence to this unique ecosystem.

Bunt *et al.*, (1982a) make the point that there are two major problems in identifying mangrove species. The first is deciding what is, and what is not, a mangrove - a problem of better defining the term. The second is a problem of scientific classification, or botanical systematics. Both will be resolved only after this community and the species are better known. This treatment addresses each problem, and presents the recent status of our knowledge, gained principally over the last decade in the Australasian region by the Australian Institute of Marine Science and Graeme Wells with the University of Sydney. The combination of these efforts arguably provide the best continental floristic record for mangroves in the world. Furthermore, these forests are mostly pristine, despite some localized vandalism in more populated areas; notably, in southern Australia where these predominantly tropical habitats are more scarce. This is not to suggest that these are any less valuable though, and recent studies of genetic variation for *Avicennia* in south-eastern Australia depict possible relict Gondwanan ancestors (Duke, 1988). This observation is supported by the special cold tolerance of these plants, seen in no other mangrove species in the world. Furthermore, the value of these forests is not just of academic interest, it also relates to their cultural and commercial value. The latter is greatly under-rated in Australia, and mangroves in this country have never been threatened by direct commercial exploitation. However, one only needs to look to the north of Australia to see quite a different situation. In Indonesia, Malaysia, Thailand, the Philippines, and just about every under-developed tropical country bordering coastal environments around the world, mangrove forests are under serious threat. This implies a greater urgency on studies of mangrove forests so we can better evaluate this resource, seeking a balance between preservation, restoration and selective utilization.

4.1.1 Mangroves Defined - Plants and Habitat

A mangrove is a tree, shrub, palm or ground fern, generally exceeding one half metre in height, and which normally grows above mean sea level in the intertidal zone of marine coastal environments, or estuarine margins. When referring to the habitat, the term 'mangroves' is used, although perhaps equally often, it is referred to as a 'mangrove forest', and sometimes, a 'tidal forest'. Another term, 'mangal', was proposed for the habitat, but this is considered redundant in view of the previously mentioned, more popular terms. In fact, it is common for individual plants to be referred to, using the term 'mangrove' as an adjective, rather than on its own. Accordingly, an individual tree in the 'mangroves', is commonly referred to as a 'mangrove tree', and so on. Hence, there is only a slight difference to how we refer to other habitats, e.g. rainforests, and the plants found in them.

A greater problem is in the definition itself. Unfortunately, it is not always clear what is a mangrove. Accordingly, species have often been categorized as 'true' mangroves and other taxa, often called 'mangrove associates'. This account makes no attempt to explore this distinction, preferring instead, to focus on those species which best fit the definition. With this in mind, all have been chosen with great care, and they were included only after field evidence of their mangrove affinity appeared undeniable. Most taxa have been commonly viewed in the field by the author, leaving only a few where the detailed descriptions of others have been necessary to gain a better understanding. One example is *Mora oleifera (=Mora megistosperma)*, which was described from the Pacific coast of Colombia by Von Prahl and colleagues (1984, 1990). Another is *Aglaia cucullata (=Amoora cucullata)*, described in an extensive study of the Bangladesh Sunderbans by Karim (1991). Clearly, there are several additional species which do occur in mangroves from time to time, but the emphasis of the definition is on those which 'normally' grow in the intertidal zone. And, 'normally' was defined on the basis of wide-ranging field observations, where systematic distinctions are known. It will be seen, however, that these distinctions are often unclear, and it is one of the aims of this work to identify such problems, by presenting also the current status of the systematics. This will enable those receptive to the remaining problems, to explore them, and to further improve our understanding of the characteristics and biogeography of each taxon.

4.1.2 Attributes of Mangroves

The combination of morphological and physiological adaptations seen in this diverse and unique group of plants have no equal (Saenger, 1982; Tomlinson, 1986). However, their chief attributes are collectively shared by others in at least three very different habitats, namely deserts, tropical rainforests and freshwater swamps. In xeric habitats, there are many plants, called halophytes, which have both physiological and anatomical adaptations for growing in salty environments. In mangroves, these attributes are necessary to grow in the marine environment, often with their roots immersed in varying concentrations of saltwater. This may change regularly with daily, monthly and annual tidal fluctuations, and with seasonal rainfall and river out-flow. In this way, salinity variations of interstitial waters are often widely and regularly variable. It is believed that this variation at particular sites influences both the types of mangroves that can become established and survive (e.g. Karim, 1991), and their morphology (e.g. Soto and Corrales, 1987; Duke, 1990). Hence, the type and condition of mangroves at particular sites reflect the physical conditions of those sites.

Furthermore, different taxa have different mechanisms for coping with high salt concentrations, and not all have salt-excreting glands on their leaves. Others exclude salt at the roots, although this creates xeric conditions for the plant. Another group also allows low concentrations of salt into their sap, but this is neutralized by its transfer into senescent leaves, or by storing it in their bark or wood. Mangroves also need to cope with growing in water-saturated, often anerobic, substrates. Some of these latter characters are shared with freshwater swamp trees. This lack of gaseous exchange in the substrate requires them to have special breathing structures on the exposed roots and/or trunk. These maybe quite different, depending on the taxon. Some, notably *Rhizophora* spp., have aerial prop roots bending down from either the trunk or branches, high above the substratum. Others have shallow, subsurface cable roots with series of vertical, stem-like breathing roots, called pneumatophores. By contrast, in some mangroves, there are no elaborate physical structures, instead numerous small air-breathing lenticels are often present on the trunk. Another essential attribute in this water-saturated environment are structures to support the above ground mass of the tree. This is very important to larger individuals which commonly attain 40 metres in height. So, where roots are unable to penetrate more than a metre or so because of the anerobic conditions, then lateral support structures are essential; i.e., beyond the mutual support provided by the neighbouring community of trees. In these cases, the root structures, already described, contribute a great deal. However, other support structures like trunk buttresses, well-known in tropical rainforests, are also common in mangrove plants.

An important attribute of mangroves is also their viviparous propagules. Again, not all mangroves share this character, and others have lesser degrees of vivipary, while a smaller group have no apparent specialization, compared to their terrestrial counterparts. In most cases, however, the propagules are buoyant for at least a short dispersal phase. The undeniable dispersal specialists, however, are Rhizophoras. These have highly-developed viviparous propagules (in fact, young seedlings), which are believed to be able to endure for several months at sea in a semi-dormant state. In general however, the different propagules, and their various dispersal ranges, reflect the diversity of plant groups found in mangroves.

Table 1. A classification of world plant families with mangrove species (chiefly, Cronquist 1981). Distributions are listed as cosmopolitan (C) or tropical (T), and either widely occurring (pan), or found in specific areas (numbers refer to six global regions in Figure 1). Habit is defined as tree (T), shrub (S), herb (H), climber (C) or ground fern (F). Families underlined have all mangrove species. Orders and families in brackets are those currently less acceptable - note authorities in footnotes.

Sub-class Order	Family	Distrib.	Habit	Name
[Division POLYPODIOPHYTA]				
	Pteridaceae	Tpan	F	Ferns
[Division MAGNOLIOPHYTA - Class Magnoliopsida]				
III. 3.Plumbaginales	Plumbaginaceae[1]	C	S,H	Sea Lavender, Thrift
	(Aegialitidaceae)[1]	T,5,6	S	Club Mangrove
IV. 2.Theales	(Theaceae)[2]	T,1,2,5	T,S	Tea, Camellia, Franklinia
	Pellicieraceae[2]	T,1,2	T	Panama Mangrove
3.Malvales	Bombacaceae	T,2	T	Baobab, Balsa, Kapok, Durian
	Sterculiaceae[3]	T,pan	T,S,H	Bottle Tree, Cocoa, Kola
12.Ebenales	Ebenaceae	T,4	T,S	Ebonies, Persimmon
13.Primulales	Myrsinaceae[4]	T,4-6	T,S	Turnip-wood, Mutton-wood
	(Aegicerataceae)[4]	T,5,6	S	River Mangrove
V. 2.Fabales	(Leguminosae)[5]	T,pan	T,S,H	Peas, Bauhinia, Wattles
	Caesalpiniaceae[5]	T,pan	T,S	Cassia, Bauhinia, Tamarind
6.Myrtales	Combretaceae	T,pan	T,S,C	Combretum, Quisqualis
	Lythraceae[6]	T,pan	T,S,H	Crepe Myrtle, Cuphea, Henna
	Myrtaceae	T,6	T,S	Eucalyptus, Bottlebrush, Guava
	Sonneratiaceae	T,4,5,6	T	Mangrove Fig
7.Rhizophorales (Myrt.)[7]	Rhizophoraceae	T,pan	T,S	Viviparous Mangrove
12.Euphorbiales	Euphorbiaceae	T,pan	T,S,H	Castor-oil, Spurges
16.Sapindales (Rut.)[8]	Meliaceae[9]	T,pan	T,S	Mahogany, Rosewood
VI. 3.Lamiales	(Verbenaceae)[10]	T,pan	T,S,H,C	Teak, Verbenas
	Avicenniaceae[10]	T,pan	T,S	Mangrove Olive
6.Scrophulariales	Acanthaceae[11]	T,pan	S,H	Black-eyed Susan, Shrimp Plant
	Bignoniaceae[12]	T,1,2	T,S,C	Tulip Tree, Jacaranda, Catalpa
8.Rubiales (Gentian.)[13]	Rubiaceae	C	T,S,C	Gardenia, Coffee, Quinine
[Division MAGNOLIOPHYTA - Class Liliopsida]				
II. 1.Arecales	Arecaceae[14]	T,pan	T,S,C	Palms
	(Nypaceae)[14]	T,5,6	S	Mangrove Palm

1. Plumbaginaceae (Heywood 1978; Takhtajan 1980; Cronquist 1981; Tomlinson 1986): Aegialitidaceae (Airy Shaw 1973). Plumbaginaceae in Primulales (Hutchinson 1973); 2. Pellicieraceae (Willis 1966; Takhtajan 1980; Cronquist 1981; Tomlinson 1986): Theaceae (Heywood 1978); 3. Sterculiaceae in Tiliales (Hutchinson 1973); 4. Myrsinaceae (Airy Shaw 1973; Heywood 1978; Cronquist 1981; Tomlinson 1986): Aegicerataceae (Hutchinson 1973), Aegiceraceae (de Candolle 1844). Myrsinaceae in Myrsinales (Hutchinson 1973); 5. Caesalpiniaceae (Hutchinson 1973; Takhtajan 1980; Cronquist 1981): Leguminosae (Airy Shaw 1973; Heywood 1978; Morley and Toelken 1983; Tomlinson 1986). Caesalpiniaceae in Leguminales (Hutchinson 1973); 6. Lythraceae in Lythrales (Hutchinson 1973); 7. Rhizophorales (Cronquist 1981; Tomlinson 1986): Myrtales (Hutchinson 1973; Heywood 1978; Takhtajan 1980; Morley and Toelken 1983): Celastrales (Dahlgren 1988; Juncosa and Tomlinson 1988a); 8. Sapindales (Heywood 1978; Cronquist 1981): Rutales (Takhtajan 1980; Muller 1981; Morley and Toelken 1983); 9. Meliaceae in Meliales (Hutchinson 1973); 10. Avicenniaceae (Airy Shaw 1973; Morley and Toelken 1983; Tomlinson 1986): Verbenaceae (Hutchinson 1973; Heywood 1978; Takhtajan 1980; Cronquist 1981). Verbenaceae in Verbenales (Hutchinson 1973); 11. Acanthaceae in Personales (Hutchinson 1973); 12. Bignoniaceae in Bignoniales (Hutchinson 1973); 13. Rubiales (Hutchinson 1973; Heywood 1978; Cronquist 1981; Morley and Toelken 1983): Gentianales (Takhtajan 1980; Muller 1981); 14. Arecaceae (=Palmae) (Hutchinson 1973; Heywood 1978; Takhtajan 1980; Cronquist 1981; Tomlinson 1986): Nypaceae (Airy Shaw 1973).

4.2 Mangrove Floristics and Higher Systematics

The diversity of mangrove plants is best seen in the array of plant orders and families presented in Table 1. There are twenty families, from two plant divisions, including the fern family in the Polypodiophyta, and the remainder in the Magnoliophyta, also known as the angiosperms. Their classification in upper ranks depends on the system used. This account generally follows Cronquist (1981) who described nineteen families with mangrove representatives as part of two classes, six subclasses and 14 orders. However, just two families are exclusively mangrove, and there are no orders or higher ranks with all mangrove taxa. These are underlined in Table 1. Clearly therefore, mangroves are not a genetic entity, but an ecological one. It is interesting to note that even the Rhizophoraceae, often referred to as the 'true mangrove' family, has only four of its sixteen genera inhabiting mangroves. Generally, these families are more commonly represented in tropical rainforests, and most are pantropic in distribution, occurring as trees and shrubs mostly. In addition, the families are better known by their common names and these are usually based on their better-known representatives, including either garden, timber, fruit or medicinal species.

In reference to other systems of classification, there are some notable differences in the status of some mangrove groups (see Table 1). Interestingly, at least three additional, exclusively mangrove families proposed earlier, are now less acceptable. In one case, Aegicerataceae, Cronquist (1981) makes the following statement. "It certainly stands apart from the rest of the Myrsinaceae, but the relationship is not in dispute. *Aegiceras* (the sole representative of the disputed family) is a mangrove, and some of the principal characters by which it differs from typical Myrsinaceae relate to the adaptation of the seed and fruit to the mangrove habit." This clearly presents the view that a plant looks like a mangrove because it grows in a mangrove habitat. It also implies that its genetic divergence from more landward ancestors was less than that required to live in this environment. For the moment, such questions of classification remain arguable, however, at least until future analyses characterize the genetic variation of respective taxa, linking various lineages.

Contrary to the problems of naming higher classification groups, there are few problems in the assignment of generic names. There are twenty-seven genera in total, seventeen are exclusively mangrove (underlined in Table 2), while nine others include non-mangrove species. Notice also that the number of mangrove genera in these families is often relatively low. This feature is also reflected in the number of mangrove species in those genera. Although 50% of genera are monospecific for the mangrove habitat, half of these include greater numbers of non-mangrove species. For the others, the number of mangrove species never exceeds eight, although the addition of putative hybrids adds to the number of taxa by one or two. Regardless, these relatively low numbers are believed to be the result of the harsh conditions imposed within the intertidal habitat. There are expected to be fewer opportunities and less flexibility for natural experimentation and genetic selection. In general, conditions under which mangroves live provide a severe test, requiring a high level of optimized efficiency in each plants utilization of resources. The critical limits on this life-dependant optimization are also believed to be reflected in the structured distribution of species, both locally in zones, and regionally in different climates. This consideration,

Table 2. A classification of all world plant genera with mangrove species (chiefly, Cronquist 1981), including numbers of taxa by family and by genus (numbers in Australia are included in parentheses; for families, Morley and Toelken 1983; for genera, Tomlinson 1986, and personal records). Taxa underlined are those with all representatives in mangrove forests. Numbers of putative hybrids are identified following (+) the number of mangrove species.

Dicot. Sub-class	Order in Mangrove	Family with Mangrove Species			Genus with Mangrove Species		
		Name	Total Number of Genera	Total Number of Species	Name	Non-Mangrove Species	Mangrove Species+'Hybrids'
[Division POLYPODIOPHYTA]							
		Pteridaceae	35	1000	Acrostichum	0	3(1)
[Division MAGNOLIOPHYTA - Class Magnoliopsida]							
III.	3.Plumbaginales	Plumbaginaceae	10(3)	560(9)			
		(Aegialitidaceae)	1(1)	2(1)	Aegialitis	0	2(1)
IV.	2.Theales	(Theaceae)	21-25(1)	550(1)			
		Pellicieraceae	1(0)	1(0)	Pelliciera	0	1(0)
	3.Malvales	Bombacaceae	31(3)	225(3)	Camptostemon	0	2(1)
		Sterculiaceae	70(23)	1200(176)	Heritiera	29	3(1)
	12.Ebenales	Ebenaceae	3(1)	500(15)	Diospyros	400	1(1)
	13.Primulales	Myrsinaceae	35(7)	1000(25)			
		(Aegicerataceae)	1(1)	2(1)	Aegiceras	0	2(1)
V.	2.Fabales	(Leguminosae)	610(172)	18000(1885)			
		Caesalpiniaceae	150(19)	>2500(85)	Cynometra	70	1(1)

Table 2 continued

					Mora	19	1(0)
	6.Myrtales	Combretaceae	20(3)	500(34)	Conocarpus	0	1(0)
					Laguncularia	0	1(0)
					Lumnitzera	0	2+1(3)
		Lythraceae	25(8)	550(24)	Pemphis	1	1(1)
		Myrtaceae	147(70)	3000(1280)	Osbornia	0	1(1)
		Sonneratiaceae	2(1)	6+(3+)	Sonneratia	0	6+3(5)
	7.Rhizophorales (Myrt.)	Rhizophoraceae	16(5)	120(13)	Bruguiera	0	6(5)
					Ceriops	0	3(3)
					Kandelia	0	1(0)
					Rhizophora	0	6+3(4)
	12.Euphorbiales	Euphorbiaceae	300(53)	7000(215)	Excoecaria	35-40	2(1)
	16.Sapindale (Rut.)	Meliaceae	50(11)	600(34)	Aglaia	100	1(0)
					Xylocarpus	1	2(2)
VI.3.Lamiales		(Verbenaceae)	75(17)	3000(62)			
		Avicenniaceae	1(1)	8(2)	Avicennia	0	8(2)
	6.Scrophulariales	Acanthaceae	250(21)	2500(40)	Acanthus	30	2(2)
		Bignoniaceae	120(5)	800(9)	Dolichandrone	9	1(1)
	8.Rubiales (Gentian.)	Rubiaceae	500(42)	7000(203)	Scyphiphora	0	1(1)
[Division MAGNOLIOPHYTA - Class Liliopsida]							
II.	1.Arecales	Arecaceae(Palmae)	210(21)	2800(52)			
		(Nypaceae)	1(1)	1(1)	Nypa	0	1(1)

therefore, would be expected to play a large part in the biogeography and evolution of particular mangrove habitats.

Mangrove forests are dominated by two orders, the Myrtales and Rhizophorales (Table 2), often combined by some authors as a larger Myrtales order (Morley and Toelken, 1983). These comprise 25% of all mangrove families and 50% of all species. They also demonstrate a high degree of specialization for the habitat, since all but one of the ten genera are exclusively mangrove. The remaining one, *Pemphis*, has only one other species, located inland as an isolated population on the island of Madacascar. In addition, the Rhizophoraceae is recognized as a highly adapted and apparently well-advanced family, although it bears some putatively primitive characters (Juncosa and Tomlinson, 1988b). These characters are somewhat subjective, however, and while Sporne (1969) ranked it with less-advanced angiosperms, this view was not endorsed by recent more detailed appraisal (Dahlgren, 1988; Juncosa and Tomlinson, 1988b). Evidence of fossil pollen is more precise, and the earliest records of Myrtales (including the Rhizophoraceae) date from the upper Cretaceous, approximately 20 million years after the first records of angiosperms (Muller, 1981). The latter suggests that mangrove habitats were occupied very early after the evolution of angiosperms, and at least these ten families have had relatively long histories of development. Orders are also characterized by the inclusion of all putative hybrids found in mangrove forests today. These are from three genera, *Lumnitzera, Sonneratia* and *Rhizophora*. Furthermore, in each of them, the ratio of species to putative hybrids is 2:1. For the two genera with six species, this suggests that only one species crosses with some of the others. This may be the result of one widely distributed species occurring sympatrically with its otherwise allopatric conspecifics. In total, there are 69 mangrove species including putative hybrids, and 39 of them, comprising representatives of all polyspecific genera, are found in Australia. This clearly indicates the key position of Australia in the phytogeography and evolution of mangroves.

4.3 Global Distribution Patterns and Species Systematics

Mangroves are distributed according to three important scales, namely their coastal range, their location within an estuary, and their position along the intertidal profile. Each of these will be discussed, along with a brief review of the problems in naming species.

On a global scale, mangrove plants are found throughout tropical regions of the world (Figure 1). More precisely, there is a conspicuous tropical distribution pattern with major deviations matching the presence of warm and cold oceanic currents. Latitudinal ranges therefore tend to be broader on eastern continental margins and more constrained on their western sides; for example, in the North and South Atlantic Oceans, the colder Canaries and Benguela Currents, each moving toward the equator from the north and south respectively, reduce the lineal extent of warmer coastline for west Africa. Conversely, on the east coasts of North and South America, the warmer poleward moving Florida and Brazil Currents extend the warmer coastline. Mangroves generally match the winter 20°C isotherm in respective hemispheres, suggesting the profound importance of water temperature to this

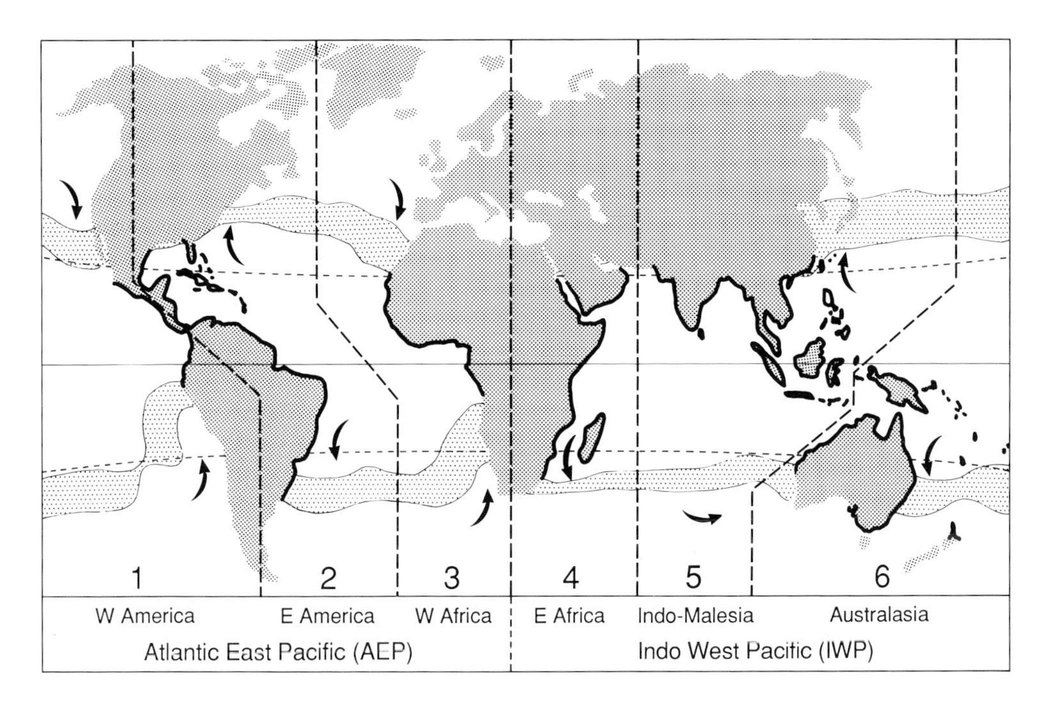

Figure 1. World distribution of mangroves showing their coastal extent (outlined), range of 20°C isotherms in January and July, major ocean currents influencing latitudinal range around the tropic lines, and six biogeographic regions grouped in two global hemispheres discussed in the text. Notice that the poleward extent of mangroves is usually associated with the winter position of the 20°C isotherm, except for, the east coast of South America (region 2), Australia, and the North Island of New Zealand (all in region 6).

habitat. There are three important exceptions to this pattern, however, and all occur in the southern hemisphere, notably along the coastlines of eastern South America, around Australia, and across the North Island of New Zealand. This could be the result of specific, small-scale extensions of warmer currents, but it appears more likely that these populations are relict, representing refuges of more poleward distributions in the past.

These exceptions notwithstanding, mangroves are found chiefly in tropical latitudes where their global dispersal by specialized water-buoyant propagules is apparently constrained by both wide bodies of water and land masses blocking the equatorial flow of tropical waters. Today, there are four major barriers, influencing the dispersal of most warm coastal marine organisms (Briggs, 1974), including the continental land masses of Africa and Euro-Asia, North and South American continents, North and South Atlantic Oceans, and the eastern Pacific Ocean. The relative effectiveness of each of these barriers differ, however, depending on geological history, and the dispersal-establishment ability and evolution of each taxon. There are two barriers which appear to have been reasonably effective during recent geological time, namely the African Euro-Asian continents, and the Pacific Ocean. Thus, mangrove species, and indeed most tropical, shallow-marine coastal habitats, are divided into two global hemispheres (Figure 1), the Atlantic East Pacific (AEP), often referred to as the New World, and the Indo West Pacific (IWP), or Old World. These more-

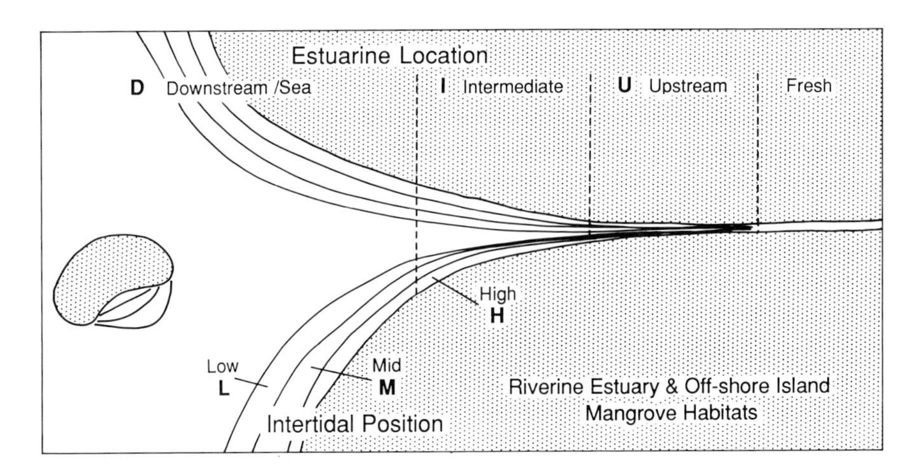

Figure 2. A sketch map delineating the two common mangrove habitats described and discussed in the text, and used in Table 3. Each is divided into three categories: estuarine location with, downstream, intermediate and upstream; and, intertidal position with, low, mid and high.

referred to as the New World, and the Indo West Pacific (IWP), or Old World. These more-or-less equal portions of the earth also have equivalent areal extents of mangrove forests (Saenger *et al.*, 1983), and represent centres of secondary radiation because they share three major mangrove genera. The AEP has fewer species, however, and fewer additional genera, although it spans two extant barriers. Meanwhile, the most diverse flora is seen in the IWP, constrained between two existing barriers. Based on species presence, this may also be divided into three regions, making a total of six for the world (Figure 1), including (1) western Americas, (2) eastern Americas, (3) western Africa, (4) eastern Africa, (5) Indo-Malesia, and (6) Australasia (Figure 1).

Mangroves are distributed according to at least two other scales, namely their location within an estuary, and their position along the intertidal profile. The latter has often been referred to as zonation. Nevertheless, both may be conveniently summarized in six categories (Figure 2), based on the two specific factors, estuarine location, and intertidal position. Each can be divided into three, including: downstream, intermediate and upstream estuarine, and, low, mid and high intertidal. Downstream also includes off-shore island communities, similarly divided into the three intertidal categories. These are essentially a simplification of the five intertidal inundation classes described by Watson (1928). Hence, 'low intertidal' represents areas inundated by medium high tides and flooded >45 times a month (Watson classes 1 & 2), 'mid intertidal' represents areas inundated by normal high tides and flooded from 20 to 45 times a month (Watson class 3), and 'high intertidal' represents areas inundated <20 times a month (Watson classes 4 & 5). By contrast, estuarine location categories are less well-defined, since the specific physical parameters are not fully understood or quantified for many tropical riverine estuaries. Hence, in this treatment, the proportional distance from the mouth of the estuary is used, where downstream represents the lower third of the estuary including off-shore islands, intermediate represents the middle third of the estuary, and upstream represents the upper third. Knowing the variability in estuarine systems, reflected mainly in intermediate locations, these categories do not adequately describe them, but for the time being, downstream and upstream divisions are useful in identifying those river systems characterized by the presence of particular mangroves. Hence, for example, a species might be known as a low intertidal, upstream specialist.

All mangrove species and putative hybrids in the world are listed in Table 3. Understandably, this compilation identifies several problems. As discussed earlier, it is one thing to have difficulty with the definition of a mangrove and deciding whether a particular species should be included in this list, but quite another, to be unable to confidently assign a name to each taxon. This is not only a problem with rare taxa, rather it applies more often to common plants. For instance, the systematic distinction between *R. mucronata* in eastern Africa and *R. stylosa* in Australia remains unclear (Duke and Bunt, 1979; Tomlinson, 1986). This becomes further confusing in places where they apparently overlap, like Malesia. And in Australia, the species referred to *R. mucronata*, is not like that in eastern Africa. This problem with Rhizophoras extends further when considering their putative hybrids. In Australia, the hybrid between *R. stylosa* and *R. apiculata* is known as *R. X lamarckii*, and its occurrence matches the overlap of its putative parents. In south-eastern India, there is a taxon that fits the description of *R. X lamarckii* (personal observation). If it is a hybrid, and the parents are *R. mucronata* and *R. apiculata*, as listed for the region, then it cannot be *R. X lamarckii*. That is, unless *R. X lamarckii* is not a hybrid, but this would conflict with recent evidence (also note Duke *et al.*, 1984). It is more likely that *R. mucronata* in India and eastern Africa is the same species as *R. stylosa* in Australasia, and that the range of *R. X lamarckii* is far greater than presently listed. Of course, this also means that the true name of the *R. mucronata* - *R. stylosa* parent must be cleared up. Then the name of the plant in Australasia, called *R. mucronata*, can also be settled. Clearly, there is an urgent need to resolve this situation before any realistic assessment of the biogeography and evolution of Rhizophoras can be confidently discussed. Furthermore, this problem with *Rhizophoras* is not restricted to the IWP, because those in the AEP have a comparable question of hybrid status for *R. X harrisonii*, with putative parents, *R. mangle* and *R. racemosa*. Other, mostly polyspecific genera, have equivalent systematic problems, and the size of these problems appear to be a function of each groups diversity. Furthermore, because specific identities are generally better-known in most countries, the main gap now appears to be more detailed comparisons between regions. This would resolve problems described for *Rhizophora*, and other taxa, including, *Sonneratia lanceolata*, *Avicennia marina* varieties, and so on (Table 3). However, this is certainly not to suggest that further intra-regional studies are not necessary. Some examples of these include, the distributions of *Acanthus ebracteatus* and species of the genus *Acrostichum*, as well as investigations of the presence of *Sonneratia X* 'merauke' in northern Australia and southern New Guinea.

The ranges of each taxa, based on their presence or absence in the six biogeographic regions, are listed also in Table 3. The greatest species diversity is found in the IWP with 58 taxa, almost six times more than the AEP with 12. There are 69 altogether, so there is only one species common between hemispheres, namely *Acrostichum aureum*, a mangrove fern. It is possible that *Rhizophora samoensis* is the same as *R. mangle* (Tomlinson, 1986), but this also requires more attention. Generic diversity is more conservative, however, with 7 genera in the AEP, 23 in the IWP, and three in common throughout the six regions. Furthermore, with respect to families, the AEP is virtually a subset of those in the IWP, with the exception of Pellicieraceae. The region with least species is western Africa (region 3) in the AEP with seven taxa, whilst Indo-Malesia (region 5) in the IWP has more than 49. In the IWP, the east African region is most depauperate with only ten taxa, notably equivalent to regions in the AEP. Their respective species compositions, however, are completely different.

The genera and families shared by the IWP and AEP, and the minimal overlap of species, provides evidence of an earlier long period of continuity or connection between regions that has been interrupted also for a long period. Furthermore, present day distributions of species do not conform with expectations either, based on their respective dispersal-establishment abilities and existing barriers. For example, note the common presence of several AEP taxa in all three regions, while others are restricted. This is shown best in the three AEP *Avicennia* which share habitat preferences and propagule characteristics, but have completely different ranges. Also note the presence of *Rhizophora samoensis* in Australasia. Clearly, such disjunctions provide clues of past distributional ranges, and hypotheses on the evolution of mangroves need to explain the presence of each species.

There is a general trend between genera and distributions, listed in Table 3, such that genera with the greatest number of species (plus putative hybrids; taxa/genera), consistently occur in the greater number of biogeographic regions ($r = 0.689$, $n = 26$, $P < 0.001$). Conversely, those with fewer species occur in fewer regions, notably seen for genera with one or two species which only occur in one or two regions (Figure 3). These make up > 50% of mangrove genera. All larger polymorphic genera (3-9 taxa/genera) are found in three or more regions, i.e. excluding the non-angiosperm family with the pantropic fern, *Acrostichum aureum*, apparently found in all regions. Nevertheless, this trend confirms the idea that more widely distributed taxa have greatest genetic diversity. Geographic isolation therefore appears to be the major mechanism promoting diversity and speciation in mangrove plants.

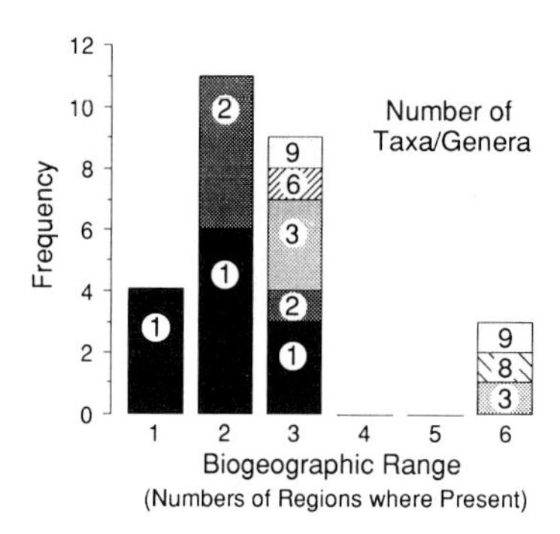

Figure 3. A frequency plot comparing biogeographic range (defined as number of regions, from Figure 1) with the number of taxa per genera for all mangrove genera in the world (Tables 2 & 3). Note that genera with one or two species are restricted mostly to two regions, while those with more than two genera always range over three or more regions.

Table 3. A classification of all mangrove species in the world (chiefly, Cronquist 1981), including their global distribution (see Figure 1), and common habitat (see Figure 2). Distribution is scored for six biogeographic regions as, present o , widely distributed © , or doubtful ?. Upstream riverine or estuarine location listed as downstream (D), intermediate (I) and upstream (U). Topographic or intertidal position listed as low (L), mid (M) and high (H). These records are based chiefly on Ding Hou (1960), Percival and Womersley (1975), Duke *et al.*, (1981), Von Prahl (1984), Tomlinson (1986), Duke and Jackes (1987), Von Prahl *et al.*, (1990), Duke (1991a), Duke (1991b), and Karim (1991). Number superscripts refer to systematic problems, outlined in the footnotes.

Dicot. Sub-class	Order in Mangrove	Family in Mangrove	Genus in Mangrove	Species in Mangrove	Global Biogeographic Regions 1	2	3	4	5	6	Common Habitat Estuary Location	Common Habitat Intertidal Position
[Division POLYPODIOPHYTA]												
		Pteridaceae	Acrostichum[1]	danaeifolium	©	©					I	H
				aureum	o	o	o	o	o	o	I	H
				speciosum					©	©	I	H
[Division MAGNOLIOPHYTA		- Class Magnoliopsida]										
III.	Plumbaginales	Plumbaginaceae	Aegialitis	rotundifolia					©		?	?
				annulata					o	©	D	M,H
IV.	Theales	Pellicieraceae	Pelliciera	rhizophorae	©	o					I,U	M,H
	Malvales	Bombacaceae	Camptostemon	philippensis					©		?	H
				schultzii						©	D,I	L,M
		Sterculiaceae	Heritiera	littoralis				©	©	©	I	H
				fomes					o		U	H
				globosa					o		U	H
	Ebenales	Ebenaceae	Diospyros[2]	ferrea						o	I,U	M,H
	Primulales	Myrsinaceae	Aegiceras	corniculatum					©	©	I,U	L
				floridum					©		?	?
V.	Fabales	Caesalpiniaceae	Cynometra	iripa					©	©	I,U	H
			Mora	oleifera	©						U	H
	Myrtales	Combretaceae	Conocarpus	erectus	©	©	©				D	H

Table 3 continued

		Laguncularia	racemosa	©	©	©				D,I	M,H
		Lumnitzera	racemosa				©	©	©	D	M,H
			X rosea					?	o	I	H
			littorea					©	©	I	M
	Lythraceae	Pemphis	acidula				©	©	©	D	H
	Myrtaceae	Osbornia	octodonta					o	©	D	M,H
	Sonneratiaceae	Sonneratia	apetala					©		U	L,M
			griffithii					o		D	L
			alba				©	©	©	D	L
			X sp.[3]					o		?	?
			ovata					©	o	D	H
			X gulngai					©	©	I	L,M
			caseolaris					©	©	U	L
			X 'merauke'[4]					?	o	I	M
			lanceolata[5]					o	©	U	L
Rhizophorales	Rhizophoraceae	Bruguiera	gymnorrhiza				©	©	©	D,I	M,H
			sexangula					©	©	I,U	M,H
			exaristata						©	I,U	H
			hainesii					©	o	I	H
			parviflora					©	©	D,I	M
			cylindrica					©	©	D,I	M
		Ceriops	tagal				©	©	©	D,I	M,H
			decandra					©	©	I	M,H
			australis[6]						©	D,I	H
		Kandelia	candel					©		D	L
Rhizophorales	Rhizophoraceae	Rhizophora	racemosa	©	©	©				D,I	M
			X harrisonii	o	o	o				D,?	L,?
			mangle	©	©	©				D,I	L,M
			samoensis[7]						o	D,I	L,M
			X selala						o	?	?
			stylosa					©	©	D,L	L,M
			X lamarckii					©	©	D,I	M
			apiculata[8]					©	©	I	M

				mucronata[9]				©	©	o	I,U	L,M
	Euphorbiales	Euphorbiaceae	Excoecaria	agallocha[10]				?	©	©	D,I,U	M,H
				indica					©		D,I	L,M
	Sapindales	Meliaceae	Aglaia	cucullata[11]					o		U	M
			Xylocarpus	granatum				©	©	©	I	M,H
				mekongensis					©	©	I	M,H
VI.	Lamiales	Avicenniaceae	Avicennia	germinans[12]	©	©	©				D,I	M,H
				bicolor	©						D	H
				schaueriana		©					D	M,H
				marina[13]	+			©	©	©	D,I	L,M,H
				alba					©	o	D	L.M
				rumphiana[14]					o	o	D	H
				officinalis					©	o	I	L
				integra						©	I	L
	Scrophulariales	Acanthaceae	Acanthus	ebracteatus[15]					©	o	I	M,H
				ilicifolius					©	©	I,U	M,H
		Bignoniaceae	Dolichandrone	spathacea					©	©	U	M
	Rubiales	Rubiaceae	Scyphiphora	hydrophyllacea					©	©	I	H
[Division MAGNOLIOPHYTA		- Class Liliopsida]										
VIII.	Arecales	Arecaceae	Nypa	fruticans[16]		+	+		©	©	U	L,M,H

1. *Acrostichum* species are poorly identified in most records (Tomlinson, 1986); 2. *Diospyros* reported by Duke *et al.* (1981) is not adequately described and the species name is provisional; 3. *Sonneratia* X sp. = putative hybrid *S. alba* X *S. ovata* (Muller and Hou Liu, 1966); 4. *Sonneratia* X 'merauke' represents the putative hybrid, *S. alba* X *S. lanceolata* (Duke and Jackes, 1987); 5. *Sonneratia lanceolata* and *S. caseolaris* in Australasia lack clear distinction in descriptions from Indonesia and SE Asia (Duke and Jackes, 1987); 6. *Ceriops australis* and *C. tagal* were shown in electrophoretic studies to be sibling species (Ballment *et al.*, 1988); 7. *Rhizophora samoensis* lacks clear morphological distinction from *R. mangle* (Tomlinson, 1978); 8 *Rhizophora apiculata* has cork warts (i.e., underleaf spots) on leaf specimens from Indo-Malesia, while these are not present in southern New Guinea and northern Australia (Duke and Bunt, 1979); 9. *Rhizophora mucronata* from E Africa and SE Asia is not clearly distinct from *R. stylosa* (compare Ding Hou, 1950, Tomlinson, 1986 and Duke and Bunt, 1979). Furthermore, *R. mucronata* in Australasia is possibly a different species (Duke and Bunt, 1979); 10. *Excoecaria agallocha* is not reliably known to occur in E Africa (Tomlinson, 1986); 11. *Aglaia cucullata* (=*Amoora cucullata*) habitat described by Karim (1991); 12. *Avicennia germinans* (=*A. africans*) was reportedly introduced to W Africa (Gunn and Dennis, 1976), although this claim cannot be substantiated (Tomlinson, 1986); 13. *Avicennia marina* var. *australasica* was introduced (+) to Mission Bay, California (pers. comm.); 14. *Avicennia rumphiana* (=*A. lanata*) habitat and distribution described by Duke (1991a); 15. *Acanthus ebracteatus* is not clearly identified from *A. ilicifolius* in most accounts (Tomlinson, 1986); 16. *Nypa fruticans* was apparently introduced (+) into W Africa and E Central America (Duke, 1991b).

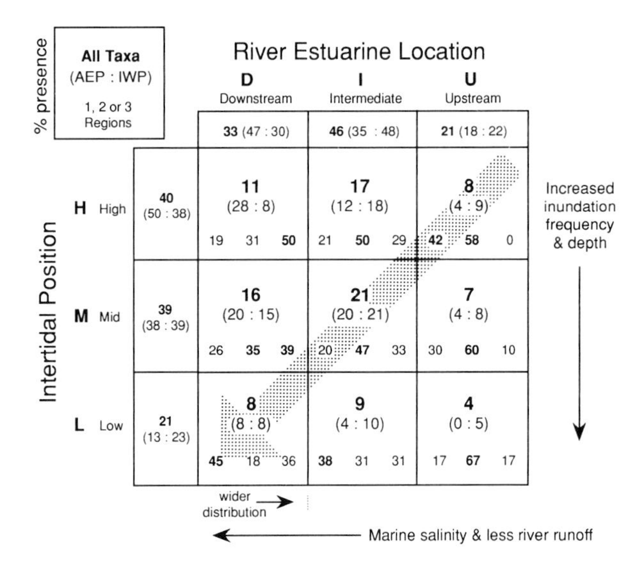

Figure 4. A two-way tabulation showing the % presence in the two common habitat categories, estuarine location and intertidal position (Figure 2), of all mangrove taxa in the world (Table 3). The categories are further divided by, presences in both the Atlantic East Pacific (AEP) and the Indo West Pacific (IWP), and for the presence in one, two or three biogeographic regions (Figure 1). This figure shows the taxa are found throughout most paired combinations of habitat types. The diagonal arrow is indicative of greater 'mangrove-ness' of taxa, notably by being more specialized for life in the intertidal environment characterized by increased inundation and higher salt concentrations.

There is also isolation at other scales of distribution, notably in the specialization of particular mangroves for certain habitats. These data, listed in Table 3, are summarized in Figure 4. It displays percent frequencies of recorded presences in the various habitats and for their paired combinations. The presence of mangrove species in the two habitat types was shown to be quite uniform, with an expected maximum frequency of 21% in mid-intertidal, intermediate combinations, and a minimum of 4% in the low-intertidal, upstream. By excluding mid-intertidal and intermediate sites, there is an apparent trend toward higher frequencies (11%) in high-intertidal downstream sites. On closer inspection, however, this was due almost entirely to species in the AEP with a maximum frequency of 28% in this habitat combination. They also have no presence in the low-intertidal, upstream. It is suggested that this difference between regions might reflect different levels of evolutionary progress of the mangrove environment. For example, it is likely that the evolution of mangroves might have progressed via the high-intertidal upstream to the low-intertidal downstream, based on their progress in adapting to both increased salinity and inundation frequency. Of course, they may also have adapted in a less balanced way, possibly by attaining halophytic attributes first, and adapting to lower intertidal positions second. If this was the case, then the taxa in the AEP could have retained ancestral characteristics, reflecting a tendency for earlier halophytic progenitors. By contrast, IWP taxa demonstrate a more mixed association with the various habitat combinations, suggesting either a more advanced stage of development for these mangrove forests, or a balanced compliment of progenitors from both upstream and downstream locations.

In Table 3, the classification is based on species, but it would also be useful to explore intra-specific variation so that inter-relationships of those in the larger polymorphic genera might be better understood. For example, it would be interesting to know whether IWP species of *Rhizophora* or *Avicennia* were more comparable with each other or their congenerics in the AEP. For *Avicennia* species, there is now a qualified answer to this question in view of the recent revision for Australasia (Duke, 1991a). Accordingly, the IWP species may be placed in two groups, with one having closer affinities with those in the AEP (Tomlinson, 1986), than with the second IWP group. These observations are provisional, however, and await further detailed studies in the AEP. Such studies are impeded by the necessity for sampling from a wide range of sites because of the notoriously variable phenotypic responses of these genera in different locations (e.g. Soto and Corrales, 1987; Duke, 1990). In such circumstances, the choice of morphological characters for systematic classification needs to be carefully re-assessed, otherwise diagnoses might be based on environmentally influenced, phenotypic characters, a major problem with some earlier treatments. In general, virtually all evidence in this respect is morphological, and most groups have been described in the treatment by Tomlinson (1986). This work identified many gaps, recently filled partly for several genera, notably *Sonneratia* (Duke and Jackes, 1987), and *Avicennia* (Duke, 1990, 1991a). There are at least two other larger groups, *Bruguiera* and *Rhizophora*, which also need to be re-assessed. Some of the problems in the latter have already been discussed. In each case, it would be useful to understand inter-specific affinities in an effort to outline possible phylogenetic relationships. The latter may then be tested in genetic studies, and this would provide the basis for models of the evolution of particular mangrove species in relation to their extant distributions.

4.4 Distribution and Discontinuities in Australasia

The Australasian region includes Australia, New Guinea, New Zealand, the south-western and central Pacific islands, as described in Figure 5. Also included in this figure is the distribution of mangroves in the region, identifying more-or-less continuous coastal distributions in the tropics, and discontinuous, isolated populations in temperate latitudes of Australia, and around the North Island of New Zealand. The region is also conveniently divided into six subregions, including: (1) south-western Australia; (2) north-western Australia; (3a) north-eastern Australia, (3b) southern New Guinea, and (3c) north-eastern Australia, New Caledonia; (4a) south-eastern Australia and (4b) New Zealand; (5) northern New Guinea, Solomon Islands and western Pacific islands north and west; and (6) western and central Pacific islands. These were based on the occurrence of mangrove taxa, and the geological history of the region (eg. Hilde *et al.*, 1977). Hence, in the latter context, Australia, southern New Guinea, New Caledonia and New Zealand (subregions 1 to 4) are ostensibly parts of the Australian Continental Plate, while northern New Guinea and the western and central Pacific Islands (subregions 5 and 6) represent the western Pacific Plate. All are recognised as floral discontinuities for mangrove taxa in this region.

The mangrove flora of Australasia is one of the richest in the world (Table 3), having around five times greater species diversity than all other regions, excepting neighbouring

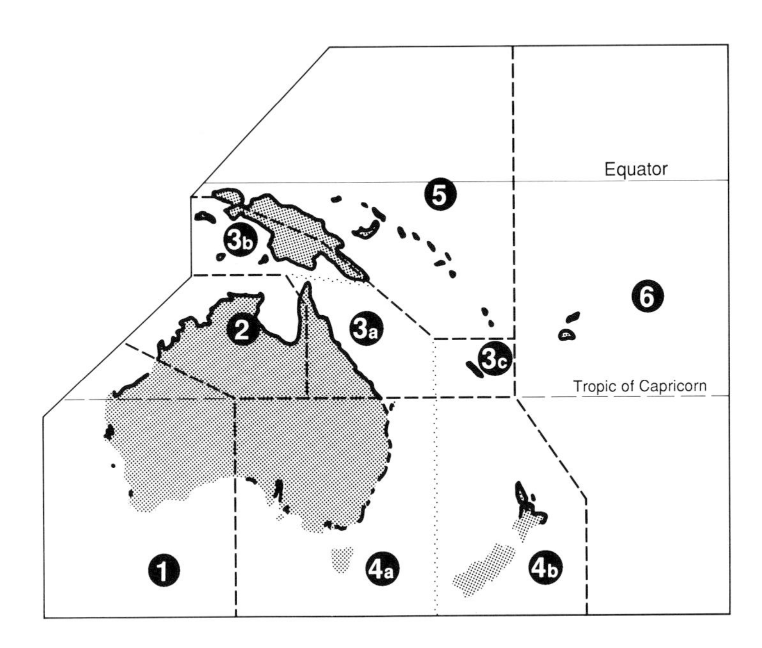

Figure 5. The distribution of mangroves in Australasia (region 6 in Figure 1), showing coastal extent (outlined), the Equator and Tropic of Capricorn, and six subregions discussed in the text. Notice that subregions three and four are further divided to isolate Australia. The extent of mangroves along tropical coastlines is notably more-or-less continuous, while those particularly in southern Australia are discontinuous and patchy (subregions 1 & 4).

Indo Malesia. The latter has just two more taxa, according to present evidence, but this is not to suggest that those in Australasia are a subset of those in Indo Malesia. In Australasia, there are 47 taxa in 21 genera, listed in Table 4. Two Indo Malesian monotypic genera, *Aglaia* and *Kandelia*, are absent but another, *Diospyros* (Duke *et al.*, 1981), is apparently unique to this region. There is a problem with the specific systematic status of this taxon (see notes with Table 3), but there are no reports of this genus in mangrove forests elsewhere. Nevertheless, if this taxon is included, then there are seven endemic taxa (including six species and one putative hybrid) in Australasia, compared with ten found in Indo Malesia. *Avicennia integra*, located in the Northern Territory (subregion 2), has the distinction of being the only endemic mangrove species in Australia. Three others, *Camptostemon schultzii, Ceriops australis,* and *Bruguiera exaristata*, are all common in northern Australia and southern New Guinea. By contrast, *Rhizophora samoensis* and *R. X selala* are located east, on several south-western Pacific islands from Fiji to New Caledonia (Tomlinson, 1978). There are also two other putative hybrids presently known only from this region, but these are expected to be found in Indo Malesia, considering the ranges of their putative parents. These include, *Lumnitzera X rosea*, located in north-eastern Queensland (subregion 3a) (Tomlinson *et al.*, 1978), and, *Sonneratia X* 'merauke', located in the Northern Territory (subregion 2) and southern New Guinea (subregion 3b) (Duke and Jackes, 1987).

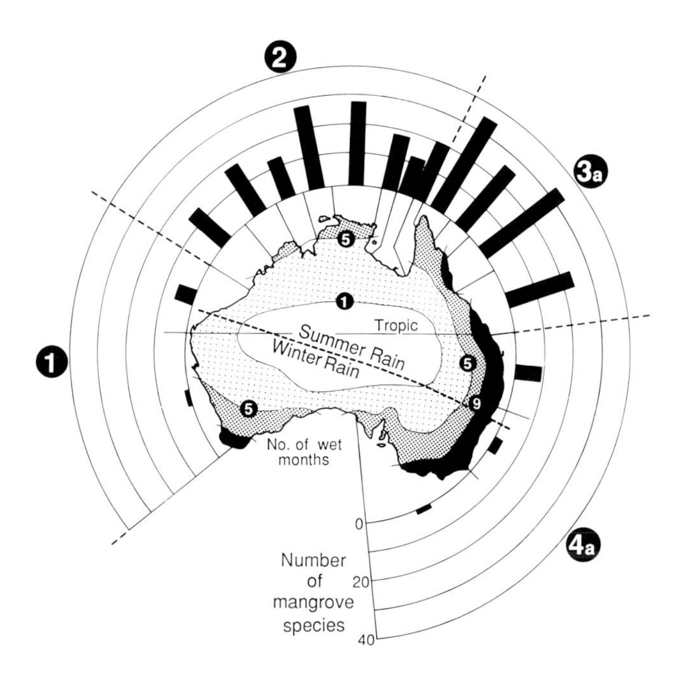

Figure 6. The distribution of mangrove species around the Australian coastline (Figure 5) are characterized by variable numbers of species along alternate coastal sections, shown in this figure. There are two important patterns, namely the latitudinal decline in species numbers in the east (subregions 3a & 4a) and west (subregions 1 & 2), and the correlation in northern tropical areas (subregions 2 & 3a) between higher species numbers and the greater number of wet months, notably those in which precipitation exceeds evaporation by one-third (climatic data taken from The Australian Environment, CSIRO and Melbourne University Press, 1970).

Generally, the greatest concentration of mangrove species in Australasia is found in north-eastern Australia and southern New Guinea (subregion 3), together sharing 45 taxa. The lowest numbers were found in the south-western Pacific (subregion 6), having only six. This range is numerically equivalent to that reported for global regions, but unlike their distribution patterns, there were no comparable major disjunctions. There are also no obvious barriers to the dispersal of propagules within this region, but there are some remarkable discontinuities of certain species.

Around mainland Australia, for example, while all 21 genera are represented, there are eight less species, i.e. 39, compared with populations just to the north in New Guinea. This is possibly due to a latitudinal effect, since most species are found in more equatorial latitudes and there is a progressive reduction in numbers to the south (Figure 6), with nine south of the Tropic of Capricorn (subregion 4). This pattern does not apply further north, however, since New Guinea has 15 species less in the north than in the south, and they are not a subset of those in the south. These forests are not simply reflecting a reduction in habitat, instead they are indicative of a floral discontinuity between the north and south coasts of New Guinea. Some notable examples of north coast taxa include, *Sonneratia caseolaris* and *Avicennia alba*, while south coast taxa include, *S. lanceolata, S. ovata, A. marina* var. *eucalyptifolia, A. officinalis, Camptostemon schultzii, Osbornia octodonta, Bruguiera exaristata* and *Ceriops australis.* Given our limited knowledge of this area, some changes are expected, but the evidence appears convincing that New Guinea marks a fusion boundary between two previously isolated and different mangrove floras.

Another discontinuity is shown by the presence of *Rhizophora samoensis* in the south-western Pacific. This, however, is particularly unusual since no other species has a similar restricted distribution, and it is virtually indistinguishable morphologically from *R. mangle* (Tomlinson, 1986), known only in the AEP. The usual trend is for species to range eastward by varying degrees, resulting in a dramatic reduction of species further east. Apparently, they are unable to disperse throughout the south-western Pacific islands today since those to the east of Tonga and Samoa are without natural stands. The discontinuity of *R. samoensis* therefore suggests an ancient connection with the AEP, possibly across the Pacific to South America via an earlier island archipelago formed during the formation of the Pacific Plate (note, Schlanger and Premoli-Silva, 1981; Schlanger *et al.*, 1981). This putative migration was apparently only one-way, and the distances between the old islands must have been small enough to allow only the dispersal specialist, *Rhizophora,* to make the crossing.

There is a similar reduction in species from south-eastern Australia to New Zealand, where there is only one species, *A. marina* var. *australasica.* Two other varieties of *A. marina* are found in Australia, and these together identify three discontinuities around the coastline of this continent. Hence, south-western Australia (subregion 1) is characterized by var. *marina,* northern Australia (subregions 2 and 3), by var. *eucalyptifolia,* and south-eastern Australia (subregion 4) by var. *australasica.* Identification of these varieties is based on genetic characters, determined by isozyme electrophoresis (Duke, 1988, plus additional unpublished data). This study also discovered that mixing of genes between varieties was apparently uninhibited, and there were zones of overlap several hundred kilometres wide. One zone on the east coast of Australia, between var. *eucalyptifolia* and var. *australasica,* possibly extends to New Caledonia. In addition, var. *marina* on the south-west coast is also found in western Thailand, Malaysia and south-eastern Africa. These findings are significant since this species has relatively short-lived propagules, unable to cross sea distances more than approximately two hundred kilometres, based on buoyancy tests (Steinke, 1975, 1986) and supported by general observations. For example, *A. officinalis*, a congeneric with similar propagules, has not crossed the 100 kilometre distance, separating southern New Guinea from north-eastern Australia. Therefore, considering that *A. marina* would have similar limited distributional abilities, how was it able to achieve such a wide distribution range? And, how have the varietal differences been maintained over such widely located sites? This evidence suggests that these areas were once in closer proximity via connecting mangrove habitats (Hilde *et al.*, 1977), and the Australian continent was divided into two or three major parts early in the evolution of the species.

The presence of *A. integra* in the Northern Territory of Australia, and the allied *A. officinalis* in southern New Guinea, also represents a discontinuity, with neither found in north-eastern Australia. As noted before, this suggests that the dispersal or established range of some common ancestor included both places, but curiously not the north-east coast of Australia where all other Australian species are found.

Table 4. A classification of all mangrove species in Australasia (chiefly Cronquist 1981), including their distribution in six subregions (see Figure 5) as, present o, widely distributed ©, or doubtful ?. In addition, those in parentheses are not known in Australia, notably subregion 3. Note, number superscripts refer to the varieties of *Avicennia marina*, 1. var. *marina*, 2. var. *eucalyptifolia* and 3. var. *australasica* (Duke, 1991a). References include, those cited in Table 3 plus, Percival and Womersley (1975), Semeniuk *et al.* (1978) and Wells (1982, 1983).

Dicot. Sub-class	**Order in Mangrove**	**Family in Mangrove**	**Genus in Mangrove**	**Species in Mangrove**	**Australasian Biogeographic Sub-Regions**					
					1	**2**	**3**	**4**	**5**	**6**
[Division POLYPODIOPHYTA]										
		Pteridaceae	Acrostichum	aureum			(?)b		©	
				speciosum		©	©	o	©	
[Division MAGNOLIOPHYTA - Class Magnoliopsida]										
III.	Plumbaginales	Plumbaginaceae	Aegialitis	annulata	o	©	©	o	o	
IV.	Malvales	Bombacaceae	Camptostemon	schultzii		©	o			
		Sterculiaceae	Heritiera	littoralis			©		©	o
	Ebenales	Ebenaceae	Diospyros	ferrea			o		?	
	Primulales	Myrsinaceae	Aegiceras	corniculatum	o	©	©	©	©	
V.	Fabales	Caesalpiniaceae	Cynometra	iripa			©		o	
	Myrtales	Combretaceae	Lumnitzera	racemosa		©	©	o	o	
				X rosea			o			
				littorea		o	©		©	
		Lythraceae	Pemphis	acidula		©	©		©	
		Myrtaceae	Osbornia	octodonta	o	©	©			
		Sonneratiaceae	Sonneratia	ovata			(o)b			
				alba		©	©		©	
				X gulngai			o		o	
				caseolaris			o		©	
				X 'merauke'		o	(o)b			
				lanceolata		©	©			
	Rhizophorales	Rhizophoraceae	Bruguiera	gymnorrhiza		©	©	o	©	©
				sexangula		o	©		©	
				exaristata	o	©	©			
				hainesii			(o)b			
				parviflora		©	©		©	
				cylindrica			©		o	
			Ceriops	tagal	o	o	©		©	
				decandra		o	©			
				australis		©	©	o	?	
			Rhizophora	samoensis			(o)c			©
				X selala			(o)c			o
				stylosa	©	©	©	©	©	o
				X lamarckii		o	©		o	
				apiculata		©	©		©	
				mucronata			©		o	
	Euphorbiales	Euphorbiaceae	Excoecaria	agallocha		©	©	o	©	©
	Sapindales	Meliaceae	Xylocarpus	granatum		o	©		©	
				mekongensis		©	©			
VI.	Lamiales	Avicenniaceae	Avicennia	marina	©[1]	©[2]	©[2,3]		©[3]	o[2]
				alba			(o)b		©	

Table 4 continued

				rumphiana		(o)b	o
				officinalis		(©)b	o
				integra	©		
	Scrophulariales	Acanthaceae	Acanthus	ebracteatus		o	?
				ilicifolius	©	©	©
		Bignoniaceae	Dolichandrone	spathacea		©	©
	Rubiales	Rubiaceae	Scyphiphora	hydrophyllacea	©	©	
[Division MAGNOLIOPHYTA - Class Liliopsida]							
II.	Arecales	Arecaceae	Nypa	fruticans	o	©	©

There is an additional discontinuity in Australia, seen in the distribution of *Sonneratia caseolaris* and *S. lanceolata* (Duke and Jackes, 1987). These species occupy exactly the same intertidal and estuarine habitats, notably low-intertidal upstream, but they have not been found in the same estuary. In this way, their respective distributions are usually exclusive along contiguous sections of coastline, as exemplified by, *S. caseolaris* on the north coast of New Guinea (subregion 5), *S. lanceolata* on the south coast (subregion 3b), and the Northern Territory coast of Australia (subregion 2). However, on the north-east coast of Australia (subregion 3a) there is some overlap of estuarine sites. This subregion is essentially dominated by *S. caseolaris*, but there are two estuaries within this range which have *S. lanceolata* instead. This overlap in Australia is curious in itself, but their respective common ranges in north-eastern Australia and the Northern Territory represent a significant discontinuity. This situation suggests that an established Australian species, probably *S. lanceolata*, was invaded by a newly introduced one, *S. caseolaris*. This corresponds with the idea of two fused floras, mentioned earlier for the New Guinea north-south discontinuity.

4.5 Environmental Factors and Regional Biogeography in Australia

Not all distributional patterns, however, are related to tectonic events as suggested by the discontinuities described above. Just as the global distribution of mangroves was seen to be influenced by temperature (Figure 1), so their distribution within regions is expected to be equally constrained by various environmental factors. This is important for the idea of species presence indicating suitability of the environment but not the reverse, since the presence of particular species depends firstly on the proximity and dispersal ability of respective propagules. Therefore, where a species exists in isolation of its conspecifics, this might be the result of either, past changes in climate, tectonic events, or both. This would apply particularly for populations distributed along continuous coastlines, or those which may have been joined during periods of lower sea levels. In such instances tectonic changes are expected to be less important. In either case, it would be extremely valuable to know how various environmental factors might influence the distribution of mangrove species.

The mangroves of Australia are distributed around most of the mainland coast, but their concentration is greater in the north, both linearly (Figure 5) and by number of species

(Figure 6). In Figure 6, the four Australasian subregions which make-up Australia are displayed as numerous smaller coastal sections with species numbers totalled for each (Wells 1982, 1983; and personal unpublished data). It is apparent that numbers of species are associated with latitude and rainfall. A latitudinal effect is seen in the north-south decline in species numbers on both west or east coasts. This is believed to reflect the influence of temperature since the cline on the west coast is more extreme, and there are many more species at equivalent latitudes in the east. The greater southward extent of mangrove species on the east coast, however, also might be related to rainfall since the east coast is appreciably wetter than the west (Figure 6). A better defined rainfall effect was suggested by higher numbers of species (ranging up to 36) in wetter tropical areas of northern Australia (subregions 2 and 3a), indicating a possible generalized distribution pattern.

In summary, two major environmental factors, temperature and rainfall, apparently explain the general distribution of mangroves, with low temperatures restricting the latitudinal extent of different species in different ways. Areas of higher rainfall usually have greater numbers of species, therefore in wet equatorial areas, mangrove forests would be expected to have their greatest levels of species diversity.

4.5.1 Species and Environmental Data from Estuaries in Northern Australia

Clearly, these general patterns in species diversity and environmental factors can be investigated further by considering data from a number of estuarine sites. Analysing these basic units of mangrove species distribution, we have learned that localized species diversity is often higher in those estuaries which are longer and/or those which occur in areas of higher rainfall (Bunt *et al.*, 1982a; Smith and Duke, 1987; also see Chapter 5, this volume). The influence of rainfall, therefore, not only comes from rain falling directly on mangrove forests, but also via runoff from the total upstream catchment area.

Table 5. Mean, (standard error), and [range] of mangrove species numbers and environmental parameters, including annual rainfall, estuary length, tidal variation and catchment area, for estuarine sites in two subregions of northern Australia (see Figure 5 for location of subregions).

	West subregion 2	**East subregion 3a**
Species Numbers per estuary	14.1 (0.4) [6-19]	17.0 (0.9) [3-30]
Annual Rainfall (mm)	1198 (24) [500-1500]	1737 (47) [1100-3000]
Estuary Length (km)	31.0 (3.4) [5-130]	9.2 (1.5) [0.2-49]
Tidal Variation (m)	4.02 (0.23) [1.08-7.94]	2.47 (0.09) [1.69-4.32]
Catchment Area (km^2)	5083 (1908) [31-91170]	799 (259) [1-12991]

Four parameters of estuaries are further investigated here, including, annual rainfall, estuary length, tidal variation and catchment area, for 102 sites in northern Australia (42 in subregion 2, and 60 in subregion 3a). Species data were taken from Messel *et al.*, (1979, 1980, 1981) for western sites (subregion 2), and from unpublished data and Duke (1985) for eastern sites (subregion 3a). These are summarized in Table 5, indentifying some of the major differences between subregions. Note particularly, the higher rainfall and greater species numbers in the east, and that western sites have longer estuaries, greater tidal variation, and larger catchment areas. Each parameter was compared to the number of species in respective estuaries, and the results are presented in four plots (Figure 7). Curiously, annual rainfall was not correlated with species numbers in either subregion. Similarly, tidal variation was not correlated with species numbers either, although the combination of data from both subregions suggests a negative correlation, with species numbers higher where tidal variation was lower. Unfortunately, without greater overlap, it is impossible to discern whether this is an environmental effect, or indicative of a floral discontinuity. By contrast, estuary length (log transformation to normalize the data) was correlated with species numbers in both the west ($r = 0.446$, $n = 42$, $P < 0.005$) and the east

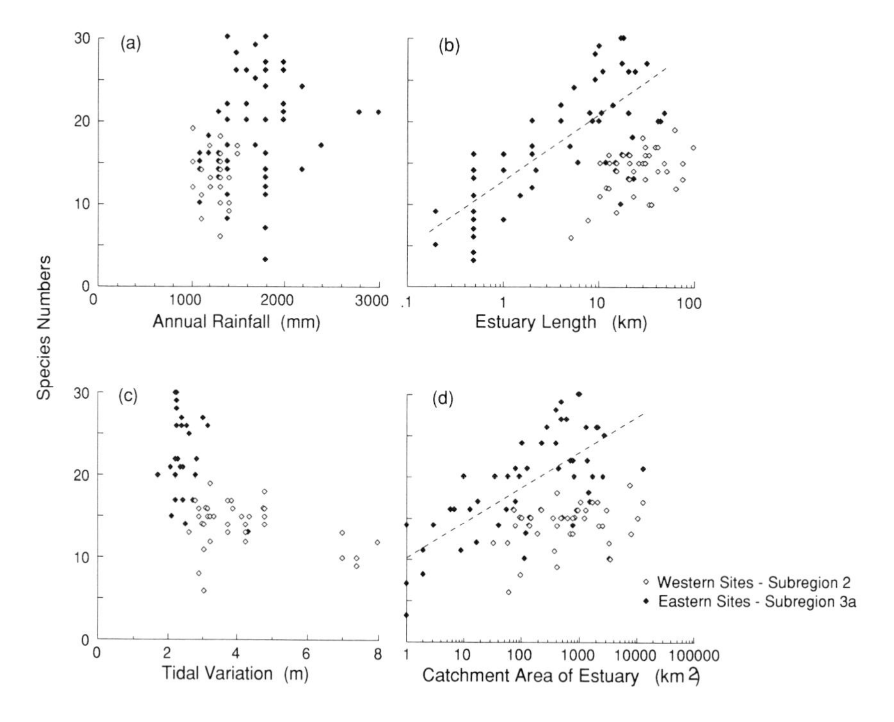

Figure 7. Four plots comparing species numbers of mangroves from 102 estuarine sites in two subregions of northern Australia (2 & 3a) with four environmental factors including, annual rainfall (a), estuary length (b), tidal variation or range (c), and catchment area of the estuary (d). The dashed lines indicate significant correlations ($P<0.005$) for eastern sites (subregion 3a). The only other significant correlation for individual subregion data was with estuary length and species numbers in western sites. Correlations for both sites combined are not presented. Notice that species numbers are highest in larger estuaries.

(r = 0.770, n = 56, P < 0.001). Accordingly, species numbers were higher in longer estuaries. However, eastern sites were characterized by having approximately ten more species than western estuaries of similar length. This also could be indicative of a floral discontinuity, since the data are not following the same trend, but it might also reflect the influence of another factor. Further investigation found eastern sites with lower annual rainfalls had below average numbers of species, placing them closer to western sites. In Figure 7b, nearly 70% of data points below the regression line represent sites with rainfall in the lower 25% of the range. All western sites fit within this same low rainfall category, notably less than or equal to 1500 mm (Table 5). The fourth factor, catchment area was highly correlated with estuary length, but nevertheless, it was only correlated with species numbers in eastern sites (r = 0.723, n = 50, P < 0.001). Multiple factor regressions (two and three factors, excluding catchment area) were no more significant in eastern sites, however, in western sites, the level of significance was improved, comparing estuary length (log transformation) and tidal variation (r = 0.563, n = 42, P < 0.001).

While single and multiple factor regressions did not indicate any direct relationship between species numbers and rainfall, a plot of estuary length and annual rainfall with sites classified by numbers of species (Figure 8a), shows sites with the greatest numbers of species (26-30) restricted to certain estuaries (dotted outline). These were moderate in length (ca. 9

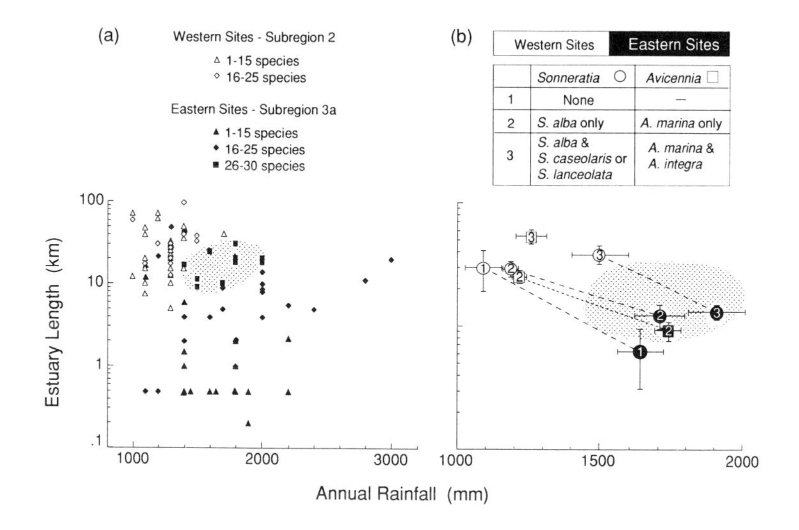

Figure 8. Two plots comparing estuary length and annual rainfall in northern Australian sites. The first (a), has sites grouped by subregion, western (subregion 2) and eastern (subregion 3a), and by three classes of species numbers per site, 1-15, 16-25, and 26-30. Notice that all sites with maximal numbers of species (all from the east) are clustered in estuaries between 9 and 30 km long, and receiving between 1400 and 2000 mm of annual rainfall. The second (b), has means and error bars for sites grouped by presence or absence of *Sonneratia* and *Avicennia* species in either western (subregion 2) or eastern (subregion 3a) sites. The area of maximal species numbers is transcribed for interest. This plot shows sites with certain species have distinctive characteristics, notably higher rainfall and longer estuaries, explaining the trends observed in species numbers comparisons.

to 30 km) and influenced by an intermediate range of annual rainfall (ca. 1400 to 2000 mm). Sites having conditions beyond these combined ranges, either higher or lower, had less species. This suggests that these parameters define some optimal range for maximal species richness. Furthermore, as sites in the west all have less than 20 species, then it is not surprising that they occur mostly outside of this zone of greatest species diversity, defined by estuary length and annual rainfall. Environmental factors, therefore, can explain major differences between east and west mangrove floras.

Although this conclusion is based on numbers of species, it tells us little about the composition of the respective floras, and it would be interesting to know how certain species, or groups of species, are influenced by these same factors. Furthermore, the presence or absence of certain mangrove species may also be indicative of local conditions, notably regarding estuary length and annual rainfall. The three species of *Sonneratia* in northern Australia have very different ranges, and *S. alba* is widespread whilst *S. caseolaris* (only found on the east coast) and *S. lanceolata* (found chiefly on the west coast) have restricted distributions within the range of the first. Another polymorphic genus, *Avicennia*, has a comparable distribution pattern with *A. marina,* a widespread species, and *A. integra* (found only in the west) restricted within the range of the other. Does this difference reflect different environmental factors? To answer this, sites were grouped according to either, absence, presence of the widespread species, or presence of both. The plot (Figure 8b) compares mean estimates (with standard error bars) of estuary length and annual rainfall for these grouped site categories. Notice that groupings reflect the greater influence of rainfall in the east, and larger estuaries in the west. Also, notice that the categories for Sonneratias in each subregion follow the same order, notably with increasing rainfall and estuary length from sites with none, to one, and to two species. Accordingly, sites with no Sonneratias have either smaller estuaries in high rainfall areas, or larger estuaries in low rainfall areas. By comparison, sites with only *S. alba* have even higher rainfall, as seen in the west, or even larger estuaries, as seen in the east. Where rainfall is limited in the west, this appears to be compensated by larger estuaries. Sites with either upstream *Sonneratia, S. caseolaris* or *S. lanceolata,* differ by having a much greater rainfall in both the east and the west. The pattern for *Avicennia* species is similar, but there is at least one important difference. Therefore, while they reflect overall site differences between the east and west, those in the west differ by having *A. integra* associated with larger estuaries rather than higher rainfall. This suggests the absence of *A. integra* in north-eastern Australia might be related to the lack of larger estuaries there, further supporting the idea that environmental factors could explain the difference between mangrove floras in the east and west.

4.5.2 Estuarine Range of Mangrove Species and Salinity in Northern Australia

These observations clearly show the influence of freshwater runoff and estuary size on the composition of mangrove floras, suggesting why particular species grow in certain estuaries and not others. This could reflect either, differing ranges of limiting conditions for establishment and growth, or differing amounts of space, or both. Smith and Duke (1987) suggested that larger estuaries have more species because they have greater amounts of

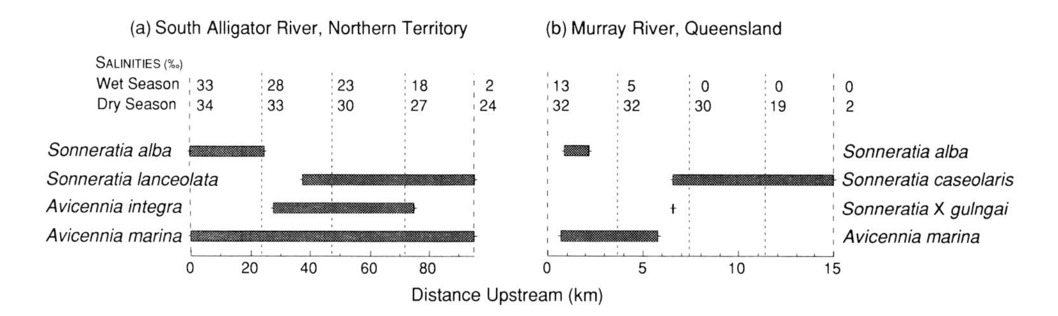

Figure 9. Two bar diagrams display the estuarine ranges of *Sonneratia* and *Avicennia* species in two riverine estuaries, (a) the South Alligator River in the Northern Territory (subregion 2), and (b) the Murray River in north eastern Queensland (subregion 3a). Salinities in wet and dry seasons are included for comparison.

habitat, eluding to the idea of island biogeography where those with greater land area also have more species. But, in the case of estuaries, it is perhaps more complex than simply more available space, because in larger estuaries, there is a much greater range of specialized habitats. These are created upstream along the length of the estuary by the complex interplay of dynamic factors and processes, such as tidal fluctuations and seasonal climatic conditions. Therefore, certain mangroves grow in particular estuaries because the special habitat conditions suitable for their establishment and survival are present. With these qualifications, estuaries can be equated to islands, and it is interesting to speculate on the process of dispersal, where some mangrove species can be isolated on continuous coastlines, as if estuaries were islands. Therefore, in the same way, dispersal success and genetic exchange are dependant primarily on suitable currents and the longevity of propagules in transit, and secondarily on suitable habitats for establishment within an estuary.

This is further influenced by the particular estuarine distribution range of each mangrove species. As noted, all have defined upstream ranges (Table 3) with some located mostly downstream, while others might be found upstream or in intermediate locations. Some may also have very limited ranges in particular portions of the estuary, while others are more widely distributed. These distributions, furthermore, are often structured as an ordered sequence of species upstream, depending on the location (e.g. Bunt *et al.* 1982b). Accordingly, species found only upstream would find it more difficult to get to similar positions in other estuaries, especially if conditions in the original river were rare elsewhere. Two genera, *Sonneratia* and *Avicennia,* include such upstream species with restricted and disjunct distributions, but they also include species which occur downstream and are more widespread. These are characterized by both upstream or downstream limits, shown (Figure 9a) in the South Alligator River (Messel *et al.,* 1979 monograph 4). This very large riverine estuary is located in the species rich area of the Northern Territory (subregion 2). Only *A. marina* extends along almost the entire length of the estuary, while by contrast, the others have certain distinct estuarine limits. *Avicennia integra* (reported as *A. officinalis*) is found midway along this range, identifying both upstream and downstream limits. *Sonneratia alba* appears only to have an upstream limit, which closely matches the downstream limit of *S. lanceolata* (reported as *S. caseolaris*). This latter species extends upstream to the estuarine limit, and sometimes beyond, thus identifying no upper

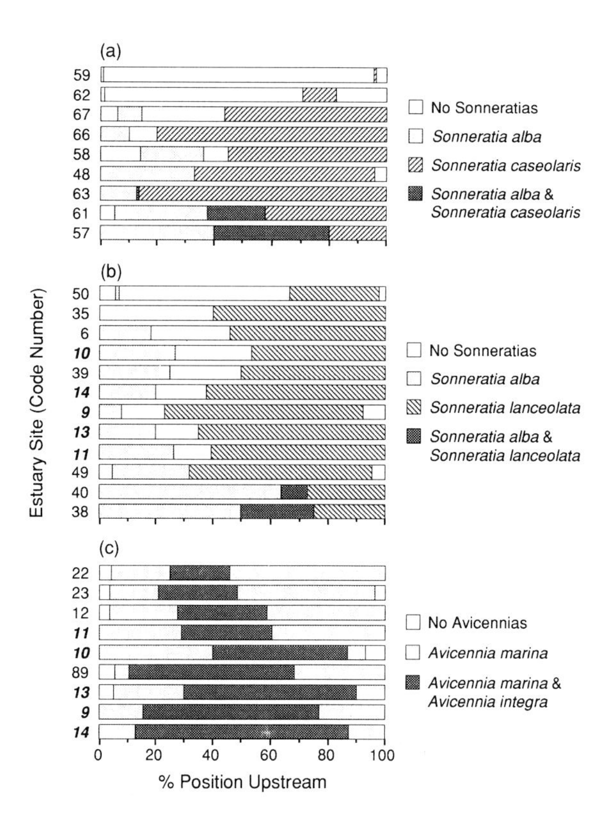

Figure 10. Three bar diagrams display the proportional estuarine ranges of all northern Australian sites with either two *Sonneratia* species, either (a) *S. alba* and *S. caseolaris*, or (b) *S. alba* and *S. lanceolata*, or two *Avicennia* species, (c) *A. marina* and *A. integra*. Sites in (a) and (b) are ordered by firstly diminishing gaps between downstream and upstream taxa, then by increasing overlap. Note that different factors must control the respective upstream and downstream limits of each species. Sites in (c) are ordered by increasing proportional range of *A. integra*.

estuarine limit where salinity profiles approach zero. This is reflected (Figure 9b) in the Murray River (Duke, 1985, and personal observations), located in a species rich area of the east (subregion 3a), in north-eastern Queensland. In this case, *S. caseolaris* replaces *S. lanceolata* in the upstream range, and *A. integra* is absent. *Sonneratia alba* is found in a similar downstream range, and a putative hybrid, *S. X gulngai,* is present at the downstream limit of *S. caseolaris*, but here the range of *A. marina* is different, since in this system it has an upstream limit.

Salinity appears to be important and undoubtably influences estuarine range, but each species appears to be affected in subtly different ways, as suggested by the different ranges of *A. marina* in the examples (Figure.9). This is further shown using all estuaries with congeneric pairs of either *Sonneratia* or *Avicennia* species (Figure 10) where there is considerable variation in the proportional estuarine limits of respective taxa. *Sonneratia* sites (Figure 10 a & b)were ranked such that upstream and downstream limits of the downstream *S. alba*, and those upstream, *S. caseolaris* or *S. lanceolata,* have relative positions ranging

Table 6. Correlation coefficients for comparisons of % estuarine position from the mouth of critical upstream and downstream limits of *Sonneratia alba*, *S. lanceolata* and *Avicennia integra*, with four environmental factors including annual rainfall, tidal variation, estuary length and the % estuarine position of the 28‰ salinity reading in wet and dry seasons. Levels of significance, * P<0.05, ** P<0.01, and *** P<0.005.

		Sonneratia alba	**S. lanceolata**	**Avicennia integra**	
		Upstream	**Downstream**	**Downstream**	**Upstream**
	n=	**23**	**9**	**8**	**9**
Annual Rainfall		-0.070	-0.497	0.277	0.820***
Tidal Variation		0.516**	0.466	-0.241	0.538
Estuary Length		0.046	-0.270	0.238	-0.414
28‰ Salinity	- Wet	0.243	-0.171	0.743*	0.639*
Upstream	- Dry	0.499*	0.436	0.807***	0.046

from ~90% of the estuary separating them (sites 59 & 50), to where they just meet (sites 48 & 49), to ~30% overlap (sites 57 & 38). As noted, the pattern for Avicennias is quite different (Figure 9a), with *A. integra* occupying an intermediate portion of the wide ranging *A. marina*. These sites were ordered (Figure 10c) by the proportional range of *A. integra* in each estuary, showing how the five sites in common (Figure 10 b & c) have different rank orders of sites; comparing the sequences 10, 14, 9, 13, 11, and 11, 10, 13, 9, 14, respectively.

These differences highlight the idea that each species reflects the range of environmental factors in different ways. Several correlations (Table 6) were made by comparing the proportional estuarine limits of three species with the range of environmental factors presented earlier. The upstream limit of *S. alba* is positively correlated with tides, not with rainfall or estuary length, and it is significantly correlated with salinity in the dry season which again suggests the importance of tides. However, the downstream position of *S. lanceolata* was not significantly correlated with any of the factors, although there was a tendency toward a negative correlation with rainfall, indicating that the species might be found further downstream where rainfall was higher. The downstream extent of *A. integra*, however, was significantly correlated with salinity in both wet and dry seasons, and more so in the dry. By contrast, the upstream extent of this species was correlated with wet season salinity and rainfall. Clearly, each species and even their respective upstream and downstream limits appear to be influenced by different factors.

It is generally believed that salinity is largely responsible for the presence and range of particular species, but the above observations provide important qualifications for any such view. All available data on salinities, separated according to wet and dry seasons, for the estuarine limits of the taxa discussed in this treatment are summarized in Figure 11. These are also divided into the two subregions, showing how interpretations can change for different places. Data from southern New Guinea (subregion 3b), included *Avicennia officinalis*, a species occupying similar habitats to *A. integra*, and having a similar estuarine

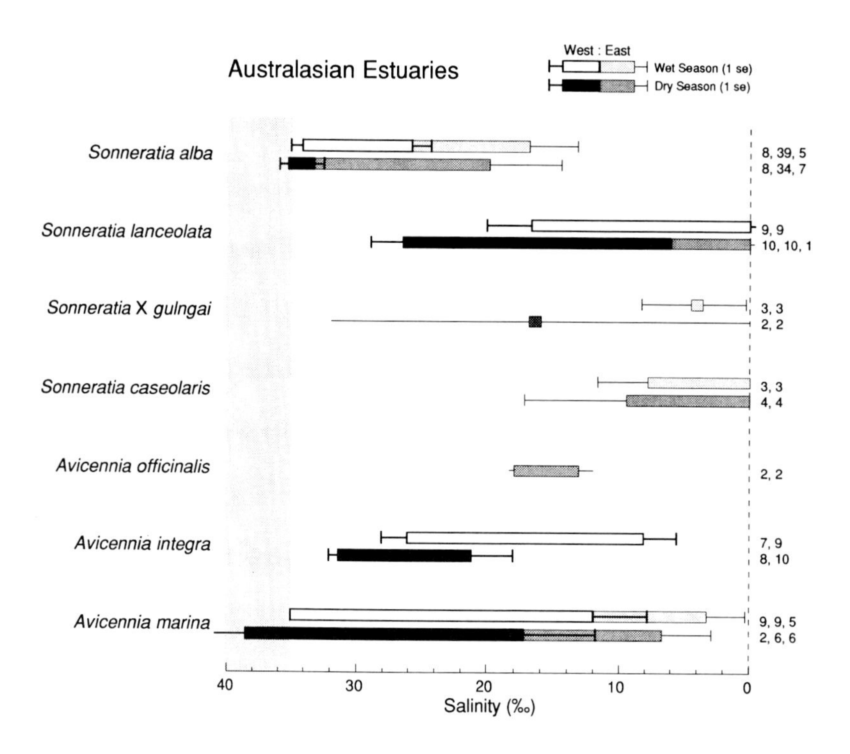

Figure 11. This bar diagram compares wet and dry season salinities associated with the upstream and downstream limits of all *Sonneratia* and *Avicennia* species in northern Australia (subregions 2 & 3a), and *A. officinalis* from southern New Guinea (subregion 3b). Notice that salinities scored in the east are generally lower, notably influencing the upstream limits of *S. alba* and *A. marina*. The shaded salinity range above 35‰ represents those found in 'reversed' rivers, discussed in the text. The numbers on the right are the number of readings used to determine the means, and standard error bars (if appropriate).

range. As all salinities were taken from the mainstream of estuaries, these are likely to be quite different than those closer to the trees, and in the substrate. However, Karim (1991) measured interstitial salinity in upstream limiting sites of *A. marina* in Bangladesh and reported values ranging from 10 to 19‰ as the lower range for this species. These results are comparable with the upstream range of the present study (Figure 11).

There are several estuaries in northern Australia which are considered 'reversed' since salinities increase upstream, becoming hypersaline. Some estuaries were 'reversed' for only the dry season, and others were reportedly hypersaline upstream all year round. The upstream distribution of species in these systems should be considered separately since they obviously demonstrate an extended and opposing range of salinities. For downstream species, like *S. alba,* their upstream distribution in a 'reversed' system will presumably be a function of the upper salinity range for that species, as opposed to a lower range in the normal system. Furthermore, those species with downstream limits in normal, non-'reversed', rivers would not be expected to be found there. Conversely, it is possible that some species might occur exclusively in these systems, preferring the higher range, although none have been identified yet.

Nevertheless, the occurrence of these two types of systems, complicates a generalized analysis of upriver distribution which does not take them into account. This could be done by making two separate analyses, however, in this study the upstream position of the 28‰ salinity level was followed in all estuarine salinity profiles for both wet and dry seasons. In this way, 'reversed' estuaries were excluded, and salinity was found to be a function of several environmental factors (Table 7), notably annual rainfall, tidal variation and catchment area during the wet season, or tidal variation and catchment area during the dry season. Notice that each correlation conforms with expectations, and the one for annual rainfall in the wet season was negative, meaning that lower salinities were found further downstream in sites with higher rainfall. Other factors, including tidal variation and catchment area, were positively correlated with the salinity profile such that for larger river systems and larger tides, higher salinities were found further upstream.

Table 7. Correlation coefficients for comparisons of three environmental factors, namely annual rainfall, tidal variation and catchment area, with the estuarine position, measured (in km) from the mouth, of the 28‰ salinity reading in 'normal' riverine estuaries in northern Australia, for wet and dry seasons. Levels of significance, * $P<0.05$, ** $P<0.01$, and *** $P<0.001$.

	Wet Season n=30	**Dry Season n=24**
Annual Rainfall	-0.377*	0.154
Tidal Variation	0.450*	0.327
Catchment Area (log)	0.489**	0.379
Rain X Tide	0.558**	0.327
Tide X Catchment	0.549**	0.504*
Catchment X Rain	0.516**	0.426*
Rain X Tide X Catchment	0.585***	0.505*

Table 8. Coefficients and constant terms for linear regressions of the estuarine position of 28‰ salinity readings and the critical limiting positions upstream and/or downstream of three mangrove species, *Sonneratia alba*, *S. lanceolata* and *Avicennia integra*. Salinities were recorded for both wet and dry seasons, and distances were all measured (in km) from the estuary mouth. Levels of significance, *$P<0.05$, ** $P<0.02$, and *** $P<0.005$.

Species	**Estuarine Limit**	**Season**	**Coefficient**	**Constant**	**r**	**n**
Sonneratia alba	Upstream	Wet	0.81 (0.18)	3.29 (3.34)	0.694***	22
		Dry	0.59 (0.12)	-2.80 (3.22)	0.768***	18
Sonneratia lanceolata	Downstream	Wet	0.64 (0.20)	10.93 (2.78)	0.764**	8
		Dry	0.49 (0.13)	5.97 (3.19)	0.803***	9
Avicennia integra	Downstream	Wet	0.49 (0.19)	8.44 (2.99)	0.688*	8
		Dry	0.30 (0.10)	5.48 (3.14)	0.729**	9
Avicennia integra	Upstream	Wet	1.33 (0.50)	22.74 (7.84)	0.687**	9
		Dry	0.64 (0.29)	21.17 (9.15)	0.593*	10

The position of the 28‰ salinity readings were then compared with the limiting ranges of Sonneratias and Avicennias; all were highly correlated (Table 8). It is possible that the shape of the salinity profile might change as it moves up and down the estuary, but even so, it is expected that the slope of the linear regression would be close to unity when the species position was proportional, and correlated, with the salinity profile. Furthermore, a positive or negative intercept, or constant term, would confirm whether the species limit was located either further upstream or downstream, respectively. As noted, it was necessary to consider both wet and dry seasons, since it was suspected that one might be more important in limiting the estuarine distribution of each species. The linear regression coefficients suggest that the wet season salinity profile appears to be the most important, because these slopes were closest to unity, for all species and limiting positions. Notice also that the upstream limit of *S. alba* was close to the 28‰ salinity position, shown by the constant term (note the units of this term are km). It appears that the downstream limits of *S. lanceolata* and *A. integra*, were relatively close, just slightly upstream of the 28‰ salinities. By contrast, the upstream limit of *A. integra* was about 20 km further upstream on average, although there is a tendency for this distance to be greater in longer estuaries. In any case, these comparisons clearly show the significance of salinity on the estuarine ranges of Sonneratias and Avicennias with their respective limits either upstream or downstream, or both.

4.6 Conclusions

In this chapter the floristics and biogeography of mangroves have been re-assessed based on recent findings, particularly in Australasia, and more so in northern Australia. Accordingly, evidence has been gleaned from several sources combined with personal records to focus on not only the complement of species found in mangrove forests, but also where these species are distributed, globally and within the Australasian region, and, both between and within estuaries of northern Australia.

The definition of mangrove plants was also revised, and all species in the world, fitting this description, were listed along with their respective regional distribution patterns and common habitat. There are several advantages with this list, but the main one is that it highlights the problems in its construction. These problems occur in two fundamental areas. First, there is difficulty in knowing precisely which species should be included, and second, there are problems with the naming of taxa even though they might be well-known in mangroves. One function of the present work was to identify the major problems and prepare the ground for those who will solve them. This is an on-going process and progress is good, but the increase in new mangrove taxa over the last two decades suggests that more will be included. In Australia, for example, there has been a steady increase in species numbers from around 19 to 39 over the last 23 years (MacNae, 1968; Jones, 1971; Lear and Turner, 1977; Dowling and MacDonald, 1982; Wells, 1983; Duke *et al.*, 1984; and the present study). The inclusion of additional species occurs in two ways, either when unknown taxa are discovered, or when presently known taxa are re-named. In Australia recently, both circumstances applied, notably with the same plant. Accordingly, the first recording of *Avicennia officinalis* in northern Australia (Wells, in Messel *et al.*, 1979 monograph 4), was later corrected and the species re-named as *A. integra* (Duke, 1988). Similarly, the discovery of *Rhizophora mucronata* in north-eastern Australia (Duke and Bunt, 1979) may also require re-classification, along with several others (see Table 3). Despite this, the discovery of additional, previously unknown taxa in Australia is unlikely considering the extensive surveys referred to in this treatment. The days are gone therefore when international botanists felt the need to publish papers like the example entitled, "Do....and...occur in Queensland or the Northern Territory?" (van Steenis, 1968). These reservations, however, do not apply elsewhere in the world.

The surveys in northern Australia provide much information on mangrove species distributions, including the location for not only particular estuarine sites, but also for particular sections of estuaries. This knowledge was used in this treatment to evaluate particular distribution patterns of some mangroves, discovering that this depended on several factors. Hence, while species numbers were generally higher in sites with larger estuaries, they were maximal in moderate-sized estuaries with only moderately high annual rainfall. In this way, species with fixed estuarine ranges, apparently dependant on salinity, are excluded from the estuary by either high or low salinities. Furthermore, salinities are significantly influenced by three main factors, including rainfall, tidal variation and catchment area of the estuary. Accordingly, these factors combine to influence the distribution of most species by restricting their presence in some estuaries and not others, where estuaries are the basic population units, or islands, for mangrove forest communities. Genetic exchange between populations via dispersal of water-buoyant propagules is ultimately controlled by climatic and geological conditions, since their fluctuations determine whether stands remain in contact, or become isolated. That is, despite the possible presence of a continuous coastline, and despite the more generalized distribution of some other mangrove species. All mangroves are not the same, and it is these differences which need to be stressed since the forests formed by them are also different. These forests change according to both the intertidal position, seen as distinct zonal bands (also see Chapter 5, this volume), their location along an estuary, and the estuary itself.

Mangrove forests are ecological, not genetic, entities. Each species within the mangrove flora has followed a different evolutionary pathway and evolved at different times and rates. Some taxa are quite old, possibly arising just after the first angiosperms, around 114mya. Fossil remains are scarce, reflecting lack of detection rather than lack of specimens. It appears that current distribution patterns of some major mangrove genera are remnants of past distributions, since it is not possible for them to disperse freely across some existing geographic barriers.

During relatively recent geological time, there have been vast changes to both the climate and the land. These included the movement and relocation of huge continental fragments by processes described in the theory of continental drift. As the southern supercontinent, Gondwanaland, was breaking up, the angiosperms evolved and diversified (White, 1990). In addition, the climate early on, in the mid Cretaceous, was much warmer and tropical, extending to latitudes as high as 60°. After this time, the extent of tropical latitudes contracted steadily. All these changes must have profoundly influenced the different mangrove ancestors, with older groups being influenced the most. Current hypotheses on the evolution of mangroves apply generously to this theory, with dispersal routes often traced-out on provisional Cretaceous maps (e.g. Chapman, 1975; McCoy and Heck, 1976; Specht, 1981; Mepham, 1983). But, in all these, few authors consider individual genera and species, assuming genetic and ecological uniformity, leading Tomlinson (1986) to conclude that none accounted for the distribution of extant taxa. The hypotheses, therefore, appear too general, and it is believed that "the cart was placed before the horse" by not seeking to know the individuals before describing the evolution of the habitat. For this reason, future hypotheses need to account for the distribution and ecological characteristics of individual taxa, their phylogenetic affinities, their specific fossil record, and their distributional disjunctions and discontinuities. In fact, it might be better to view the latter as particularly valuable evidence, since they represent the residual boundaries of global footprints from which we might track the ancestral paths of all taxa and discover more about their respective origins, including the development of mangrove habitats.

4.7 Acknowledgments

During the preparation of this article, the author was supported jointly by the Minerals Management Service of the U.S. Government (Contract No. 14-12-0001-30393) and the Smithsonian Tropical Research Institute. I am also grateful for the advice and encouragement of several colleagues and friends, near and far, particularly John Bunt, Betsy Jackes, Barry Tomlinson, Al Robertson, Tom Smith, Kevin Boto and Dan Alongi.

4.8 References

Ballment, E.M., Smith III, T.J., and Stoddart, J.A., 1988. Sibling species in the mangrove genus *Ceriops* (Rhizophoraceae), detected using biochemical genetics. *Australian Systematic Botany* **1**:391-7.

Briggs, J.C., 1974. *Marine Zoogeography*, McGraw-Hill, New York, 475pp.

Bunt, J.S., Williams, W.T., and Duke, N.C., 1982a. Mangrove distributions in north-east Australia. *Journal of Biogeography* **9**:111-20.

Bunt, J.S., Williams, W.T., and Clay, H.J., 1982b. River water salinity and the distribution of mangrove species along several rivers in North Queensland. *Australian Journal of Botany* **30**:401-12.

Candolle, A.C. de, 1844. *Prodromus systematis naturalis*, part 8, Paris, Fortin Masson et Soc.

Chapman, V.J., 1976. *Mangrove Vegetation*, Cramer, Vaduz, 447pp.

Chapman, V.J., 1977. Introduction. In: Chapman, V.J., (Ed.), *Wet Coastal Ecosystems: Ecosystems of the World*, pp.1-29, Elsevier, Amsterdam.

Cronquist, A., 1981. *An Integrated System of Classification of Flowering Plants*, Columbia University Press, New York, 1262pp.

Dahlgren, R.M.T., 1988. Rhizophoraceae and Anisophylleaceae: summary statement, relationships. *Annals of the Missouri Botanical Gardens* **75**:1259-77.

Ding Hou, 1960. A review of the genus *Rhizophora. Blumea* **10**:625-34.

Dowling, R.M., and MacDonald, T.J., 1982. Mangrove communities of Queensland. In: Clough, B.F., (Ed.), *Mangrove Ecosystems in Australia: Structure, Function and Management,* pp. 79-93, Australian National University Press, Canberra.

Duke, N.C., 1985. The mangrove genus, *Sonneratia* L.f. (Sonneratiaceae), in Australia, New Guinea and the south-western Pacific region, MSc Thesis, 245pp, Botany Department, James Cook University of North Queensland.

Duke, N.C., 1988. An endemic mangrove species *Avicennia integra* sp. nov. (Avicenniaceae) in Northern Australia. *Australian Systematic Botany* **1**:177-80.

Dukc, N.C., 1988. The mangrove genus *Avicennia* (Avicenniaceae) in Australasia, Ph.D Thesis, 195 pp., Botany Department, James Cook University of North Queensland.

Duke, N.C., 1990. Morphological variation in the mangrove genus *Avicennia* in Australasia: systematic and ecological considerations. *Australian Systematic Botany* **3**:221-39.

Duke, N.C., 1991a. A systematic revision of the mangrove genus *Avicennia* (Avicenniaceae) in Australasia. *Australian Systematic Botany* **4**:299-324.

Duke, N.C., 1991b. *Nypa* the mangrove palm in Central America: introduced or relict? *Principes*, **35**:127-132.

Duke, N.C., and Bunt, J.S., 1979. The genus *Rhizophora* (Rhizophoraceae) in north-eastern Australia. *Australian Journal of Botany* **27**:657-78.

Duke, N.C., and Jackes, B.R., 1987. A systematic revision of the mangrove genus *Sonneratia* (Sonneratiaceae) in Australasia. *Blumea* **32**:277-302.

Duke, N.C., Birch, W.R., and Williams, W.T. 1981. Growth rings and rainfall correlations in a mangrove tree of the genus *Diospyros* (Ebenaceae). *Australian Journal of Botany* **29**:135-42.

Duke, N.C., Bunt, J.S., and Williams, W.T., 1984. Observations on the floral and vegetative phenologies of north-eastern Australian mangroves. *Australian Journal of Botany* **32**:87-99.

Gunn, C.R., and Dennis, J.V., 1976. *World Guide to Tropical Drift Seeds and Fruits*, New York Times Book Co., New York, 240pp.

Heywood, V.H., 1978. *Flowering Plants of the World,* Mayflower Books, New York, 336pp.

Hilde, T.W.C., Uyeda, S., and Kroenke, L., 1977. Evolution of the Western Pacific and its margin. *Tectonophysics* **38**:145-65.

Hutchinson, J., 1973. *The Families of Flowering Plants*, 3rd edn., Oxford University Press, London, 968pp.

Jones, W.T., 1971. The field identification and distribution of mangroves in eastern Australia. *Queensland Naturalist* **20**:35-51.

Juncosa, A.M., and Tomlinson, P.B., 1988a. A historical and taxonomic synopsis of Rhizophoraceae and Anisophylleaceae. *Annals of the Missouri Botanical Gardens* **75**:1278-95.

Juncosa, A.M., and Tomlinson, P.B., 1988b. Systematic comparison and some biological characteristics of Rhizophoraceae and Anisophylleaceae. *Annals of the Missouri Botanical Gardens* **75**:1296-318.

Karim, A., 1991. Environmental factors and the distribution of mangroves in Sunderbans with special reference to *Heritiera fomes* Buch.-Ham., Ph.D Thesis, 230pp, Botany Department, University of Calcutta.

Lear, R., and Turner, T., 1977. *Mangroves of Australia,* University of Queensland Press, St. Lucia, 84pp.

Macnae, W., 1968. A general account of the fauna and flora of mangrove swamps and forests in the Indo-West Pacific region. *Advances in Marine Biology* **6**:73-270.

Mepham, R.H., 1983. Mangrove floras of the southern continents, Part 1, The geographical origin of Indo-Pacific mangrove genera and the development and present status of Australian mangroves. *South African Journal of Botany* **2**:1-8.

Messel, H., Wells, A.G., and Green, W.J., 1979. *Surveys of Tidal River Systems in the Northern Territory of Australia and Their Crocodile Populations, Monographs 2-8*, Pergamon Press, Sydney.

Messel, H., Vorlicek, G.C., Wells, A.G., and Green, W.J., 1980. *Surveys of Tidal River Systems in the Northern Territory of Australia and Their Crocodile Populations, Monographs 9-14,* Pergamon Press, Sydney.

Messel, H., Vorlicek, G.C., Wells, A.G., and Green, W.J., 1981. *Surveys of Tidal River Systems in the Northern Territory of Australia and Their Crocodile Populations, Monographs 15-17*, Pergamon Press, Sydney.

McCoy, E.D., and Heck Jr., K.L., 1976. Biogeography of corals, seagrasses, and mangroves: an alternative to the center of origin concept. *Systematic Zoology* **25**:201-10.

Morley, B.D., and Toelken, H.R., (Eds.), 1983. *Flowering Plants in Australia*, Rigby Publishers, Adelaide, 416pp.

Muller, J., 1981. Fossil pollen records of extant angiosperms. *Botanical Review* **47**:1-142.

Percival, M., and Womersley, J.S., 1975. Floristics and ecology of the mangrove vegetation of Papua New Guinea, Botanical Bulletin, No. 8., Department of Forestry, Division of Botany, Lae, Papua New Guinea, 96pp.

Prahl, H. Von, 1984. Notas sistematicas de las diferentes especies de mangle del Pacifico Colombiano. *Cespedesia* **13**:222-38.

Prahl, H. Von, Cantera, J.R., and Contreras, R., 1990. *Manglares y hombres del Pacifico Colombiano,* Fondo Fen Colombia, Bogata.

Saenger, P., 1982. Morphological, anatomical and reproductive adaptations of Australian mangroves. In: Clough, B.F., (Ed.), *Mangrove Ecosystems in Australia: Structure, Function and Management,* pp. 153-91, Australian National University Press, Canberra.

Saenger, P., Hegerl, E.J., and Davie, J.D.S., 1983. Global Status of Mangrove Ecosystems. *The Environmentalist,* **3**:1-88.

Schlanger, S.O., Jenkyns, H.C., and Premoli-Silva, I., 1981. Volcanism and Vertical Tectonics in the Pacific Basin Related to Global Cretaceous Transgressions. *Earth and Planetary Science Letters* **52**:435-49.

Schlanger, S.O., and Premoli-Silva, I., 1981. Tectonic, Volcanic and Paleogeographic Implications of Redeposited Reef Faunas of Late Cretaceous and Tertiary Age from the Nauru Basin and Line Islands. In: Initial Reports of the Deep Sea Drilling Project **61**:817-27.

Semeniuk, V., Kenneally, K.F., and Wilson, P.G., 1978. *Mangroves of Western Australia,* Western Australian Naturalists Club, Perth, 92pp.

Shaw, H.K. Airy, 1973. *A Dictionary of the Flowering Plants and Ferns by J.C. Willis,* 8th edn. revised by Shaw, H.K. Airy, Cambridge University Press, London, 1245pp.

Smith III, T.J., and Duke, N.C., 1987. Physical determinants of inter-estuary variation in mangrove species richness around the tropical coastline of Australia. *Journal of Biogeography* **14**:9-19.

Soto, R., and Corrales, L.F., 1987. Variación de algunas características foliares de *Avicennia germinans* (L.) L. en un gradiente climático y de salinidad. *Revista de Biologia Tropical* **35**:245-56.

Specht, R.L., 1981. Biogeography of halophytic angiosperms (salt-marsh, mangrove and sea-grass). In: Keast, A., (Ed.), *Ecological Biogeography of Australia*, pp. 575-590, Dr. W. Junk, The Hague.

Sporne, K.R., 1969. The ovule as an indicator of evolutionary status in angiosperms. *New Phytologist* **68**:555-66.

Steenis, C.G.G.J. van, 1968. Do *Sonneratia caseolaris* and *S. ovata* occur in Queensland or the Northern Territory? *North Queensland Naturalist* **35**:3-6.

Steinke, T.D., 1975. Some factors affecting dispersal and establishment of propagules of *Avicennia marina* (Forsk.) Vierh. In: Walsh, G., Snedaker, S., and Teas, H., (Eds.), *Proceedings of the International Symposium on the Biology and Management of Mangroves,* **2**:402-14, University of Florida, Gainesville.

Steinke, T.D., 1986. A preliminary study of buoyancy behaviour in *Avicennia marina* propagules. *South African Journal of Botany* **52**:559-65.

Takhtajan, A., 1980. Outline of the Classification of Flowering Plants (Magnoliophyta). *Botanical Review* **46**:225-359.

Tomlinson, P.B., 1986. *The Botany of Mangroves*, Cambridge University Press, Cambridge, 413pp.

Tomlinson, P.B., 1978. *Rhizophora* in Australasia - some clarification of taxonomy and distribution. *Journal of the Arnold Arboretum* **59**:156-69.

Tomlinson, P.B., Bunt, J.S., Primack, R.B., and Duke, N.C., 1978. *Lumnitzera X rosea* (Combretaceae) - its status and floral morphology. *Journal of the Arnold Arboretum* **59**:342-51.

Watson, J.G., 1928. Mangrove forests of the Malay Peninsula. *Malayan Forest Records* **6**:1-275.

Wells, A.G., 1982. Mangrove Vegetation in Northern Australia. In: Clough, B.F., (Ed.), *Mangrove Ecosystems in Australia: Structure, Function and Management*, pp. 57-78, Australian National University Press, Canberra.

Wells, A.G., 1983. Distribution of mangrove species in Australia. In: Teas, H.J., (Ed.), *The Biology and Ecology of Mangroves*, pp. 57-76, 8, Dr. W. Junk, The Hague.

White, M.E., 1990. *The Flowering of Gondwana*, Princeton University Press, Princeton, pp.256.

Willis, J.C., 1966. *A Dictionary of the Flowering Plants and Ferns* (7th edn. revised by Shaw, H.K. Airy,), Cambridge University Press, Cambridge, pp.1214.

5

Forest Structure[1]

Thomas J. Smith III

5.1 Introduction

Description of a forest's "structure" may include measures of species composition, diversity, stem height, stem diameter, basal area, tree density, and the age-class distributions and spatial distribution patterns of the component species in the forest. The most noted feature of mangrove forest structure is the often conspicuous zonation of tree species into monospecific bands parallel to the shoreline (Snedaker, 1982). Zonation has been a dominant theme in the voluminous literature on mangroves (Frith, 1977; Rollet, 1981), as well as in that on other vegetation types (Whittaker, 1967). Zonation, however, is not the only manifestation of "structure" in mangroves. Lugo and Snedaker (1974) described six types of mangrove forests in Florida, a region with only three mangrove species. Their classification of forests into riverine, overwash, fringe, basin, scrub, and hammock was based on differences in size, productivity, and composition of Florida mangroves which were caused by differing geomorphic and hydrologic factors. Janzen (1985) recently commented on the apparent lack of a distinct understory in mangrove forests. Other forest types (e.g., oak- hickory, pine, and tropical rainforest) have a suite of species that have adapted to life in the lower strata of the forest. These grasses, herbs, and shrubs are absent from most mangroves. The seedling and sapling size-classes are often absent from the understory as well. Several hypotheses have been advanced to account for this missing understory (Janzen, 1985; Corlett, 1986; Lugo, 1985; Snedaker and Lahmann, 1988). These observations all describe aspects of forest structure.

In this review I will concentrate on those factors that influence the species composition within a mangrove estuary; the distribution of the component forest species across the intertidal zone and along the length of tidal rivers; and the measures of physical attributes of the forest such as stem density and height. Particular attention will be given to comparisons between regional and continental differences of mangrove forests and between mangroves

---

[1] This chapter is dedicated to the memory of the late William E. Odum, Professor of Environmental Sciences at the University of Virginia, USA. Bill conducted pioneering research on the ecological dynamics of coastal wetlands, including mangroves. But he was more than a scientist. He was a teacher, a mentor, a colleague, and most of all, a friend. He will be sorely missed and always remembered.

and other forest types. I believe that "mangrove" ecologists have all too often considered mangroves to be very different from other forest ecosystems, and so they have felt that any ecological concepts derived from other ecostystems do not apply to mangroves.

5.2 Mangrove Species Richness

Smith and Duke (1987) examined the influence of tidal amplitude, average temperature (hottest and coldest months), annual rainfall, rainfall variability, runoff, catchment area, frequency of tropical cyclones, and estuary length on the tree species richness of 92 mangrove-dominated estuaries in northern Australia. Their analysis indicated that the suite of environmental parameters which had the greatest influence on species richness were different for estuaries in eastern versus western Australia. In both regions, temperature and tidal amplitude affected species richness. Increasing temperatures led to greater species richness. Species richness decreased with increasing tidal amplitude. In eastern, but not in western, Australian mangroves, the size of the surrounding catchment, the variation in rainfall, and the frequency of tropical cyclones all influenced species richness in the forest. Estuaries which are long and have large catchments tend to have more species than estuaries which are shorter and have small catchments. High interannual rainfall variability and frequent cyclones tended to decrease species richness in eastern Australian mangrove forests but had no effect in western Australia. Interestingly, the amount of freshwater runoff did not appear to be important in either region. This result did not support the hypothesis of Saenger and Moverly (1985), who felt that runoff was a key factor in controlling mangrove species richness.

A factor not considered by Smith and Duke (1987) was freshwater seepage into the intertidal zone. In many mangrove forests worldwide, the highest intertidal zone terminates rather abruptly at a hill or ridge. At this topographic juncture, fresh water often seeps into the intertidal area and reduces salinity. This results in what Seminiuk (1983, 1985) terms "hinterland fringe" mangrove communities. These areas tend to be species rich in comparison to other portions of the forest. Along the dry western and northwestern Australian coast, freshwater seepage is an important determinant of species richness in mangroves. Estuaries that receive freshwater seepage have more mangrove species than those which do not (Semeniuk, 1983). The process also operates along the northeast Queensland coast but does not appear to be as important because of the generally higher rainfall there (This author, pers. obs.). The importance of freshwater seepage to the maintenance of species richness in mangrove forests elsewhere in the world is unknown. However, Thomson (1945) provides evidence that in Sierra Leone, freshwater seepage may actually decrease species richness. Forests with pronounced freshwater seepage contained only *Rhizophora*; those without freshwater seepage had *Rhizophora* and *Avicennia* (Thomson, 1945).

Species richness within an estuary is probably not a result of the dispersal properties of mangrove propagules. The long-distance dispersal ability and propagule longevity of many mangrove species is a paradigm in the literature (Gunn and Dennis 1973, Rabinowitz 1978c). Viable *Rhizophora mangle* propagules are routinely dispersed to the beaches and estuaries of

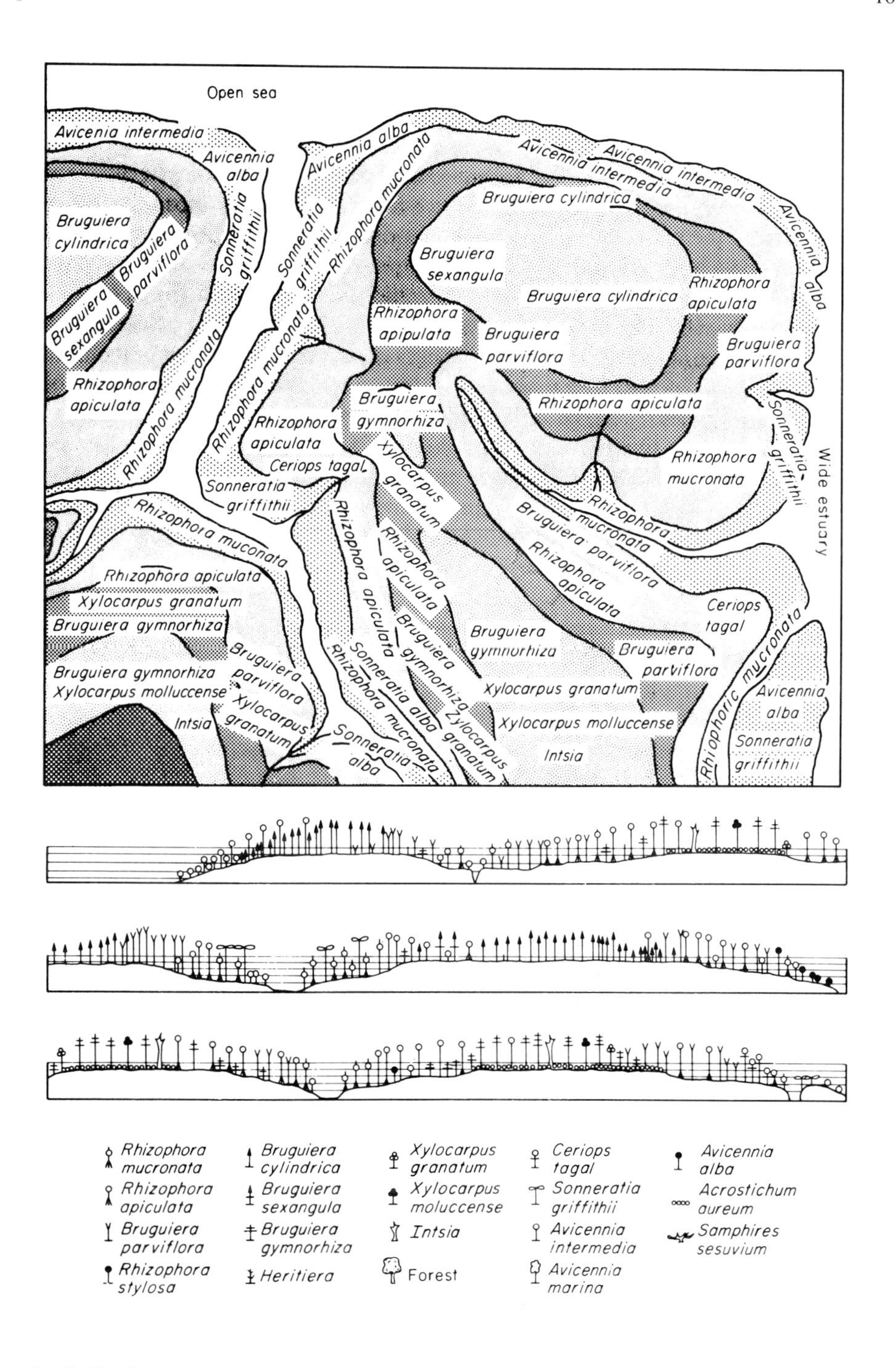

Figure 1. Stylized representation of mangrove zonation in Malaysia (from Watson, 1928). The figure does not represent actual transects through a forest but rather, Watson's synthesis of zonation based on his extensive surveys in peninsular Malaysia.

south Texas (Sherrod and McMillan, 1985; Sherrod *et al.*, 1986). These propagules are from populations several hundred kilometers to the south in Mexico. Although some may become established, regular, severe winter freezes will eventually kill them all (Sherrod *et al.*, 1986). Thus, the local species richness is limited to mangroves that are freeze tolerant. At a biogeographic scale, however, dispersal properties of mangrove propagules may play a very important role in determining the species richness of a region (Tomlinson, 1986; and see Chapter 4, this volume). In an interesting recent paper, Clarke and Myerscough (1991) reported that very few *Avicennia marina* propagules were dispersed away from the parent tree. Most propagules stranded and established near the parent. This is interesting as *A. marina* has the largest geographic range of all mangroves (Duke, 1990).

5.3 Species Zonation Patterns

Zonation patterns have been described for Malaysia (Watson, 1928), east Africa (Walter and Steiner, 1936; Grewe, 1941, Macnae, 1968), Australia (Macnae, 1969; Semeniuk 1980; Elsol and Saenger, 1983), Papua New Guinea (Johnstone, 1983), Indonesia (Van Steenis, 1957; Prawiroatmodjo *et al.*, 1985), India (Sidhu, 1963), Burma (Stamp, 1925), Florida (Bowman, 1917; Davis, 1940), west Africa (Thomson, 1945), and Panama (Rabinowitz, 1978a-c) to name but a few. Typical zonation patterns from the Indo-Pacific region show *Aegiceras, Avicennia,* and *Sonneratia* occupying the lowest intertidal zones; various species of *Bruguiera* and *Rhizophora* in the mid-intertidal areas; and *Heritiera, Xylocarpus,* and numerous other species in the higher intertidal regions (Figure 1). Walter and Steiner (1936) found *Avicennia* in the highest intertidal (Figure 2). Macnae (1969) and Johnstone (1983) have reported "double distributions." These are situations in which a species may be abundant in two different zones of the forest. For example, *Avicennia marina* is often the dominant species in both the lowest and highest intertidal zones and is rare or absent in the middle intertidal (Figure 3). Observations such as these make interpreting zonation patterns difficult. Bunt and Williams (1981) concluded that "generalizations from relatively local observation may be expected to continue as a source of needless debate."

In addition to describing zonation patterns across the intertidal, early workers also noted patterns of distribution along the length of an estuary (e.g., Grewe, 1941; Figure 4). Some species which are common at the seaward mouth of an estuary are not present nearer the fresher, more riverine, headwater regions of the estuary (Bunt *et al.*, 1982a).

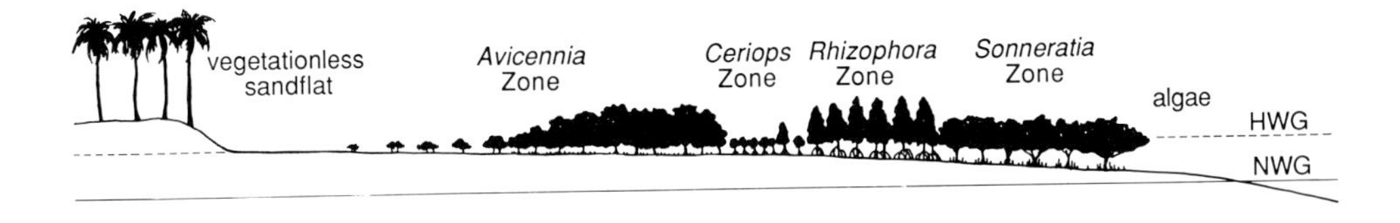

Figure 2. Mangrove zonation in east Africa (after Walter and Steiner, 1936).

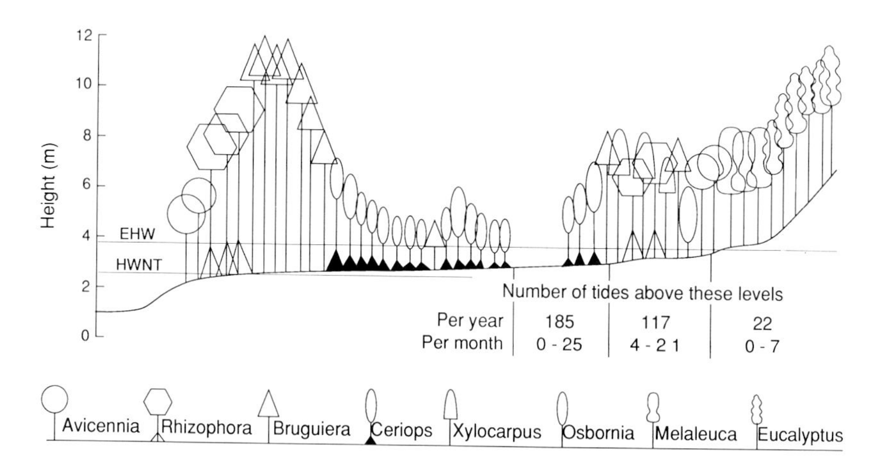

Figure 3. Zonation along a transect through mangroves near Townsville, north Queensland, Australia (from Macnae, 1969). Note "double" distribution of *Rhizophora* and *Avicennia.*

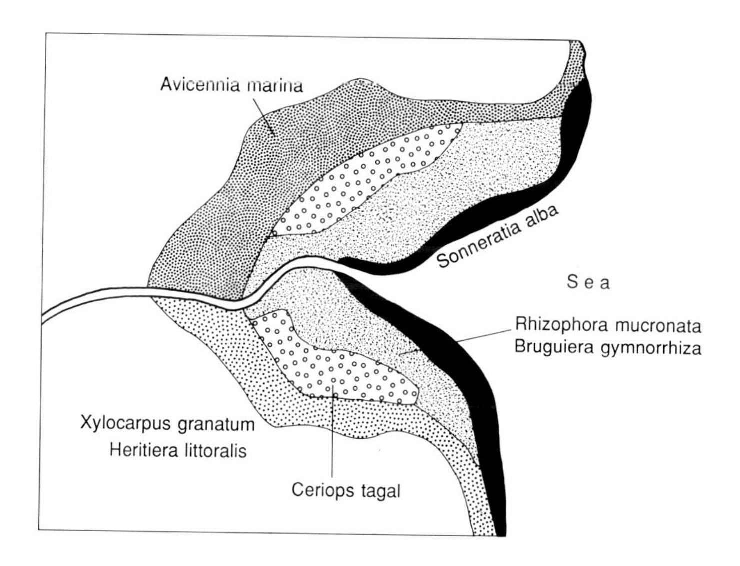

Figure 4. Mangrove zonation in Madagascar showing both across intertidal and upstream - downstream patterns (after Grewe, 1941).

Chapman (1976) provides an extensive synthesis of the early literature describing mangrove forest zonation. These observations led to the paradigm that zonation was the classical feature of mangrove forests and was present in almost all mangroves worldwide (Chapman, 1976).

Not all researchers reported this "classical" view of mangrove zonation, however. Thom (1967) and Thom *et al.,* (1975) describe spatial patterns of occurrence that are not in accordance with the classical view. West (1956) was unable to describe zones in the mangroves of Colombia. In Tanzania, mangroves have been reported as both zoned (Chapman, 1976) and unzoned (Macnae and Kalk, 1962). Bunt and colleagues performed extensive surveys in Australian mangrove forests and reported some 29 species associations ("communities"), based on a species pool of only 35 (Bunt and Williams, 1980, 1981; Bunt *et al.*, 1982b). Classical zonation patterns in Australian forests tend to be the exception rather than the rule.

Several hypotheses have been advanced to date, including the following: 1) plant succession due to land building (Davis, 1940), 2) response to geomorphological factors (Thom 1967; Woodroffe, Chapter 2, this volume), 3) physiological adaptation to gradients across the intertidal zone (Macnae, 1968), 4) differential dispersal of propagules (Rabinowitz, 1978a), 5) differential predation on propagules across the intertidal zone (Smith, 1987a), and 6) interspecific competition (Clarke and Hannon, 1971). Unfortunately, there appear to be many papers which give specific examples of mangrove zonation and few papers which provide rigorous experimental tests of the hypotheses which attempt to explain why mangrove zonation occurs.

5.3.1 Land Building and Plant Succession

The view that zonation in mangroves represents a successional sequence from pioneer colonizers to mature climax forest is by far the most popular and most often invoked mechanism (Snedaker, 1982). The idea is that species which grow in the lowest intertidal zone successfully trap sediments. Over time, the sediment builds up and new mangroves are able to invade and outcompete the colonizers. The process continues until the land is no longer intertidal. The key to this explanation is the ability of the colonizer to trap and hold sediment and thus build land.

Curtiss (1888) makes one of the earliest claims regarding the ability of mangroves to build land, specifically for *Rhizophora mangle* in Florida. Davis (1940) expanded the supposed land-building role of *Rhizophora* into a complete successional sequence in which-seagrasses colonized bare, subtidal areas and trapped sediments to the point that *R. mangle* would colonize the area and trap more sediment; *Rhizophora* would then be replaced by *Avicennia germinans*, which in turn would give way to a tropical forest climax association. Chapman (1976) provided a synthesis of the "zonation represents succession" theory and provided examples from around the world. Although Chapman (1976) himself noted numerous exceptions and variations to this theme, he attributed them to differing local environmental factors.

Criticism of the "zonation represents succession" hypothesis appeared early in the literature. Watson (1928) claimed that mangroves responded to depositional processes rather than causing them. In Watson's, view frequency of tidal inundation, salinity, and soil type were the important determinants of mangrove zonation. Egler (1950) presented evidence that each mangrove zone behaved differently in terms of its development and control. He emphasized the roles of disturbance from fire and hurricanes as factors influencing the distribution of *Rhizophora, Laguncularia,* and *Avicennia* in Florida mangroves. Egler (1950) also stated that the idea of land building by mangroves was "part of arm-chair musings of air-crammed minds."

The idea of succession in mangroves still appears in the literature. Elsol and Saenger (1983) and Johnstone (1983) discuss zonation patterns as successional sequences. Johnstone (1983) does not make the claim that mangroves will succeed to dry land, rather he finds a "climax" in forests dominated by *Bruguiera gymnorrhiza.* Putz and Chan (1986) analyzed over 60 years of forest composition and growth data from permanent plots in the Matang mangroves of Malaysia. They reported increased species diversity of the forest over time, as shade-tolerant species invaded the understory. *B. gymnorrhiza,* one of the most shade-tolerant mangrove species, increased most in abundance (Putz and Chan, 1986). It is obvious that within a mangrove forest, classical ecological succession can and does occur, as it does in every other of the world's forest types (Shugart, 1984). This succession, however, is not the result of mangroves building land.

5.3.2 Geomorphological Influences

It is now widely recognized that mangroves respond to geomorphological changes rather than cause the changes themselves. Detailed studies by Thom, Woodroffe, and coworkers have established that mangrove vegetation is directly dependent on the dynamics of topography. Mangroves do not override abiotic land-building processes (Thom, 1967; Thom *et al.,* 1975; Woodroffe 1981, 1982, see Woodroffe, Chapter 3, this volume). Stoddart (1980) has expanded these results to include mangroves associated with coral reef environments. Detailed analyses of long-term stratigraphic records from peat deposits also show the dependence of mangrove forest development on geomorphic factors, in particular on relatively stable sea level. During periods of rapid sea-level rise, the size and extent of mangrove forests decrease (Woodroffe *et al.,* 1985; Ellison and Stoddart, 1991). Results of these studies, however, leave unanswered questions regarding explanations of zonation based in terms of different biological adaptations of individual species to contrasting physiographic factors within the intertidal environment.

5.3.3 Physico-chemical Gradients and Zonation

A dominant theme in vegetation ecology is the idea that a species adapts physiologically to physico-chemical gradients in the environment (Watt, 1947; Whittaker, 1967). Two flavors of the "gradient" hypothesis exist: the distinct-preference hypothesis and the same-preference hypothesis (Pimm, 1978). The distinct-preference hypothesis (Pimm, 1978) states that each species has its own optimum along the gradient which controls where that species occurs. Because different species have different optima, zonation results. An alternate view is that many species share the same optimum and that other factors (e.g., competition, seed dispersal, predation) cause zonation (Vince and Snow, 1984; Ball, 1988a; Figure 5). The idea of physiological adaptation has been used to explain the zonation patterns observed in a variety of plant communities, including mangroves (Watson, 1928; Macnae, 1968; Clarke and Hannon, 1970). In this section I briefly review the types of data which have been used to make inferences concerning mangrove physiology and forest structure. The data are of three general varieties: field observations, field experiments, and laboratory experiments. First, however, we need to look at the types of physico-chemical gradients which occur in the intertidal zone.

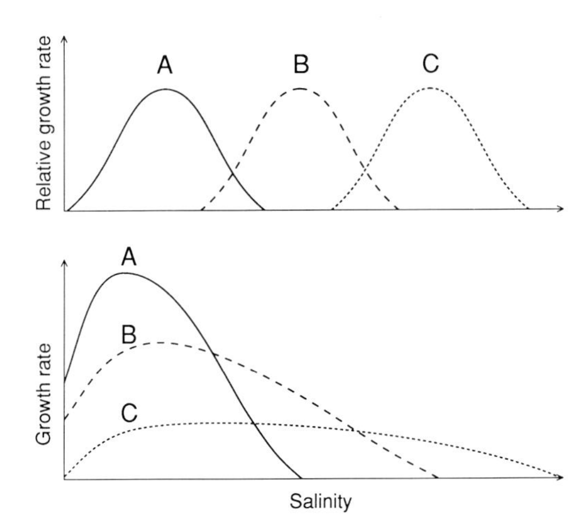

Figure 5. Hypothetical ecological distribution of three mangrove species along a salinity gradient as a result of their physiological response to salinity (from Ball, 1988).

Which gradients occur in the intertidal zone?

Frequency of tidal inundation is the most obvious parameter which varies across the intertidal zone, and is most often cited as a cause of zonation. Low intertidal areas are inundated much more frequently than high intertidal regions. Tidal action, however, introduces two other gradients: soil pore water salinity and soil waterlogging (Giglioli and Thornton, 1965; Clarke and Hannon, 1967). These two gradients may not vary in the same way as frequency of inundation. The pattern of soil pore water salinity across the intertidal zone is influenced by the salinity of the flooding tidal water, rainfall, and freshwater runoff and seepage. Pore water salinity in the lowest intertidal area approximates the salinity of the flooding water: 35‰ near the ocean and <1‰ at the upstream end of riverine mangrove systems (Bunt *et al.*, 1982b). The pattern of salinity variation in the high intertidal zone is complex and usually site specific. In arid regions, pore and surface water salinities in the high intertidal zone may exceed 90‰ (Wells, 1982). High intertidal zone salinities are often lower than that of the flooding water in regions with abundant rainfall, freshwater runoff and / or seepage (Semeniuk, 1983).

Other factors that vary across the intertidal zone include nutrients such as nitrogen and phosphorus (Boto and Wellington 1983, 1984), oxidation-reduction potential (Nickerson and Thibodeau, 1985; McKee *et al.*, 1988), pH (Thornton and Giglioli, 1965), pore water sulfide concentrations (Carlson *et al.*, 1983) and soil texture (Watson, 1928). These gradients are often intercorrelated. For example, fine- grained, clay sediments are often the most highly reduced, whereas, coarser sands are more oxidized (Giglioli and Thornton, 1965). An almost unstudied aspect of mangrove forest ecology is the influence of the fauna on physico-chemical gradients. In particular, organisms which burrow have the potential for modifying chemical and physical factors (see Chapter 3, this volume). For example, burrowing by crabs has been shown to alter the topography and textural properties of mangrove soils (Warren and Underwood, 1986). Recently burrowing has been shown to influence soil nutrient and redox characteristics and hence forest productivity. In the absence of crab burrowing redox potentials increased and forest productivity decreased (Smith *et al.*, 1991).

Field observations of zonation and physico-chemical gradients

Numerous authors have used field surveys to make inferences concerning the tolerances of mangroves to various environmental parameters. Extensive reviews can be found in Chapman (1976) and Hutchings and Saenger (1987). Based on this voluminous literature the conclusion could be drawn that most mangroves have extremely wide tolerances to many factors including: salinity, pH, nutrients, redox potential and soil texture. Data for two factors, salinity and pore water sulfide concentration, illustrate this point.

Wells (1982) conducted extensive field work in the mangroves of northern Australia. He found seedlings of a many species were found growing in soils with salinities over 65‰ (eg. *Avicennia marina, A. officianalis, B. exaristata, Rhizophora stylosa*, see Table 1). There were a few species, however, which appeared to be restricted to soils with salinities less than 40‰ (e.g. *B. sexangula, R. mucronata, Lumnitzera racemosa, Sonneratia caseolaris*, see Table 1). Wells (1982) also examined the texture of the soils and found that virtually every species could be found in sand, silt or clay soils. Jimenez and Soto (1985) reported similar observations for mangroves in the eastern Pacific and Caribbean. Most species where found over an extremely broad range of soil salinities, some at salinities in excess of 90‰ (Table 1). Only three species appeared to be restricted to soil salinities less than 40‰ (*Pelliciera rhizophorae, A. tonduzii, R. racemosa*, see Table 1). These data may indicate two groups of mangroves: one group having extremely broad salinity tolerances and another with slightly narrower tolerances.

Field measurements of soil redox potential and pore water sulfide concentration have been used to speculate about mangrove zonation in the Caribbean. Nickerson and Thibodeau (1985) and Thibodeau and Nickerson (1986) correlated the distribution of *A. germinans* and *R. mangle* to pore water sulfide concentrations. They hypothesized that these species oxidized anaerobic substrates differently, which explained their differing distribution patterns in the field. They found that the substrate around *A. germinans* roots had much less pore water H_2S and was less reduced than substrates away from *Avicennia* roots. They found no differences in these parameters around *Rhizophora* roots. From these results Nickerson and Thibodeau (1985) and Thibodeau and Nickerson (1986) concluded that *Avicennia* is able to exploit lower intertidal, more highly reduced substrates than is *Rhizophora*. McKee *et al.*, (1988) reexamined the issue because many species of mangrove are known to have well developed aerenchyma, which reportedly allow effective gas transport from the air to the rhizosphere (Scholander *et al.*, 1955; Saenger, 1982). McKee *et al.*, (1988) found that redox potential and pore water sulfide concentrations were significantly correlated with the presence of roots of both species. Their results suggested that *Rhizophora* and *Avicennia* were equally capable of exploiting highly reduced sediments as long as their respective pathways for root aeration remained functional. This suggests that soil redox potential might not be a determinant of zonation between these species.

Table 1. Ecological characteristics of various mangrove species. Under "Shade": T = Tolerent, I = Intolerent; "Salinity" (in ‰): MS = Maximum Porewater Salinity measured in the field at sites where the species was growing, OG = salinity for Optimum Growth based on culture studies. ?? = unknown at present time. Data have been extracted from Clarke and Hannon (1970), Clough (1984), Downton (1982), Jimenez (1984, 1990), Jimenez and Soto (1985), Macnae (1968), MacMillan (1971), Putz and Chan (1984), Rabinowitz (1978a), Saenger (1982), Sidhu (1975), Smith (1987a, 1988b), Steinke (1975), Watson (1928), Wells (1982).

Species	**Shade**		**Salinity**	
	T	**I**	**MS**	**OG**
Acanthus ilicifolius	X		65	8
Aegialitis annulata		X	85	??
Aegiceras corniculatum		X	67	8-15
Avicennia marina		X	85	0-30
A. officianalis		X	63	??
A. germinans		X	100	<40
A. bicolor		X	90	??
Bruguiera exaristata		X	72	8
B. gymnorrhiza	X		50	8-34
B. sexangula	X		33	??
B. parviflora		X	66	8-34
B. cylindrica	X		??	??
Ceriops decandra	X		67	15
C. australis		X	80	15-30
C. tagal		X	45	0-15
Rhizophora mangle		X	70	??
R. racemosa		X	40	??
R. apiculata		X	65	8-15
R. stylosa		X	74	8
R. mucronata		X	40	8-33
R. harrisonii		X	65	??
Camptostemon schultzii	X		75	??
Excoecaria agallocha	X		85	??
Lumnitzera littorea		X	35	??
L. racemosa		X	78	??
Laguncularia racemosa		X	90	??
Pelliciera rhizophorae	X		37	??
Sonneratia alba		X	44	??
S. caseolaris		X	35	??
Xylocarpus granatum	X		34	8
X. mekongensis	X		76	8
Heritiera littoralis	X		??	??
Osbornia octodonta	??	??	56	??
Scyphiphora hydrophyllacea	??	??	63	??

Inferences concerning mangrove physiological adaptation based on field observations and measurements must be made with care. Firstly, many published observations are based on the distribution of adult individuals. The physiological tolerances of seedlings may be much narrower than those of adults (Ball, 1988a; McKee *et al.*, 1988). Secondly, the environmental conditions at a site may change over time such that adults persist, but seedlings can no longer become established. Reports of long-term observations from mangrove forests are rare in the literature (but see Putz and Chan, 1986). Thirdly, the entire suite of physico-chemical parameters is rarely measured. Therefore it is virtually impossible to separate the influence of a single factor (e.g. salinity) from other variables (e.g. redox potential). Finally, and most importantly, the data are correlative, and correlation does not prove causation. Correlative data are useful, even necessary, for developing hypotheses. These hypotheses must then be tested with controlled experiments in order to make strong inferences regarding underlying causal mechanisms (Platt, 1947).

Field experiments

Transplant experiments have also been used to examine the question of mangrove zonation. Rabinowitz (1978a), working in Panama, planted seedlings of four species (*R. mangle, A. germinans, P. rhizophorae, and Laguncularia racemosa*) in forests dominated by conspecific adults and forests dominated by each of the other three species. In general, she found that all of the species could grow in any of the "zones" in the forest. In fact, most species grew best away from the "parent" zone of the forest. Recently, Jimenez and Sauter (1991), working in Costa Rica, found that *A. bicolor* grew best in a lower intertidal zone which was dominated by *R. racemosa. R. racemosa* grew best in its home zone, but it did survive and grow in higher intertidal forests dominated by *A. bicolor*. In both of these studies the authors concluded that physiological adaptation could not explain the observed distributional patterns of the species across the intertidal zone and that some other mechanism must be operating.

In Australia, Smith (1987b) planted propagules of four species (*A. marina, B. gymnorrhiza, C. australis* and *R. stylosa*) into both high and low intertidal forests that differed in both frequency of inundation and salinity. The high intertidal forest was characterized by low frequency of inundation and high soil salinity, whereas soil salinity was low and inundation frequency high in the low intertidal forest. All four species had their greatest survival in the high intertidal compared to low intertidal zones. Relative growth rates of *R. stylosa, C. australis* and *A. marina* were also greater in the high intertidal zone. Relative growth rates for *B. gymnorrhiza* did not vary between sites. Although *R. stylosa, B. gymnorrhiza,* and *A. marina* survived best in the high intertidal sites, they reached their greatest natural densities in lower intertidal forests. *C. australis* was the only species that survived and grew best in the zone in which it naturally occurs, but even there it was outperformed by the other three species. The results of this experiment also appear to not support the physiological adaptation hypothesis.

Osborne (1988) examined the influences of salinity (upstream versus downstream river location) and intertidal position (high versus low) on the survival and growth of *Aegiceras corniculatum* seedlings in the Murray River estuary of northeast Queensland. Her results indicated that survival and growth were generally higher in the low intertidal zone in both the upstream, low-salinity (<5‰) portions of the estuary and in the higher salinity estuary mouth (>35‰). Her results are partially supportive of the physiological adaptation hypothesis for *A. corniculatum*. While salinity did not appear to influence where this species grew best, frequency of tidal inundation did.

Criticisms of the field experimental approach include the lack of an adequate control and that most of the experiments are not complete (or even partial) factorial designs. For example, Rabinowitz (1978a) had no controls and in my own study (Smith 1987b) two factors (salinity and frequency of inundation) covaried so it was not possible to separate them. Osborne (1988) selected her field plots such that the factors of interest (salinity and inundation frequency) were not confounded. Additionally, most published field experiments have not reported the cause of death of the seedlings. It is assumed to be the physico-chemical environment in which the seedling has been planted. An examination of Rabinowitz's field notes (on file in the library of the Smithsonian Tropical Research Institute) indicates that many propagules were actually consumed by crabs. Smith (1987b) did record cause of death. In the low intertidal zone 100% of the *Rhizophora* propagules were killed by larvae of a scolytid beetle, a biological vector having nothing to do with a seedling's physiological tolerance.

Laboratory experiments

Laboratory culture studies provide the best data with which to examine the tolerance of mangroves to various physico-chemical parameters. Salinity has been well studied. Clarke and Hannon (1970) found that *Avicennia marina* and *Aegiceras corniculatum* seedlings survived and grew at salinities from 0-35‰, but that maximum growth occurred between 7-14‰. Downton (1982) reported a larger optimal growth range for *A. marina* of 3-20‰. Clough (1984) tested the hypothesis that *A. marina* had a broader salinity tolerance than *R. stylosa*. Both species had growth optima at 9‰. Biomass accumulation in *R. stylosa* fell sharply at salinities over 18‰, whereas *A. marina* showed extended growth responses up to 26‰. Both of these species had the least growth at 0 and 35‰ (Clough, 1984). Clough (1984) attributed the broader growth response of *A. marina* to its ability to excrete salt via salt glands in its leaves. This may account for Bunt *et al.*'s (1982a) observation that *R. stylosa* was restricted to river mouth situations, but *A. marina* was likely to be encountered almost anywhere along an estuary.

Smith (1988b) made detailed comparisons between the observed seedling distributions of *Ceriops tagal* and *C. australis* (see Ballment *et al.*, 1988) in the field and their survival and growth performance along a laboratory salinity gradient. The salinity at which seedlings of both species reached their maximum abundance in the field did not correspond to the laboratory salinity at which maximal growth or survival were measured. Both species grew best at 15‰ in the laboratory. In the field, seedlings of *C. tagal* were most abundant between 20-35‰, whereas *C. australis* reached greatest abundance between 50-60‰.

Ball and colleagues have conducted a series of elegant laboratory studies on carbon gain and water use efficiency of several mangrove species in Australia (Ball and Cowan, 1984; Ball, 1988b; Ball *et al.*, 1988). Ball (1988b) reported that *Aegiceras corniculatum* had a less conservative water use strategy than did *A. marina.* The growth rate of *Aegiceras* was high at low salinities and dropped rapidly as salinity increased. *Avicennia* however, had a lower growth rate, but one that did not drop sharply as salinity increased (Ball, 1988b). She interpreted this to account for the dominance of *Aegiceras* in low salinity areas where it purportedly would be able to outcompete the slower growing *Avicennia. Avicennia* in turn would dominate higher salinity areas because *Aegiceras* simple can't tolerate high salinities. In a second study, Ball *et al.*, (1988) showed that water use efficiency was related to salinity tolerance and leaf size in several mangroves. *Ceriops tagal* var. *australis* had the most conservative water use, the smallest leaves and was most salinity tolerant. *Bruguiera gymnorrhiza* had larger leaves, was least efficient at water use and was least salinity tolerant (Ball *et al.*, 1988). These results also support Ball's view that zonation along salinity gradients is a result of differing water use efficiencies between species.

A problem with the above studies is that they examined a single potential causal agent individually. As noted earlier, there are a number of factors which vary across the intertidal zone. Controlled experiments describing the growth responses of mangroves to two or more factors simultaneously are virtually non-existent in the mangrove literature. The possibility of interactive effects between variables cannot be discounted.

In fact, McMillan (1975) has clearly demonstrated this phenomena for two mangroves. He found that the salinity tolerances of *A. germinans* and *L. racemosa* were highly modified by soil texture. Seedlings grown in hypersaline conditions in sand failed to survive. Seedlings grown in soil composed of 90% sand and 10% clay had 100% survival in hypersaline conditions, but showed some leaf discoloration. At 75% sand and 25% clay there was 100% survival with no observable effect on the leaves (McMillan, 1975).

Reconciling experimental results with field observations

The multi-factored, intercorrelated nature of the environmental gradients found in mangrove forests makes the deduction of causal agents from field observations impossible. Furthermore, the extrapolation of single factor laboratory experiments (e.g. Ball, 1988b), no matter how well controlled or elegantly performed, to the multiple factor field situation is tenuous at best. This is because the plant's physiological response to one factor often varies depending on the level of other factors present in the environment (e.g. McMillan, 1975). Multi-factor, controlled experiments are necessary to fully elucidate the physiological mechanisms involved in zonation. In particular the interaction of salinity, soil texture, and sediment redox potential on seedling establishment and growth deserves study (McKee and Mendelssohn, 1987; McKee *et al.*, 1988).

The studies to date clearly demonstrate that many mangroves can grow over the broad range of conditions found across the intertidal zone (Table 1). Data relating species distributions to soil salinities suggest that two groups of mangroves exist (Table 1). The first has very broad tolerances and can grow and survive in salinities two to three times that of seawater. The second group appears to be restricted to salinities less than 40‰. This latter group is composed of species that have predominately upstream distributions in river dominated estuaries (e.g., *R. mucronata*) or those restricted to geographic areas of abundant rainfall (e.g., *P. rhizophorae*). In this regard it seems that some adaptation to salinity gradients may have occurred which influences distributions within and between estuaries. This view is supported by very limited experimental results for less than 10 mangrove species (Ball, 1988b; Ball *et al.*, 1988). Much more data for other physico-chemical factors (e.g., soils, nutrients, redox potential) are required before the physiological adaptation hypothesis can be fully tested as an underlying cause of mangrove zonation patterns.

5.3.4 Propagule Dispersal and Zonation

Rabinowitz (1978b) hypothesized that zonation in Panamanian mangrove forests was controlled by the influence of tidal action on mangrove propagules. She observed that the species were distributed from the low to high intertidal zone in a manner inversely related to the size of their propagules (Rabinowitz, 1978b). *Avicennia* and *Laguncularia* were restricted to high intertidal zones because they had small propagules that high tides would carry the farthest inland. Large propagules, such as those of *Rhizophora* and *Pelliciera*, would become snagged and not get carried into higher intertidal areas. Thus, tidal action "sorted" the propagules across the intertidal inversely according to their size.

Rabinowitz (1978c) also attempted to relate zonation to the dispersal properties of mangrove propagules, such as floating and rooting time. Her experiments indicated that *Avicennia* and *Laguncularia* required 5-7 days to take root in mangrove soils, whereas *Rhizophora* and *Pelliciera* needed 11-15 days to become rooted. Based on these results, one would expect the species with smaller propagules to be more abundant in the lower intertidal zone because that zone experiences periods of inundation at shorter intervals. The high intertidal, with long periods between inundations, should be favorable to all species. The pattern of zonation reported by Rabinowitz (1978a-c) is exactly the opposite of what would be expected based on her own results.

The "tidal sorting" hypothesis has recently been resurrected by Jimenez and Sauter (1991) to explain the zonation of *R. racemosa* and *A. bicolor* in Costa Rica. They interpreted the results of their reciprocal transplant studies as being supportive of tidal sorting. They observed that the high intertidal *A. bicolor* dominated forest they studied was being invaded by *R. racemosa*. They attributed this to a rise in sea level which would permit more high tides to penetrate the forest.

Observations of species distributions in Australia and elsewhere, however, indicate clearly that tidal sorting is not a mechanism which influences zonation patterns. Species in the genus *Sonneratia* routinely colonize the lowest intertidal zone (Watson, 1928; Duke 1984). The seeds of *Sonneratia* are only some 10-15mm in length, which is small in comparison to most other mangroves (Tomlinson, 1986). In discussing *Sonneratia*, Rabinowitz (1978b) mistakenly referred to the entire seed capsule (which may contain >500 seeds) as the unit of dispersal. The capsule sinks very quickly and then releases individual seeds, which are then dispersed. The genera *Aegiceras* and *Avicennia* also have small propagules and are typically abundant in low intertidal areas (Watson, 1928; Bunt and Williams, 1981; Wells, 1982). They are also common in the highest intertidal areas (Wells, 1982; Johnstone, 1983; Smith, 1987c; Osborne and Smith, 1990). Saenger (1982) provides data on seedling recruitment in mangrove forests at Port Curtis, on the central Queensland coast of Australia. He found that seedlings of all species were found in all plots. *Rhizophora stylosa*, which had the largest propagules, was found across the entire intertidal gradient. It is obvious that tidal action delivers propagules of all species to all portions of the intertidal zone. The question is not so much does dispersal take place?, as much as it is, which factors regulate post-dispersal establishment, survival, and growth?

5.3.5 Seed Predation and Forest Structure

Predation on seeds has been recognized as an important process in a variety of ecosystems (Janzen, 1971; Whelan *et al.*, 1990). Watson (1928) and Noakes (1955) commented on the role of crabs as consumers of mangrove propagules, particularly in the managed forests of west Malaysia. (Because most mangroves are viviparous the unit of dispersal is a propagule, not a true seed.) Watson (1928) stated, "The most serious enemies to mangroves are crabs" and "It is doubtful whether these pests do much damage under natural conditions, but they can, and do, cause great trouble in plantations." Noakes (1955) claimed that "...crabs are a major pest and may entirely prevent regeneration or planting by their attacks on seedlings." He went on to say, "... nothing is known of their effect on natural regeneration, the presence of crabs being no proof that it is likely to fail." The crabs to which Watson and Noakes referred belong to the family Grapsidae. This group is a ubiquitous feature of mangrove forests, especially in the Indo-Pacific region. Crabs are the dominant macrofauna of mangrove forest soils in terms of both numbers (Jones, 1984) and biomass (Golley *et al.*, 1962).

Recent experimental evidence has revealed that consumption of mangrove propagules by grapsid crabs greatly effects natural regeneration and influences the distribution of certain species across the intertidal zone. Smith and colleagues (Smith, 1987a,c, 1988a; Smith *et al.*, 1989; Osborne and Smith, 1990; Smith and Duke, In review) conducted a series of experiments in which mangrove propagules were tethered in the forest and then the amount of consumption was determined over time. The initial experiments were conducted in northeastern Queensland, Australia. For *A. marina*, *R. stylosa*, *B. gymnorrhiza*, and *B. exaristata* there appeared to be an inverse relationship between the dominance of the species in the canopy and the amount of predation on its propagules (Figure 6, Smith 1987a). This relationship was not found for *C. australis*, however. Caging experiments were used to study the establishment and growth of *A. marina* in middle intertidal forests (Smith 1987b). *A. marina* is usually absent from this region of the intertidal zone (Macnae, 1969; Johnstone, 1983; see Figure 3). The results indicated that when protected from crabs, *A. marina* propagules survived and grew. The conclusion was that virtually 100% of the *A. marina* propagules that were

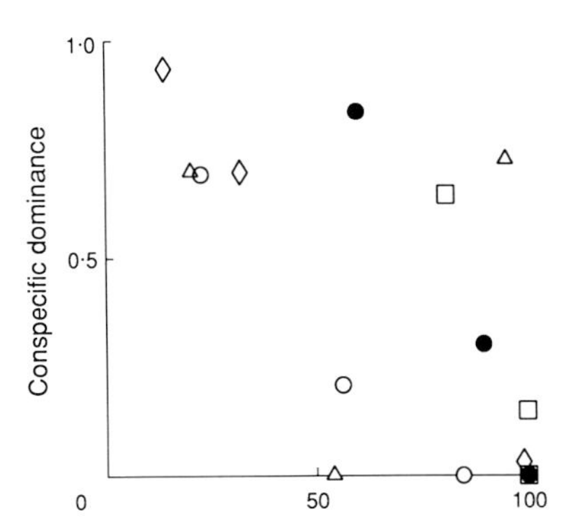

Figure 6. Relationship between conspecific dominance and cumulative amount of predation on five mangrove species from north Queensland, Australia (from Smith, 1987a, reproduced with permission from *Ecology*). Square = *Avicennia marina*, Solid circle = *Bruguiera exaristata*, Open circle = *B. gymnorrhiza*, Triangle = *Ceriops australis*, Diamond = *Rhizophora stylosa*.

dispersed into middle intertidal forests were consumed by crabs; hence, seed predation was an important determinant of the forest's species composition and structure (Smith, 1987b).

Subsequent studies indicated that seed predation was important over a much larger geographic region than northeast Queensland. Data from Malaysia and Florida revealed high levels of predation on the propagules of *A. officianalis* and *A. alba* in Malaysia and on *A. germinans* in Florida (Smith *et al.*, 1989). For all three species, predation was higher where the species was absent from the canopy, and it was lower in forests where conspecific adults were present. For *Rhizophora* and *Brugiuera,* however, equivocal results were obtained. In Malaysia, results for *B. cylindrica* supported the predation hypothesis, whereas results from *B. gymnorrhiza* did not. No predation on *R. mangle* in Florida was observed, but in Panama more *R. mangle* were consumed in a forest where the species was present in the canopy than were consumed in a forest where it was absent (Smith *et al.*, 1989).

More extensive propagule predation experiments have now been conducted in Panama. These studies utilized some of the same forests that Rabinowitz used some 15 years ago and were carried out along both the Pacific and Caribbean coasts (Smith and Duke, In Review). These new results indicate that predation on propagules may effectively preclude the establishment of *A. germinans* and *Laguncularia racemosa* in forests dominated by *R. mangle* and *Pelliciera rhizophorae*. However, the reverse is not true. The amount of predation on *R. mangle* and *P. rhizophorae* propagules in forests dominated by *Avicennia* and *Laguncularia* was not high. Therefore, it seems that predation is not sufficient to account for all of the species distribution patterns observed in Panamanian forests, however, it does account for some (Smith and Duke, In Review).

Predation on propagules has also been proposed as an influence on succession in north Queensland mangrove forests (Smith, 1988a). For example, no *A. marina* saplings were observed in a forest in which the canopy size-class was dominated by this species. The sapling size-class was composed of *B. gymnorrhiza, B. exaristata,* and *C. australis.*

Predation studies showed that >95% of the *A. marina* propagules were consumed in this forest, but <25% of the propagules of the other species were eaten (Smith 1988a). The question arises, how did this forest become dominated by *Avicennia?* Did the predators move into the forest after *Avicennia* became dominant? Long-term studies of the crab populations and their food sources are needed (Whelan *et al.*, 1990).

Crabs are not the only consumers of mangrove propagules. Robertson *et al.*, (1990) have recently shown that insects attack and kill a substantial number of the seeds and propagules of some mangroves. *Heritiera littoralis, Xylocarpus granatum,* and *X. australasicus* all have seeds with hard pericarps that are highly resistant to attack by crabs. More than 55% of the seeds of these species were attacked by insect predators (Robertson *et al.*, 1990). Growth and survival of insect damaged and non-damaged control seeds from seven mangrove species were compared by these authors. Insect attack reduced survival and growth in *X. granatum* and *X. australasicus. B. parviflora* had decreased survival but no differences in growth. *A. marina* and *B. exaristata* had no differences in survival, but insect damage resulted in decreased growth. *R. stylosa* and *B. gymnorrhiza* showed no differences in survival or growth between control and insect damaged propagules. Robertson *et al.*, (1990) concluded that insects are a major determinant of seed survival and possibly of seedling distribution for these north Queensland mangrove species.

The role of insects as seed predators in mangrove forests elsewhere in the world is equivocal. In Florida and the Caribbean, conflicting observations have been published. Onuf *et al.*, (1977) found that infestations of a scolytid beetle in *R. mangle* propagules significantly reduced their growth and survival. In Panama, Rabinowitz (1977) found no effect from insect borers on the propagules of *R. harisonii.* Detailed experimental analyses appear to be lacking.

Seed predation studies have also revealed both local- and biogeographic-scale patterns in the process. Consumption of propagules appears to be least in the lowest intertidal zone and increases to maximum amounts in the high intertidal zone (Smith, 1988a; Smith *et al*,. 1989; Osborne and Smith, 1990; Smith and Duke, In review). Grapsid crab populations tend to be greatest in high intertidal areas (Frusher *et al.*, unpub. data). Additionally, there is often a marked zonation in the crab fauna both across the intertidal zone and upstream - downstream along the length of the estuary (Verwey, 1930; Tweedie, 1950; Snelling, 1959; Berry, 1964; 1972; Hartnoll, 1965, 1973, 1975; Barnes, 1967; Warner, 1969; Sasekumar, 1974; Icely and Jones, 1978; Jones, 1984). An understanding of what determines crab zonation is almost totally lacking at this time. Preliminary observations from northeastern Queensland suggest that salinity is not as important as soil textural properties such as organic matter content and percentages of sand, silt, and clay (Frusher *et al.*, unpub. data; Smith *et al.*, unpub data).

Biogeographic patterns have also been noted in the consumption of mangrove propagules (Smith and Duke, In review). Rates of predation are highest in the Indo-Pacific region, decrease towards the east across the Pacific Ocean to Panama, and are least in the western Atlantic. A latitudinal gradient also exists from Panama northward to Florida (unfortunately, the potential of latitudinal gradients in predation has not been addressed in Australia). Similar patterns have been observed in a number of other tropical marine communities. In coral reef ecosystems, higher rates of predation by fish on invertebrates were observed in the Pacific than in the Atlantic (Bakus, 1966, 1969). Palmer (1978) observed higher predation on molluscs in the eastern Pacific than in the western Atlantic. The scarcity and lower profile of algae on eastern Pacific reefs relative to that on Caribbean reef systems was attributed by both Earle (1972) and Glynn (1972) to higher levels of predation. Vermeij (1976, 1978) provides extensive data, both experimental and observational, that indicates higher levels of predation in the Indo-Pacific than in the Caribbean. He showed that gastropods of the Indo-Pacific were much more highly evolved in their predator defenses than were gastropods in the western Atlantic.

It is interesting that the predator guilds change across this same broad region. The grapsid crabs are most diverse in the Indo-Pacific, with diversity steadily decreasing eastward across the Pacific to the western Atlantic (Jones, 1984). The grapsid fauna changes latitudinally as well (Jones, 1984). Only five species are found in the mangroves of southwest Florida (Smith, pers. obs.). Unfortunately, accurate measures of abundance and/or biomass of the crab fauna in mangroves have not been made. This author's personal experience indicates that both biomass and abundance follow the same pattern as diversity. In both Australia and Malaysia the grapsid crabs composed >95% of the predators on propagules. In south Florida, however, they accounted for <6% of propagule consumption. The snails *Melampus coeffeus* and *Cerithidea scalariformis* consumed >70% of the propagules in Florida (Smith *et al.*, 1989). These predators are only capable of consuming *Avicennia* and possibly *Laguncularia*, which was not tested (Smith *et al.*, 1989).

5.3.6 Competition and Forest Structure

Competition has been studied in a variety of wetland plant communities (e.g., Grace and Wetzel, 1981; Silander and Antonivics, 1982), but few studies have examined the role of competitive interactions in mangrove forests. Ball (1980) examined the colonization of high intertidal habitats in south Florida by *R. mangle* and *L. racemosa*. Based on historical aerial photographs and measurements of living and dead tree densities and the densities of saplings and seedlings, she inferred that *Laguncularia* was being replaced by *Rhizophora*. Competition was the mechanism invoked by Ball (1980) to account for the replacement. Unfortunately, Ball's study was observational, not experimental, so other possible alternatives (e.g., seed predation or changing environmental conditions) for the species replacement were not examined.

Smith (1988b) tested for possible competitive interactions between *C. tagal* and *C. australis* along an experimental salinity gradient. Seedlings were grown in mono- and polycultures at salinities from 0-60‰. *C. tagal* grew better than *C. australis* did at lower salinities, whereas the reverse was true at high salinities. Competition was gauged by comparing the reduction in growth of each species in the presence of the other to the growth of that species alone. Growth of *C. tagal* was reduced less at 0 and 15‰ than was *C. australis* at all densities. The effect of *C. tagal* on *C. australis* was some two to four times greater than *C. australis's* effect on *C. tagal* (Smith 1988b). For salinities >45‰, however,

Table 2. Ecological characteristics of pioneer- and mature-phase terrestrial forest communities with mangrove forest species and communities. Modified from Tomlinson (1986).

Specific Characters	**Pioneer Phase**	**Mature Phase**	**Mangroves**
Propagule size	Small	Large	Variable
Propagule number	Numerous	Few	Numerous
Propagule production	Continuous	Discontinuous	Continuous
Propagule dormancy and viability	Long	Short	Long
Dispersal agent	Often abiotic (e.g. wind)	Usually biotic (e.g. birds)	Always abiotic (water)
Dispersibility	Wide	Limited	Wide
Seedlings dependent on	Light-demanding, dependent on seed seed reserves	Not light demanding, and many dependent reserves	Light demanding on seed reserves
Reproductive age	Early	Late	Most early
Geographic range	Broad	Narrow	Variable
Life span	Short	Long	Vivparous - long Nonvivparous - short
Leaf palatability	High	Low	Most low
Wood	Soft, light	Heavy, dense,	Most heavy, dense
Crown shape	Uniform	Varied	Uniform
Competitiveness	For light	For many resources	Mainly for light
Pollinators	Not specific	Highly specific	Rarely specific
Flowering period	Prolonged	Short	Prolonged
Breeding mechanism	Usually inbreeding	Usually outbreeding	Inbreeding favored
Community Characters			
Species richness	Poor	Rich	Poor
Stratification	Few or no strata	Many strata	Few or no strata
Size distribution	Even-sized	Uneven-sized	Mainly even-sized
Large stems	Absent	Present	Present only in old, undisturbed stands
Undergrowth	Dense	Sparse	Usually absent
Climbers	Few	Many	Few
Epiphytes	Few	Many	Few

this result was strongly reversed, suggesting that *C. australis* was the superior competitor at higher salinities. In the field, however, it was observed that both species were shifted to salinities higher than their growth optima salinities in the laboratory. Smith (1988b) hypothesized that both may be outcompeted at lower salinities by species such as *Heritiera littoralis, Xylocarpus granatum,* or *Bruguiera gymnorrhiza.* Additional experimental analyses and long-term studies of permanent forest plots would be very helpful at unravelling the role of competition in mangrove forests.

5.4 Stand Structure in Mangroves

Stand structure in mangrove forests is relatively simple when compared to that of other forest types (Table 2). The number of strata is often reduced to one: the main canopy. In some forests a carpet of seedlings may form a second layer, but the abundant lianas and subcanopy trees and shrubs common to most tropical forests are largely absent in mangrove forests. Janzen (1985) commented on this "missing" understory. Subsequent hypotheses have postulated that the combination of salinity-stress and the need for light is enough to prohibit the development of understory vegetation and therefor poses an evolutionary hurdle which has not been crossed (Lugo, 1985; Snedaker and Lahmann, 1988). There are mangrove forests with understories, however. These tend to be in areas with abundant year-round rainfall and freshwater runoff (Corlett, 1986). In this situation a number of smaller tree and shrub species can be found in the forest as mangrove associates, but these species are much more common in freshwater swamp or rainforest environments (Tomlinson, 1986).

The age- (or size-) class structure of mangrove forests is also characteristic of pioneer formations (Table 2). Most mangrove forests have an even-aged size-class structure. The question of how this arises in mangroves has not been addressed. The possibility exists that large-scale disturbance will destroy large tracts of forest, which then regenerate at approximately the same time. It has been hypothesized that mangroves in Florida have adapted to a 25 year disturbance cycle, the approximate return time for major hurricanes (Odum *et al.,* 1982).

Stand height, density, and biomass accumulation appear to be related to climatic factors, particularly rainfall. Pool *et al.,* (1977) combined measures of species richness, stem density, canopy height and basal area into a complexity index to make geographic scale comparisons across the Caribbean region. They found that the least complex stands were in arid regions. These stands were marked by high stem density, but low species richness, height, and basal areas. Complex stands, characterized by tall canopies, high basal areas, and lower stem densities, were common in wet, high rainfall areas (Pool *et al.,* 1977). Complementary results that are based on different methods are available from the Indo-Pacific region (e.g., Boto *et al.,* 1984; Putz and Chan, 1986). Rainfall and freshwater runoff appear to be major determinants of stand structure.

5.5 Mangroves and Recent Theories of Forest Ecology

Over the past 40 years ecologists have developed the view that pattern in vegetation is the result of dynamic processes operating over a continuum of spatial and temporal scales: from days and weeks to centuries and from square meters to hundreds of square kilometers. In particular, the influence of natural disturbances on vegetation structure has been the subject of intense interest (Watt, 1947; White, 1979; Pickett and White, 1985; Whitmore, 1989). Forests and other ecosystems are now seen in the context of "gap dynamics" and "patch phase mosaics" (Shugart, 1984; Pickett and White, 1985). In this context the landscape is viewed as a patchwork quilt in which the individual patches are different ages or stages of development. This view of forest ecosystems has proven especially amenable to the development of ecosystem level models to explore successional patterns, nutrient cycling, and other system dynamics (Shugart, 1984). Almost all types of forest systems have been examined in light of this "gap-dynamic" or "patch phase mosaic" paradigm, with the exception of mangrove forests (e.g., Barden, 1989; Brokaw, 1985; Christensen, 1985; Lorimer, 1989; Spies and Franklin, 1989; Runkle, 1985; Veblen, 1985). But are mangrove forests really different from other forest types? Processes such as primary production, decomposition, herbivory, and competition, which operate in other forest systems, certainly operate in mangrove forests. So must the processes of natural disturbance that generate canopy gaps and forest mosaics.

This modern view of forest ecology began with the realization that forest trees can be grouped into two classes based on their reproductive strategies (e.g., Swaine and Whitmore, 1988; Whitmore, 1989). The climax class contains those species which have seeds that can germinate under the forest canopy and which have seedlings that can become established in shade. The pioneer group consists of those species that become established in the full sunlight of canopy gaps.

Mangrove species and the mangrove community have characteristics of both pioneer- and mature-phase forest communities (Table 3). For example, they produce a copious seed rain, a pioneer-phase trait. Jimenez (1990) estimated that >2,000,000 propagules/ha were produced in an *A. bicolor* forest in Costa Rica. Other species may be as productive (Duke *et al.*, 1981). Mangrove propagules, however, are often rather large and have a very long period of dispersal and longevity. These are mature-phase traits. On balance, it seems that mangroves have more pioneer-phase characteristics and therefor they should be viewed as pioneer communities (Tomlinson, 1986). Pioneer species have adapted to natural disturbance.

A number of authors have alluded to the importance of disturbance and gap dynamics in mangrove forests (e.g., Watson, 1928; Macnae, 1968; Rabinowitz, 1978a; Wells, 1982; Putz and Chan, 1986; Smith, 1987b,c; Jimenez, 1988, 1990), but no detailed analysis has been made to date. Watson (1928) commented on the shade intolerance of the seedlings of many mangrove species in Malaysia. He also remarked on the regular occurrence of gaps in the canopy, which provided the habitat needed for these species to regenerate. Macnae (1968) provided a partial classification of species into shade tolerant and shade intolerant based on his observations in the Indo-Pacific region. Wells (1982) classed Australian mangrove species as shade intolerant and shade tolerant based on his extensive observations (Table 1). Only a few experimental studies have been published that relate to gap dynamics in mangroves.

In Panama, Rabinowitz (1978a) related rates of seedling mortality to initial propagule size. She noted that mortality was inversely related to propagule size. Species with smaller propagules (*Avicennia* and *Laguncularia*) established cohorts on the forest floor every year, and these cohorts died relatively rapidly. *Rhizophora* and *Pelliciera,* however, which have larger propagules, had cohorts which overlapped; that is, seedlings were always present, but there was a constant turnover of the seedling pool (Rabinowitz 1978a). She also reported that *Pelliciera* grew better under a closed canopy than did *Rhizophora*. Rabinowitz (1978a)

Table 3. Current status of several hypotheses proposed to explain mangrove zonation.

Hypothesis	**Status**
1. Zonation represents land building and plant succession	Not supported by the data.
2. Geomorphological control	Geomorphological factors that regulate sediment supply, soil type, texture, accretion and erosion all play an important role in setting the framework within which mangrove forests develop. Climatic factors, particularly rainfall and freshwater runoff, are also important.
3. Physiological adaptation to gradients	Application of the results of single factor experiments to the field situation is tenuous at best. Extensive controlled, multi-factor, experiments are needed to fully test this hypothesis. Based on salinity tolerances, two groups of mangroves can tentatively be identified: one with an extremely broad range and the other with a narrower range of tolerance.
4. Tidal sorting of propagules	Not supported by the data.
5. Differential predation on propagules	More important for some mangroves (e.g, *Avicennia*) and in certain regions (e.g., the Indo-Pacific) than for other groups or regions.
6. Interspecific competition	Very limited data indicate that competitive interactions occur which could influence zonation.

suggested that if light gaps were important, then *Rhizophora* and *Pelliciera* would probably havc an advantage in colonizing them.

Putz and Chan (1986) reported a relationship between the abundance of mangrove seedlings and the illumination of the forest floor in Malaysian mangrove forests. The forest canopy was very dense in the 1920s and seedlings were scarce. As the canopy matured and individual trees began to die, seedling abundance increased (Putz and Chan, 1986).

5.5.1 Natural Disturbance in Mangroves

A variety of natural disturbance regimes affect mangrove forests. These may be relatively local-scale events such as breakage of branches during wind storms (Putz *et al.*, 1984), lightning strikes (Paijmans and Rollet, 1977), frost damage (Lugo and Patterson-Zucca, 1977) that may be very patchy but may extend over large areas, and whole-scale destruction of the forest by hurricanes (Craighead and Gilbert, 1962). Gradients in the types and frequency of disturbance are also present across the geographic range of mangrove forests. For example, the mangroves of Panama are not subjected to frost or hurricanes; the predominant natural disturbance is lightning strikes. To the north, in Belize, frost is again unimportant, but hurricanes and lightning strikes are common. In south Florida, disturbances from frost, hurricanes, and lightning are common (Odum *et al.*, 1982). The influence of disturbance on the structure and function of mangrove forests is poorly investigated and most reports are anecdotal.

Smith and Duke (1987) found a positive relationship between large- scale disturbance (cyclones) and species richness in the mangrove forests of northeastern Queensland, Australia. Forests that were impacted, on average, by one cyclone every 5 years had more species than forests affected by fewer storms. Species in the Rhizophoraceae often dominate these forests (Bunt *et al.*, 1982b). In the Sunderbans mangroves of Bangladesh, the Rhizophoraceae are minor components of the forest community (Blasco *et al.*, 1975). The Sunderbans are struck by up to 40 cyclones a year. The Rhizophoraceae's inability to coppice, in comparison to other groups (e.g., *Avicennia, Laguncularia, Excoecaria* and *Xylocarpus*), may account for their vunerability to cyclones.

5.5.2 Gap Dynamics in Mangroves

Canopy gaps are common in mangrove forests. In addition to what most forest ecologists would recognize as a canopy gap, Smith (1987c) observed that low intertidal, accreting mudbanks also act as "light gaps." Individuals in these areas are exposed to more light than are individuals under the nearby canopy. Most gaps (the traditional kind) are probably created by lightning strikes. Lightning strikes create relatively circular patches in the forest from the top of the canopy to the forest floor. An interesting aspect of lightning strikes is that a number of trees are usually killed rather than a single individual, and those dead trees often remain standing for several years (Duke *et al.*, 1991). Seedlings that are present under the canopy are often killed as well. Smith (unpub. data) surveyed 391 gaps in northeast Queensland; they ranged from $<10m^2$ to $>500m^2$ in size (Figure 7), with the modal size 40- $60m^2$. Gaps were evenly distributed across the forest from low to high intertidal zones and from upstream to downstream along the length of the estuaries. Saplings of several species, including *A. marina, B. parviflora, B. exaristata,* and *R. apiculata,* were found to be significantly more abundant in these gaps than under the surrounding canopy (Smith, 1987a,c).

The physical environment in light gaps is substantially different from that under the surrounding canopy (Figure 8). Measurements made in light gaps in high and low intertidal areas in Australia indicated differences in photosynthetically active radiation (PAR), pore water salinity, and soil temperature (Smith, 1987c). The canopy was so dense that it dampened the annual cycle in PAR, which was pronounced in nearby gaps. There were consistent differences in soil temperature. Sediments in light gaps were 3-5°C warmer than were soils under the canopy. Pore water salinity was also lower in gaps, by 1-2‰, than under nearby canopies.

The entire Murray River estuarine system in north Queensland was surveyed for gaps using recent and historical aerial photographs. It was determined that between 4-15% of the

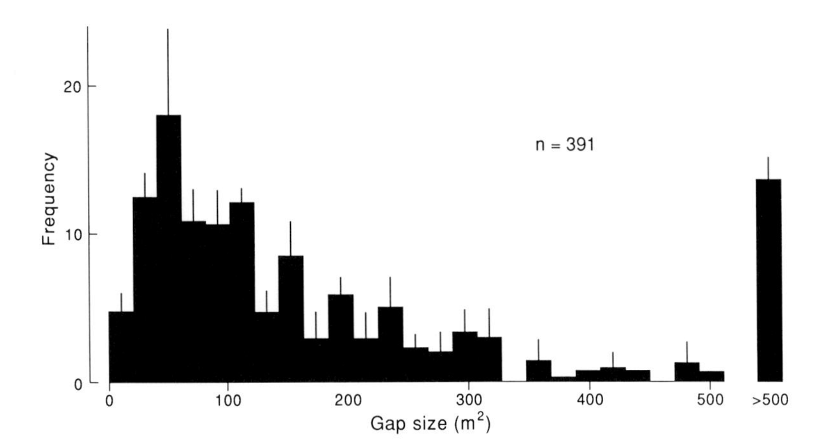

Figure 7. Frequency distribution of canopy gaps by size class from northern Australia and southern Papua New Guinea. Data based on surveys of 391 gaps. Gaps were randomly assigned to three groups and then the mean (± 1SD) was calculated. Data are previously unpublished from the author.

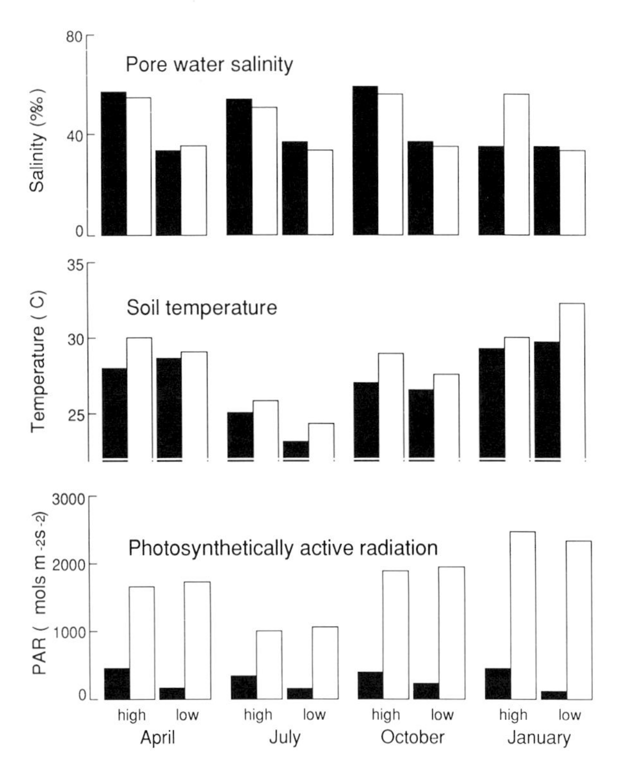

Figure 8. Seasonal variation in some physical characteristics of gap (open bars) and understory (solid bars) environments for both high and low intertidal habitats in north Queensland, Australia (from Smith, 1987b, used with permission). Mean ± 1SD, for n=5.

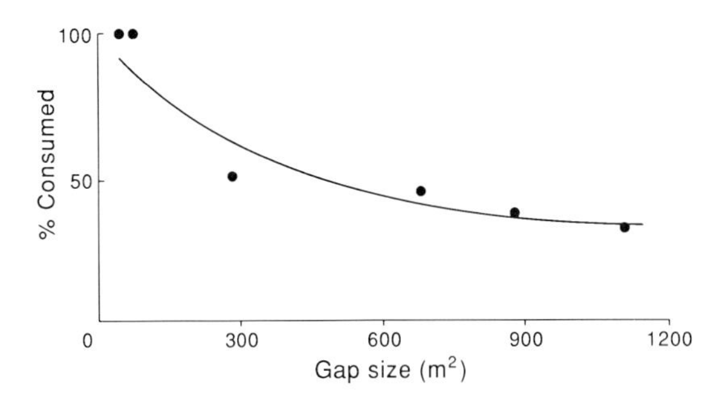

Figure 9. Amount of predation (%) on *Avicennia marina* propagules located in canopy gaps as a function of gap size (m^2), from Osborne and Smith (1990). Actual data points and their best fit nonlinear regression equation as shown. The regression equation is given by: %Consumed $=100 \cdot e^{-0.001275 \cdot \text{gap size}}$ and is highly significant ($F_{1,4}=73.1$, $p \leq 0.01$, $r^2=0.96$). Reproduced with permission of *Vegetatio*.

forest was in the gap- phase at any one time and that the forest "turned over" approximately every 150-170 years (Smith, unpub. data). Comparable data are not available for mangroves elsewhere in the world. However, given the very high frequency of thunderstorms, and hence lightning strikes, in south Florida (Michaels *et al.*, 1987) and visual observation of the forests, it appears that a larger percentage of Florida mangroves are in the gap-phase stage than are forests in northern Australia (Smith, pers. obs.).

Seedling survival and growth for several mangrove species have been examined in gap and understory habitats in both high and low intertidal zones. Smith (1987b,c) found that survival of *A. marina, R. stylosa, B. gymnorrhiza,* and *C. australis* was higher in gaps than under the canopy and greater in high intertidal gaps than in low intertidal gaps. Relative growth rates for all species except *B. gymnorrhiza* were also greater in light gaps. Osborne (1988) found that *A. corniculatum* survived and grew best on open (unshaded) accreting mudbanks. Within high intertidal forests, survival and growth was greatest in canopy gaps, but was still lower than on accreting mudbanks. Duke (unpub. data) tagged a number of seedlings of *C. australis, B. gymnorrhiza, X. granatum,* and *H. littoralis*, all of which were growing under a shaded canopy. Growth of all species was minimal, <1cm/yr, measured over 10 years of observation. Survival, however, was >80% for *B. gymnorrhiza, X. granatum,* and *H. littoralis*. For *C. australis* survival was <20%.

In Australia canopy gaps may provide some mangroves with a refuge from seed predators. Osborne and Smith (1990) observed that predation on propagules of *A. marina* was higher in small gaps and decreased with increasing gap size (Figure 9). Visual observations indicated that the crab fauna in gaps was dominated by ocypodids (Ocypodidae, primarily *Uca*), whereas grapsids dominated under the canopy. Ocypodids are not known to consume mangrove propagules, but grapsids do (Smith, 1987b). The increase in soil temperatures that accompany gap formation may underlie this shift in the crab fauna, as *Uca* appear to prefer warmer sediments (Jones, 1984).

5.6 Conclusions

Despite several thousand publications concerning mangrove forests (Frith, 1977; Rollet, 1981), a clear understanding of the dynamics in mangrove ecosystems is just beginning to emerge. Of the hypotheses advanced to account for species zonation, several warrant further attention, but others should be laid to rest (Table 3). In particular, hypotheses concerning zonation as plant succesion and the tidal sorting of propagules clearly are not supported by the available data. Geomorphological factors establish much, but not all, of the framework within which mangrove forests develop. Climatic factors, particularly rainfall, are important determinants of species richness, stand structure, and growth dynamics in mangrove forests. Two groups of mangroves can be tentatively identified based on salinity tolerance data; one has a very broad range (0-80‰) and the other has a narrower range (<40‰) of tolerance. Extensive controlled experimentation is required to fully understand how mangrove physiological responses to other environmental gradients (e.g., soil texture, redox potential, nutrients) may influence observed zonation patterns. In particular experiments which address potential interactions between variables are needed. Biotic factors such as predation on propagules, are important influences on the distributional patterns of some groups of mangroves and in certain geographic regions (Table 3). Competitive interactions may be important in determining some aspects of forest structure, but much more experimental and long- term observational work is needed.

A more important consideration is that the dynamics of mangrove forest systems fit within current theories and paradigms developed for other vegetation systems. Ideas of gap-phase dynamics, natural disturbance, and forest mosiacs are applicable to mangrove ecosystems and will provide a fruitful avenue for further research.

5.7 Acknowledgements

Numerous persons have provided assistance, advice, and encouragement in my mangrove research over the years, including Drs. D.M. Alongi, K.G. Boto, J.S. Bunt, B.F. Clough, N.C. Duke, N.Q. Lager, C.C. McIvor, M.B. Robblee, and A.I. Robertson. I thank them for their patience, understanding and insights. Preparation of this manuscript was supported by the Florida Department of Natural Resources and the U.S. National Oceanic and Atmospheric Administration (Grant #NA90AA-H-CZ717). Earlier drafts of the manuscript benefited from comments by D. Crewz, I. Feller, J. Lieby, I. Mendelssohn, L. Nall, F. Putz, and especially, K. McKee. M. Ball is thanked for so succinctly demonstrating to me how easy it is for a scientist to be blinded by their pet theories.

5.8 References

Bakus, G.J., 1966. Some relationships of fish to benthic organisms on coral reefs. *Nature* **210**:280-284.

Bakus, G.J., 1969. Energetics and feeding in shallow marine waters. *International Review General Experimental Zoology* **4**:275-369.

Ball, M.C., 1980. Patterns of secondary succession in a mangrove forest in southern Florida. *Oecologia* **44:**226-235.

Ball, M.C., 1988a. Ecophysiology of mangroves. *Trees* **2**:129-142.

Ball, M.C., 1988b. Salinity tolerence in the mangroves Aegiceras corniculatum and Avicennia marina I. Water use in relation to growth, carbon partitioning, and salt balance. *Australian Journal of Plant Physiology* **15**:447-464.

Ball, M.C., Cowan, I.R., and Farquhar,G.D., 1988. Maintenance of leaf temperature and the optimisation of carbon gain in relation to water loss in a tropical mangrove forest. *Australian Journal of Plant Physiology* **15**:263-276.

Ballment, E.R., Smith, III, T.J., and Stoddart, J.A., 1988. Sibling species in the mangrove genus *Ceriops* (Rhizophoraceae), detected using biochemical genetics. *Australian Systematic Botany* **1**:391-397.

Barden, L.S., 1989. Repeatability in forest gap research: studies in the Great Smokie Mountains. *Ecology* **70**:558-559.

Barnes, R.S.K., 1967. The osmotic behaviour of a number of grapsoid crabs with respect to their differential penetration of an estuarine system. *Journal of Experimental Biology* **47**:535-551.

Berry, A.J., 1964. Faunal zonation in mangrove swamps. *Bulletin of the National Museum of Singapore* **32**:90-98.

Berry, A.J., 1972. The natural history of west Malaysian mangrove faunas. *Malaysian Naturalists Journal* **25**:135-162.

Blasco, F., Caratini, C., Chanda, S., and Thanikaimoni, G., 1975. Main characteristics of Indian mangroves. In: Walsh, G., Snedaker, S.C. and Teas, H. (Eds.), *Proceedings of the International Symposium on the Biology and Management of Mangroves*, pp.71-87, University of Florida, Gainesville.

Boto, K.G., and Wellington, J.T., 1983. Phosphorus and nitrogen nutritional status of a northern Australian mangrove forest. *Marine Ecology Progress Series* **11**:63-69.

Boto, K.G., and Wellington, J.T., 1984. Soil characteristics and nutrient status in a northern Australian mangrove forest. *Estuaries* **7**:61-69.

Boto, K.G., Bunt, J.S., and Wellington, J.T., 1984. Variations in mangrove forest productivity in northern Australia and Papua New Guinea. *Estuarine, Coastal and Shelf Science* **19**:321-329.

Bowman, H.H.M., 1917. Ecology and physiology of the red mangrove. *Proceedings of the American Philosophical Society* **56**:589-672.

Brokaw, N.V.L., 1985. Treefalls, regrowth, and community structure in tropical forests. In: Pickett, S.T.A., and White, P.S., (Eds.), *The ecology of natural disturbance and patch dynamics*, pp. 53-71, Academic Press, New York.

Bunt, J.S., and Williams, W.T., 1980. Studies in the analysis of data from Australian tidal forests ('mangroves'). I. Vegetational sequences and their graphical representation. *Australian Journal of Ecology* **5**:385-390.

Bunt, J.S., and Williams, W.T., 1981. Vegetational relationships in the mangroves of tropical Australia. *Marine Ecology Progress Series* **4**:349-359.

Bunt, J.S., Williams, W.T., and Clay, H.J., 1982a. River water salinity and the distribution of mangroves species along several rivers in north Queensland. *Australian Journal of Botany* **30**:401-412.

Bunt, J.S., Williams, W.T., and Duke, N.C., 1982b. Mangrove distributions in north-east Australia. *Journal of Biogeography* **9**:111-120.

Bunt, J.S., Williams, W.T., Hunter, J.F., and Clay, H.J., 1991. Mangrove sequencing: analysis of zonation in a complete river system. *Marine Ecology Progress Series* **72**:289-294.

Carlson, P.R., Yarbro, L.A., Zimmerman, C.F., and Montgomery, J.R., 1983. Pore water chemistry of an overwash mangrove island. *Florida Scientist* **46**:239-249.

Chapman, V.J., 1976. *Mangrove vegetation*. J.Cramer, Vaduz, Germany. 447pp.

Christensen, N.L., 1985. Shrubland fire regimes and their evolutionary history. In: Pickett, S.T.A., and White, P.S., (Eds.), *The ecology of natural disturbance and patch dynamics,* pp.86-100, Academic Press, New York.

Clarke, L.D., and Hannon, N.J., 1967. The mangrove and salt marsh communities of the Sydney district. I. Vegetation, soils and climate. *Journal of Ecology* **55**:753 771.

Clarke, L.D., and Hannon, N.J., 1969. The mangrove and salt marsh communities of the Sydney district. II. The holocoenotic complex with particular reference to physiography. *Journal of Ecology* **57**:213-234.

Clarke, L.D., and Hannon, N.J., 1970. The mangrove and salt marsh communities of the Sydney district. III. Plant growth in relation to salinity and waterlogging. *Journal of Ecology* **58**:351-369.

Clarke, L.D., and Hannon, N.J., 1971. The mangrove and salt marsh communities of the Sydney district. IV. The significance of species interaction. *Journal of Ecology* **59**:535-553.

Clarke, P.J., and Myerscough, P.J., 1991. Bouyancy of *Avicennia marina* propagules in south-eastern Australia. *Australian Journal of Botany* **39**: 77-83.

Clough, B.F., 1984. Growth and salt balance in the mangroves *Avicennia marina* (Forsk.) Vierh. and *Rhizophora stylosa* Griff. in relation to salinity. *Australian Journal of Plant Physiology* **11**:419-430.

Corlett, R.T., 1986. The mangrove understory: some additional observations. *Journal of Tropical Ecology* **2**:93-84.

Craighead, F.C., and Gilbert, V., 1962. The effects of Hurricane Donna on the vegetation of southern Florida. *Quarterly Journal of the Florida Academy of Sciences* **25**:1-28.

Curtiss, A.H., 1888. How the mangrove forms islands. *Garden and Forest* **1**:100.

Davis, J.H., 1940. The ecology and geologic role of mangroves in Florida. Publications of the Carnegie Institute, Washington, D.C. Publication #517.

Downton, W.J.S., 1982. Growth and osmotic relations of the mangrove, *Avicennia marina*, as influenced by salinity. *Australian Journal of Plant Physiology* **9**:519 528.

Duke, N.C., 1990. Phenological trends with latitude in the mangrove tree *Avicennia marina. Journal of Ecology* **78**:113-133.

Duke, N.C., and Jackes, B.R., 1987. A systematic revision of the mangrove genus *Sonneratia* (Sonneratiaceae) in Australasia. *Blumea* **32**:277-302.

Duke, N.C., Bunt, J.S., and Williams, W.T., 1981. Mangrove litterfall in northeastern Australia. I. Annual totals by component in selected species. *Australian Journal of Botany* **29**:547-553.

Duke, N.C., Pinzon, Z.S., and Prada, M.C., 1991. Recovery of tropical mangrove forests following a major oil spill: a study of recruitment and growth, and the benefits of planting. In: Yanez-Arancibia, A., (Ed.), *Mangrove ecosystems in tropical America: structure, function and management.* EPOMEX Serie Cientifica, In press.

Earle, S.A., 1972. A review of the marine plants of Panama. *Bulletin of the Biological Society of Washington* **2**:69-87.

Egler, F.A., 1950. Southeast saline Everglades vegetation, Florida, and its management. *Vegetatio* **3**:213-265.

Ellison, J.C., and Stoddart, D.R., 1991. Mangrove ecosystem collapse during predicted sea-level rise: Holocene analogues and implications. *Journal of Coastal Research* **7**:151-165.

Elsol, J.A., and Saenger, P., 1983. A general account of the mangroves of Princess Charlotte Bay with particular reference to zonation along the open shoreline. In: Teas, H.J. (Ed.), *Biology and ecology of mangroves, Tasks for vegetation science,* pp. 37-48, Dr. W. Junk Publishers, The Hague.

Frith, D.W., 1977. A selected bibliography of mangrove literature. Research Bulletin No. 19. Phuket Marine Biological Center, Phuket, Thailand. 142pp.

Giglioli, M.E.C., and Thornton, I., 1965. The mangrove swamps of Keneba, lower Gambia River basin. I. Descriptive notes on the climate, the mangrove swamps, and the physical condition of their soils. *Journal of Applied Ecology* **2**:81-103.

Glynn, P.W., 1972. Observations on the ecology of the Caribbean and Pacific coasts of Panama. *Bulletin of the Biological Society of Washington* **2**:13-30.

Golley, F., Odum, H.T., and Wilson, R.F., 1962. The structure and metabolism of a Puerto Rican red mangrove forest in May. *Ecology* **43**:9-19.

Grace, J.B., and Wetzel, R.G., 1981. Habitat partitioning and competitive displacement in cattails (*Typha*): Experimental field studies. *American Naturalist* **118**:463-474.

Grewe, F., 1941. Afrikanische mangrovelandschaffen. Wissenschaftliche Veroffentlichungen. *Museum Landerkind Leipzig* **9**:103-177.

Gunn, C.R., and Dennis, J.V., 1973. Tropical and temperate stranded seeds and fruit from the Gulf of Mexico. *Contributions in Marine Science* **17**:111-121.

Hartnoll, R.G., 1965. Notes on the marine grapsid crabs of Jamaica. *Proceedings of the Linnaen Society of London* **176**:113-147.

Hartnoll, R.G., 1973. Factors affecting the distribution of *Dotilla fenestrata* on East African shores. *Estuarine, Coastal, and Shelf Science* **1**:137-152.

Hartnoll, R.G., 1975. The Grapsidae and Ocypodidae (Decapoda: Brachyura) of Tanzania. *Journal of Zoology, London* **177**:305-328.

Hegerl, E.J., and Davie, J.D.S., 1977. The mangrove forests of Cairns, Northern Australia. *Marine Research in Indonesia* **18**:23-57.

Icely, J.D., and Jones, D.A., 1978. Factors affecting the distribution of the genus *Uca* (Crustacea: Ocypodidae) on an East African shore. *Estuarine, Coastal and Shelf Science* **6**:315-325.

Janzen, D.H., 1971. Seed predation by animals. *Annual Review of Ecology and Systematics.* **2**:465-491.

Janzen, D.H., 1985. Mangroves: where's the understory? *Journal of Tropical Ecology* **1**:89-92.

Jimenez, J.A., 1984. A hypothesis to explain the reduced distribution of the mangrove *Pelliciera rhizophorae* Tr. & Pl. *Biotropica* **16**:304-308.

Jimenez, J.A., 1988. The dynamics of *Rhizophora racemosa* Meyer, forests on the Pacific coast of Costa Rica. *Brenesia* **30**:1-12.

Jimenez, J.A., 1990. The structure and function of dry weather mangroves on the Pacific coast of Central America, with emphasis on *Avicennia bicolo*r forests. *Estuaries* **13**:182-192.

Jimenez, J.A., Lugo, A.E., and Cintron, G., 1985. Tree mortality in mangrove forests. *Biotropica* **17**:177-1185.

Jimenez, J.A., and Sauter, K., 1991. Structure and dynamics of mangrove forests along a flooding gradient. *Estuaries* **14**:49-56.

Jimenez, J.A., and Soto, R.S., 1985. Patrones regionales en la estrucutra y composicion floristica de los manglares de la Costa Pacifica de Costa Rica. *Revista de Biologia Tropical* **33**:25-37.

Jones, D.A., 1984. Crabs of the mangal ecosystem. In: Por, F.D,. and Dor, I., (Eds.), *Hydrobiology of the mangal*, pp. 89-110, Dr. W. Junk Publishers, The Hague.

Johnstone, I.M., 1983. Succession in zoned mangrove communities: Where is the climax? In: Teas, H.J., (Ed.), *Biology and ecology of mangroves, Tasks for vegetation science,* pp. 131-139, Dr. W. Junk Publishers, The Hague.

Lorimer, C.G., 1989. Relative effects of small and large disturbances on temperate hardwood forest structure. *Ecology* **70**:565-566.

Lugo, A.E., 1985. Mangrove understory: an expensive luxury? *Journal of Tropical Ecology* **2**:287-288.

Lugo, A.E., and Patterson-Zucca, C., 1977. The impact of low temperature stress on mangrove structure and growth. *Tropical Ecology* **18**:149-161.

Lugo, A.E., and Snedaker, S.C., 1974. The ecology of mangroves. *Annual Review of Ecology and Systematics* **5**:39-64.

Macnae, W., 1968. A general account of the flora and fauna of mangrove swamps in the Indo-Pacific region. *Advances in Marine Biology* **6**:73-270.

Macnae, W., 1969. Zonation within mangroves associated with estuaries in north Queensland. In: Lauff, G.H., (Ed.), *Estuaries*, pp.432-441, American Association for the Advancement of Science, Washington, D.C.

Macnae, W., and Kalk, M., 1962. The ecology of the mangrove swamps of Inhaca Island, Mozambique. *Journal of Ecology* **50**:19-34.

McKee, K.L., and Mendelssohn, I., 1987. Root metabolism in the black mangrove (*Avicennia germinans* (L.) L): Resposne to hypoxia. *Environmental and Experimental Botany* **27**:147-156.

McKee, K.L., Mendelssohn, I., and Hester, M.K., 1988. Reexamination of pore water sulfide concentrations and redox potentials near the aerial roots of *Rhizophora mangle* and *Avicennia germinans. American Journal of Botany* **75**:1352 1359.

McMillan, C., 1971. Environmental factors affecting seedling establishment of the black mangrove on the central Texas Coast. *Ecology* **52**:927-930.

McMillan, C., 1975. Interaction of soil texture with salinity tolerences of black mangrove (*Avicennia*) and white mangrove (*Laguncularia*) from North America. In: Walsh, G.E., Snedaker, S.C., and Teas, H.J., (Eds.), *Proceedings of the International Symposium on the Biology and Management of Mangroves,* pp. 561-566, University of Florida, Gainesville.

Michaels, P.J., Pielke, R.A., McQueen, J.T., and Sappington, D.E., 1987. Composite climatology of Florida summer thunderstorms. *Monthly Weather Review* **115**:2781 2791.

Nickerson, N.H., and Thibodeau, F.R., 1985. Association between pore water sulfide concentrations and distribution of mangroves. *Biogeochemistry* **1**:183-192.

Noakes, D.S.P., 1955. Methods of increasing growth and obtaining natural regeneration of the mangrove type in Malaya. *Malaysian Forester* **18**:23-30.

Odum, W.E., McIvor, C.C., and Smith, T.J., III, 1982. The ecology of the mangroves of south Florida: A community profile. U.S. Fish & Wildlife Service, Office of Biological Services. Washington, D.C. FWS/OBS - 81/24.

Onuf, C.P., Teal, J.M., and Valiela, I., 1977. Interactions of nutrient, plant growth and herbivory in a mangrove ecosystem. *Ecology* **58**:514-526.

Osborne, K., 1988. A distribution study of the mangrove, *Aegiceras corniculatum* (L.) Blanco, in some northern Australian estuaries. Bachelor of Science Honors Thesis, Department of Geography, James Cook University of North Queensland. Townsville, Australia. 82pp.

Osborne, K., and Smith, T.J., III, 1990. Differential predation on mangrove propagules in open and closed canopy forest habitats. *Vegetatio* **89**:1-6.

Paijmans, K., and Rollet, B., 1977. The mangroves of Galley Reach, Papua New Guinea. *Forest Ecology and Management* **1**:119-140.

Palmer, A.R., 1978. Fish predation as an evolutionary force molding gastropod shell form: a tropical - temperate comparison. *Evolution* **33**:697-713.

Pickett, S.T.A., and White, P.S., 1985. Natural disturbance and patch dynamics: An introduction. In: Pickett, S.T.A., and White, P.S., (Eds.), *The ecology of natural disturbance and patch dynamics,* pp.3-13, Academic Press, New York.

Platt, J.R., 1964. Strong inference. *Science* **146**: 347-353.

Pimm, S.L., 1978. An experimental approach to community structure. *American Zoologist* **18**:797-808.

Pool, D.J., Snedaker, S.C. and Lugo, A.E., 1977. Structure of mangrove forests in Florida, Puerto, Mexico, and Costa Rica. *Biotropica* **9**:195-212.

Prawiroatmodjo, S., Sapulete, D., Pratignyo, S.E. and Budiman, A., 1985. Structural analysis of mangrove vegetation in Elpaputih and Wailale, Ceram, Indonesia. In: Bardsley, K.N., Davie, J.D.S., and Woodroffe, C.D., (Eds.), *Coasts and tidal wetlands of the Australian monsoon region.* North Australia Research Unit, Australian National University, Mangrove Monograph #1. pp.153-165.

Putz, F.E., and Chan, H-T., 1986. Tree growth, dynamics and productivity in a mature mangrove forest in Malaysia. *Forest Ecology and Management* **17**:211-230.

Putz, F.E., Parker, G.G. and Archibald, R.M., 1984. Mechanical abrasion and intercrown spacing. *American Midland Naturalist* 112:24-28.

Rabinowitz, D., 1977. Effects of mangrove borer, *Poecilips rhizophorae*, on propagules of *Rhizophora harrisonii. The Florida Entomologist* **60**:129-134.

Rabinowitz, D., 1978a. Mortality and initial propagule size in mangrove seedlings in Panama. *Journal of Ecology* **66**:45-51.

Rabinowitz, D., 1978b. Early growth of mangrove seedlings in Panama, and an hypothesis concerning the relationship of dispersal and zonation. *Journal of Biogeography* **5**:113-133.

Rabinowitz, D., 1978c. Dispersal properties of mangrove propagules. *Biotropica* **10**:47-57.

Robertson, A.I., Giddins, R.L. and Smith, T.J., III, 1990. Seed predation by insects in tropical mangrove forests: extent and affects on seed viability and growth of seedlings. *Oceologia* **83**:213-219.

Rollet, B., 1981. *Bibliography on mangrove research, 1600-1975*. UNESCO, Paris. 479pp.

Runkle, J.R., 1985. Disturbance regimes in temperate forests. In: Pickett, S.TA., and White, P.S., (Eds.), *The Ecology of natural disturbance and patch dynamics*, pp.17-34, Academic Press, New York.

Saenger, P., 1982. Morphological, anatomical and reproductive adaptations of Australian mangroves. In: Clough, B.F., (Ed.), *Mangrove ecosystems in Australia*, pp.153-191, Australian National University Press, Canberra.

Saenger, P., and Moverly, J., 1985. Vegetative phenology of mangroves on the Queensland coast. *Proceedings of the Ecological Society of Australia* **13**:257-265.

Sasekumar, A., 1974. Distribution of macrofauna on a Malaysian mangrove shore. *Journal of Animal Ecology* **43**:51-69.

Scholander, P.F., Van Dam, L., and Scholander, S.I., 1955. Gas exchange in the roots of mangroves. *American Journal of Botany* **42**:92-98.

Semeniuk, V., 1980. Mangrove zonation along an eroding coastline in King Sound, north-western Australia. *Journal of Ecology* **68**:789-812.

Semeniuk, V., 1983. Mangrove distribution in northwestern Australia in relationship to regional and local freshwater drainage. *Vegetatio* **53**:11-31.

Semeniuk, V., 1985. Development of mangrove habitats along ria shorelines in north and northwestern tropical Australia. *Vegetatio* **60**:3-23.

Sherrod, C.L., and McMillan, C., 1985. The distributional history and ecology of mangrove vegetation along the northern Gulf of Mexico coastal region. *Contributions in Marine Science* **28**:129-140.

Sherrod, C.L., Hockaday, D.L., and McMillan, C., 1986. Survival of red mangrove, *Rhizophora mangle*, on the Gulf of Mexico coast of Texas. *Contributions in Marine Science* **29**:27-36.

Shugart, H.H., 1984. *A theory of forest dynamics*. Springer-Verlag, New York.

Sidhu, S.S., 1963. Studies on the mangroves of India. I. East Godavari region. *Indian Forester* **89**:337-351.

Sidhu, S.S., 1975. Culture and growth of some mangrove species. In: Walsh, G.E., Snedaker, S.C., and Teas, H.J., (Eds.), *Proceedings of the International Symposium on Biology and Management of Mangroves*, pp.394-401, University of Florida, Gainesville.

Silander, J.A., and Antonovics, J., 1982. Analysis of interspecific interactions in a coastal plant community - a perturbation approach. *Nature* **298**:557-560.

Smith, T.J., III, 1987a. Seed predation in relation to tree dominance and distribution in mangrove forests. *Ecology* **68**:266-273.

Smith, T.J., III, 1987b. Effects of light and intertidal position on seedling survival and growth in tropical, tidal forests. *Journal of Experimental Marine Biology and Ecology* **110**:133-146.

Smith, T.J., III, 1987c. Effects of seed predators and light level on the distribution of *Avicennia marina* (Forsk.) Vierh. in tropical, tidal forests. *Estuarine, Coastal and Shelf Science* **25**:43-51.

Smith, T.J., III, 1988a. Structure and succession in tropical, tidal forests: the influence of seed predators. *Proceedings of the Ecological Society of Australia* **15**:203-211.

Smith, T.J., III, 1988b. Differential distribution between subspecies of the mangrove *Ceriops tagal:* Competitive interactions along a salinity gradient. *Aquatic Botany* **32**:79-89.

Smith, T.J., III, and Duke, N.C., 1987. Physical determinants of inter estuary variation in mangrove species richness around the tropical coastline of Australia. *Journal of Biogeography* **14**:9-19.

Smith, T.J., III, and Duke, N.C., in review. The distribution of mangroves along an intertidal gradient: A comparison of seed predation, tidal sorting, and physiological adaptation in the mangrove forests of Panama. *Journal of Ecology.*

Smith, T.J., III, Boto, K.G., Frusher, S.D., and Giddins, R.L., 1991. Keystone species and mangrove forest dynamics: the influence of burrowing by crabs on soil nutrient status and forest productivity. *Estuarine, Coastal and Shelf Science* **33**:419-432.

Smith, T.J., III, Chan, H-T., McIvor, C.C., and Robblee, M.B., 1989. Comparisons of seed predation in tropical, tidal forests on three continents. *Ecology* **70**:146-151.

Snedaker, S.C., 1982. Mangrove species zonation: Why? In: Sen, D.N. and Rajpurohit (Eds.), *Tasks for vegetation science*, Vol. 2. Dr. W. Junk Publishers, The Hague. pp.111-125.

Snedaker, S.C., and Lahmann, E.J., 1988. Mangrove understory absence: a consequence of evolution? *Journal of Tropical Ecology* **4**:311-314.

Snelling, B., 1959. The distribution of intertidal crabs in the Brisbane River. *Australian Journal of Marine and Freshwater Research* **10**:67-83.

Spies, T.A., and Franklin, J.F., 1989. Gap characteristics and vegetation response in coniferous forests of the Pacific Northwest. *Ecology* **70**:543-545.

Stamp, L.D., 1924. The aerial survey of the Irrawaddy delta forests. *Journal of Ecology* **13**:262-276.
Sternberg, L. da SL., and Swart, P.K., 1987. Utilization of freshwater and ocean water by coastal plants of southern Florida. *Ecology* **68**:1898-1905.

Steinke, T.D., 1975. Some factors affecting dispersal and establishment of propagules of *Avicennia marina* (Forsk.) Vierh. In: Walsh, G.E., Snedaker, S.C., and Teas, H.J., (Eds.), *Proceedings of the International Symposium on Biology and Management of Mangroves,* pp.402-414, University of Florida, Gainesville.

Stoddart, D.R., 1980. Mangroves as successional stages, inner reefs of the northern Great Barrier Reef. *Journal of Biogeography* **7**:269-284.

Swaine, M.D., and Whitmore, T.C., 1988. On the definition of ecological species groups in tropical rainforests. *Vegetatio* **75**:81-86.

Thibodeau, F.R., and Nickerson, N.H., 1986. Differential oxidation of mangrove substrate by *Avicennia germinans* and *Rhizophora mangle*. *American Journal of Botany* **73**:512-516.

Thom, B.G., 1967. Mangrove ecology and deltaic geomorphology: Tabasco, Mexico. *Journal of Ecology* **55**:301-343.

Thom, B.G., Wright L.D., and Coleman J.M., 1975. Mangrove ecology and deltaic-estuarine geomorphology: Cambridge Gulf - Ord River, Western Australia. *Journal of Ecology* **63**: 203-232.

Thomson, R.C.M., 1945. Studies on the breeding places and control of *Anopheles gambiae* and *A. gambiae* var. *melas* in coastal disticts of Sierra Leone. *Bulletin of Entomological Research* **36**:185-252.

Tomlinson, P.B., 1986. *The botany of mangroves.* Cambridge University Press. Cambridge, United Kingdom. 413pp.

Tweedie, M.W.F., 1950. Notes on the grapsoid crabs from the Raffles Museum: II. On the habits of three Ocypodid crabs. *Bulletin of the Raffles Museum* **23**:317-324.

Van Steenis, C.J.J.G., 1957. Outline of vegetation types in Indonesia and some adjacent regions. *Proceedings of the Pacific Science Congress* **8**:61-97.

Veblen, T.B., 1985. Stand dynamics in Chilean Nothofagus forests. In: Pickett, S.T.A., and White, P.S., (Eds.), *The ecology of natural disturbance and patch dynamics,* pp. 35-52, Academic Press, New York.

Vermeij, G.J., 1976. Interoceanic differences in vulnerability of shelled prey to crab predation. *Nature* **260**:135-136.

Vermeij, G.J., 1978. *Biogeography and adaptation, patterns of marine life.* Harvard University Press, Cambridge. 332pp.

Verwey, J., 1930. Einiges uber die biologie Ost-Indischer mangrove krabben. *Treubia* **12**:167-261.

Vince, S.W., and Snow, A.A., 1984. Plant zonation in an Alaskan saltmarsh. I. Distribution, abundance and environmental factors. *Journal of Ecology* **72**:651-667.

Von Hagen, H.O., 1977. The tree climbing crabs of Trinidad. *Studies on the Fauna of Curacao & other Caribbean Islands* **175**:26-59.

Walter, H., and Steiner, M., 1936. Die Oekologie der ost-afrikanischen mangroven. *Zeitschrift fur Botany* **30**:65-193.

Warner, G.F., 1969. The occurrence and distribution of crabs in a Jamaican mangrove swamp. *Journal of Animal Ecology* **38**:379-389.

Warren, J.H., and Underwood, A.J., 1986. Effects of burrowing crabs on the topography of mangrove swamps in New South Wales. *Journal of Experimental Marine Biology and Ecology* **102**:223-235.

Watson, J.G., 1928. Mangrove forests of the Malay peninsula. *Malayan Forest Records* **6**:1-275.

Watt, A.S., 1947. Pattern and process in the plant community. *Journal of Ecology* **35**:1-22.

West, R.C., 1956. Mangrove swamps of the Pacific coast of Columbia. *Annals of the Association of American Geographers* **46**:98-121.

Wells, A.G., 1982. Mangrove vegetation of northern Australia. In: Clough, B.F., (Ed.), *Mangrove ecosystems in Australia: Structure, Function and Management*, pp.57-78, Australian National University Press, Canberra.

Whelan, C.J., Willson, M.F., Tuma, C.A., and Souza-Pinto, I., 1990. Spatial and temporal patterns of postdispersal seed predation. *Canadian Journal of Botany* **69**:428-436.

White, P.S., 1979. Pattern, process, and natural disturbance in vegetation. *The Botanical Review* **45**:229-299.

Whitmore, T.C., 1989. Canopy gaps and the two major groups of forest trees. *Ecology* **70**:536-538.

Whittaker, R.H., 1967. Gradient analysis of vegetation. *Biological Reviews* **42**:207-264.

Woodroffe, C.D., 1981. Mangrove swamp stratigraphy and Holocene transgression, Grand Cayman Island, West Indies. *Marine Geology* **41**:271-294.

Woodroffe, C.D., 1982. Geomorphology and development of mangrove swamps, Grand Cayman Island, West Indies. *Bulletin of Marine Science* **32**:381-398.

Woodroffe, C.D., Thom, B.G., and Chappell, J., 1985. Development of widespread mangrove swamps in mid-Holocene times in northern Australia. *Nature* **317**:711-713.

6

Benthic Communities

D.M. Alongi and A. Sasekumar

6.1 Introduction

Studies of the ecology of infaunal and above ground benthic biota in tropical mangroves have been few compared with the number of investigations of benthic communities in temperate intertidal habitats. Most of the early (pre-1975) mangrove studies focused on description of new species (e.g. Gerlach, 1957) and on changes in species composition of macroepifauna with tidal height (e.g. Macnae and Kalk, 1962). Various reviews are available, but most have either summarized the benthic fauna along with the entire mangrove dependent biocoenoses (Macnae, 1968; Milward, 1982) or have provided an overview of only one specific component, such as crabs (Jones, 1984). An earlier review by Alongi (1989) provided a critical assessment of soft bottom benthic communities in mangroves and coral reefs.

This chapter will focus on the abundance and composition of nearly all biotic groups which constitute the benthos of tropical mangrove forests, including those organisms living on prop roots and on (and in) timber resting on the forest floor. Edaphic and physicochemical factors that affect their ecology will also be examined. The role of mangrove benthos on nutrient flow will be considered in chapters 9 and 10 of this volume, but some facets of benthic energetics (secondary production, growth, reproduction) will be detailed in this chapter.

6.2 Characteristics of Mangrove Sediments

Tropical mangrove sediments possess reduction and oxidation properties, and other physicochemical characteristics (pH, grain size, etc) typical of other estuarine and marine intertidal deposits (Boto, 1984). Macnae (1968) and many other workers (e.g. Hesse, 1963) have generalized that mangrove sediments are highly anaerobic, sulphidic muds. More recent work in Indian, Southeast Asian and Australian mangroves suggest that soil texture and subsequent physicochemical properties vary greatly among forest types and with tidal elevation (Limpsaichol, 1978; Boto and Wellington, 1983, 1984; Sahoo *et al.*, 1985; Alongi, 1987a; and see Chapter 8, this volume).

Large deltaic forests possess sediments that are predominantly well sorted silts and clay with large quantities of fine, fibrous root matter (Boto and Wellington, 1984). These muds have the highest concentrations of organic carbon and nitrogen, ranging from 0.5 to 15.0% and 0.2 to 0.5% by sediment dry wt, respectively. In contrast to temperate salt marsh muds, mangrove muds are less anaerobic, rarely more negative than 300 mV, and usually within the range of 200 to +100 mV over a one meter profile (Limpsaichol, 1978; Boto and Wellington, 1984; Alongi, 1987a). Factors that prevent the buildup of entirely anoxic conditions in mangrove muds are not known, but this characteristic has been ascribed to several factors such as crab burrowing, poor quality of organic matter, high bacterial activity and translocation of oxygen by mangrove trees to their roots.

In riverine systems where fringing forests occur, sediment facies are predominantly fine sand with a mud veneer. These deposits are more oxidised and compacted than the finer deposits. pH readings in all deposits are consistently within the range of 6.2 to 7.2 (Alongi, 1987a), but interstitial salinities vary greatly with frequency of tidal inundation, rainfall and the extent of groundwater seepage. Concentrations of organic carbon and nitrogen in sandier sediments are naturally lower than in muds, ranging typically from 0.1 to 0.5% and from 0.01 to 0.1% by sediment dry wt, respectively.

Table 1. Bacterial densities, productivity and specific growth rates in surface sediments (0-2 cm) in the low, mid and high intertidal zones of four northeastern Australian mangrove forests during the austral winter dry (W) and summer wet season (S) (after Alongi, 1988a). Values are means ±1 S.D.

Estuaries	Bacteria (cells.g^{-1} DW x 10^{10})		Productivity (gC.m^{-2}.d^{-1})		Growth rate (d^{-1})	
	W	S	W	S	W	S
Morgan/McIvor						
Low	3.4 ± 2.2	2.3 ± 0.3	0.8 ± 0.2	5.1 ± 0.9	0.7	5.5
Mid	3.8 ± 0.6	1.6 ± 0.6	0.8 ± 0.2	3.2 ± 1.3	0.6	5.5
High	3.1 ± 1.6	0.2 ± 0.08	0.6 ± 0.2	3.6 ± 0.8	0.2	4.2
Lockhart						
Low	35.9 ± 10.9	2.9 ± 0.5	1.3 ± 0.2	2.1 ± 0.7	0.2	3.8
Mid	25.0 ± 6.3	12.8 ± 3.9	1.7 ± 0.2	1.4 ± 0.1	0.3	0.6
High	14.5 ± 1.7	8.4 ± 2.3	1.7 ± 0.6	1.0 ± 0.2	0.4	0.5
Claudie						
Low	28.1 ± 9.4	6.1 ± 0.5	1.4 ± 0.2	1.7 ± 0.2	0.2	0.8
Mid	20.3 ± 4.7	34.4 ± 0.8	2.4 ± 0.4	2.0 ± 0.1	0.6	0.3
High	4.2 ± 1.1	1.6 ± 0.08	1.0 ± 0.5	0.2 ± 0.1	0.6	0.3
Escape						
Low	9.1 ± 3.6	8.3 ± 3.4	0.9 ± 0.2	1.6 ± 0.1	0.4	1.0
Mid	31.3 ± 6.3	10.6 ± 2.7	0.9 ± 0.1	1.1 ± 0.2	0.2	0.2
High	1.1 ± 4.1	2.9 ± 1.1	1.0 ± 0.1	1.3 ± 0.1	1.6	0.3

Monsoonal rains can drastically alter sediment characteristics. For example, in the Claudie River in northern Australia, grain size of low intertidal deposits increased from coarse silt in the winter dry season to coarse sand during the summer wet season (Alongi, 1987a). Such erosion is probably more drastic in higher rainfall areas of Southeast Asia, although this remains to be documented.

Two features of mangrove sediments are typical of other tropical marine and estuarine deposits: (1) low (μM) concentrations of dissolved porewater nutrients such as ammonium, nitrate and phosphate, and (2) the presence in the intertidal water of soluble and condensed tannins derived from leaching roots and litter on the forest floor (see chapters 9 and 10 in this volume). In mangrove soils in northern Australia, soluble tannin concentrations typically range from 0.02 to 0.3% of sediment dry weight (Alongi, 1987b; Boto *et al.*, 1989).

The low dissolved nutrient levels in mangrove porewaters are similar to concentrations in other tropical sedimentary environments, but the reasons for this phenomenon are not wholly understood (Alongi, 1990a). Several factors that may explain the low concentrations include uptake by the mangroves themselves, high rates of bacterial uptake, moderately high redox potential and low quality of deposited organic matter. It is conceivable that most of the essential nutrients are tied up in the aboveground litter and living tree biomass and in bacterial biomass, as in tropical savannahs and rainforests. In any event, it is clear that mangrove sediments possess characteristics that, in toto, constitute a unique habitat for benthic organisms.

6.3 Microbes

6.3.1 Bacteria

Most of the early bacteriological work in tropical mangrove sediments utilized a variety of plate counting techniques that provide severe underestimates of true density (Ayyakkannu and Chandramohan, 1971; Matondkar *et al.,* 1980a, 1981). Some workers still use most probable number techniques, mainly to isolate and to test for the ability of specific strains to degrade or produce particular organic compounds (e.g. phenols, Gomes and Mavinkurve, 1982).

Despite the availability of direct counting techniques by epiflourescence microscopy since the mid1970's, studies providing reasonably accurate estimates of bacterial densities in mangrove sediments or on litter are rare (Dye, 1983a, Alongi, 1988a). As in other aquatic sediments, densities in mangrove soils vary greatly among different sediment types (higher in mud than in sand) and among seasons (higher at warmer temperatures). In mangrove muds of southern Africa, Dye (1983a) utilized his own direct count method and found surface (0-1 cm) densities ranging from 0.8-1.0 x 10^9 cells.g^{-1} DW of sediment in winter. Hoppe *et al.*, (1983) counted abundances within the same range (0.4-1.3 x 10^9 cells.g^{-1} DW) during winter in mangrove lagoonal muds in Colombia.

Seasonality and intertidal zonation of bacterial abundances has been extensively examined only along the Cape York peninsula of Australia (Stanley *et al.*, 1987; Alongi,

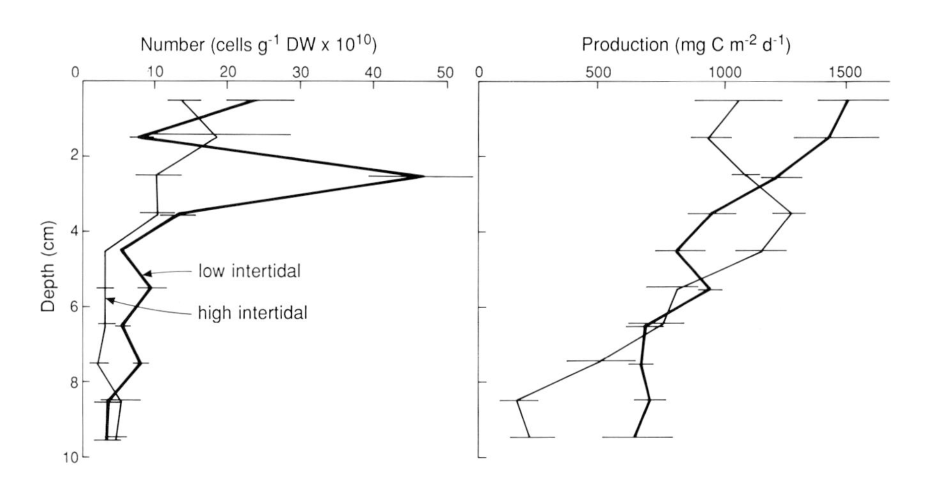

Figure 1. Vertical distribution of bacterial densities and productivity in mangroves of the low and high intertidal zones on Hinchinbrook Island (adapted from Stanley *et al.*, 1987 and unpubl. data).

1988a, b; Boto *et al.*, 1989). In four wet tropical estuaries north of 17°S latitude, bacterial densities were generally higher in the winter dry season than in the summer when scouring of surface sediments occurred (Table 1). In the mangroves of Goa, India, Matondkar *et al.*, (1980a) found similar results using plate counting methods. In the dry tropics, Alongi (1988b) found that bacterial densities exhibit little or no seasonality, with variations being related mainly to short term (weekly) changes in rainfall that affect salinities, rather than to changes in sedimentary organic carbon or nitrogen.

Intertidal zonation of bacterial abundances has been found, but patterns are inconsistent among estuaries in different seasons (Alongi, 1988a). For instance, densities in the Morgan/McIvor estuary of northern Australia were not significantly different between the low and mid intertidal forests and between the mid and high zones, but were significantly higher in the lower littoral zone compared with the upper intertidal forest. In the Lockhart estuary, zonation patterns differed seasonally where winter trends (low = mid > high) were not the same as in summer (low < mid = high). Despite these variations, bacterial densities in surface (0-2 cm) mangrove sediments in Australia are among the highest yet found (mean and range: 1.1(0.02-3.6) x 10^{11} cells.g^{-1} DW) in intertidal deposits, with seasonality explaining most of the variation. Among Australian estuaries, highest bacterial densities are generally found in the large deltaic systems (Hinchinbrook Island, Lockhart River) whereas lower abundances are usually observed in sediments of the smaller, fringing riverine forests (Morgan/McIvor River).

Vertical distribution of bacteria in tropical mangrove sediments has been examined only at one site to date (Stanley *et al.*, 1987; Boto *et al.*, 1989). In the mangroves on Hinchinbrook Island in northeastern Australia, density variations with sediment depth are erratic over 0 to 10 cm depth profiles (Figure 1). This lack of a clear increase or decrease in numbers is contrary to the decline in numbers usually observed with depth in temperate sediments. Large populations of nitrogen-fixing bacteria associated with subsurface roots and rhizomes may be partly responsible for the lack of decline with depth, although this remains to be substantiated.

Other mangrove components such as litter, harbor large populations of bacteria. For example, Robertson (1988) observed bacterial counts in excess of 10^{10} cell. g^{-1} DW of *Rhizophora stylosa* litter in litter bag decomposition experiments in tropical Australia. Bacterial densities initially increased rapidly on leaves as tannins were leached, but after 40 days densities either fluctuated widely, declined or increased slowly, depending on habitat and species of leaf. Previous studies have observed similar rates of litter decomposition (Fell and Master, 1973; Fell *et al.*, 1975; Cundell *et al.*, 1979; Boonruang, 1984) ascribed to microbial activity, but direct estimates of bacterial numbers were not assessed. Other microhabitats such as bark must have quantifiable numbers of bacteria as most of these components exhibit some form of microbial activity. For instance, Uchina *et al.*, (1984) found measurable nitrogen-fixation rates from lenticellate bark of *Bruguiera gymnorhiza,* although bacterial numbers were not estimated.

Identification of bacterial species in mangrove sediments is rare, but use of biochemical markers such as fatty acids (e.g. see Gillan and Hogg, 1984) would offer relative abundances of specific chemotypes (sulphate-reducers vs denitrifiers) which may be more meaningful ecologically than compilation of species lists.

Measurements of various facets of bacterial activity (e.g. nitrogen fixation, respiration; see Chapters 9 and 10, this volume) in mangroves have been made, but estimates of bacterial cell production are few. The only available studies have been made in mangroves of tropical Australia (Stanley *et al.*, 1987; Alongi, 1988a, b; Boto *et al.*, 1989) using the tritiated thymidine methodology wherein thymidine uptake is equated to cell production via rates of DNA synthesis. The method has its problems, but offers the only reasonable way to estimate bacterial cell production in sediments.

In the same mangrove estuaries of the Australian wet tropics where bacterial densities were examined, Alongi (1988a) found consistently high rates of bacterial productivity in surface (0-2 cm) sediments. Converted to carbon, rates ranged from 0.2-5.1 $gC.m^{-2}.d^{-1}$ with an average of 1.6 $gC.m^{-2}.d^{-1}$. Specific growth rates (production divided by standing stock) ranged from 0.2-5.5 d^{-1} and on average, were fast (mean = 1.1) indicating actively dividing communities. At most sites, production rates were higher in summer than in winter (Table 1) as were daily specific growth rates. Trends in intertidal zonation generally followed those as exhibited by bacterial numbers. That is, varying greatly among estuaries with season. In the dry tropics, rates of bacterial productivity are not as high (45-1725 $mgC.m^{-2}.d^{-1}$; mean of 475), but rates of production and specific growth were faster in warmer months, although seasonality was not distinct (Alongi, 1988b).

Rates of bacterial cell production in these mangrove habitats are closely coupled to the porewater concentrations and fluxes across the sedimentwater interface of DOC and DON (Stanley *et al.*, 1987; Boto *et al.*, 1989). Vertical rates of production (Figure 1) relate well to dissolved nutrient concentrations, at least within the top 10 cm of sediment.

Flux chamber experiments conducted by Stanley *et al.*, (1987) and Boto *et al.*, (1989) in the mangrove forests on Hinchinbrook Island revealed negligible DON (as amino acids) and

DOC flux in or out of the sediments in untreated chambers, but large and significant (27-69 $mgN.m^{-2}.d^{-1}$ and $0.4\text{-}2.4gC.m^{-2}.d^{-1}$) effluxes in chambers where poisons were added to kill the benthic biota. Fluxes accounted for between 9-38% and 5-19% of the nitrogen and carbon required to support the measured rates of bacterial production (Stanley *et al.*, 1987). Similarly, DOC fluxes provided, on average, 35% of bacterial productivity requirements at the sediment-water interface. These experiments indicate that bacterial populations in surface sediments utilize all or nearly all of the DOM flux to the sediment-water interface in tropical mangroves.

The distribution, abundance and dynamics of bacteria in mangroves are controlled by a variety of chemical, biological and physical factors including nutrients, sediment type, temperature, metazoan and protozoan grazers, and frequency of tidal inundation, as they are in temperate intertidal sediments. The same major regulatory factors thus apply to sedimentary bacterial communities worldwide.

6.3.2 Cyanobacteria

Cyanobacteria are abundant in the tropics, particularly along shoal margins of reefs and coastal lagoons that are often lined with mangroves (Alongi, 1989, 1990a). To date, only the studies by Potts (1979, 1980) and Potts and Whitton (1980) have adequately documented their distribution and standing stocks in mangroves. Potts and his coworkers studied the bluegreen algal mats within the hypersaline pools and shallow lagoons of the Red Sea and Aldabra atoll. These mats are composed mainly of filamentous forms such as *Lyngbya*, interwoven with coccoid bluegreens (e.g. *Synechococcus*) and on the surface with pennate diatoms and in the deeper layers with a variety of phototrophic bacteria.

In Aldabra lagoon, cyanobacteria are sparse in sediments, but growth is considerably more abundant on prop roots and pneumatophores. In the forests, a number of Rivilariaceae species and heterocystous forms such as *Scytonema* sp. form conspicuous growths on pneumatophores, with nonheterocystous species restricted to the sediment surface. Potts and Whitton (1980) postulated that the low standing stocks in the mangroves are due to heavy grazing by gastropods, low light intensity and competition from microheterotrophs.

Cyanobacteria exhibit sharp horizontal zonation on pneumatophores and on prop roots. Potts (1979) noted that an exclusive nitrogen-fixing flora of Rivilariaceae occurred on pneumatophores in the Sinai mangroves, and cited sulfide, desiccation, salinity and frequency of tidal wetting as the major factors controlling their zonation on mangrove roots. Cyanobacterial standing stocks are low in mangroves which belies their capacity for very high rates of nitrogen fixation and photosynthesis.

Table 2. Microalgal standing stocks (μg chl *a*.g^{-1} sediment DW) in surface sediments (0-2 cm) in four northeastern Australian mangrove estuaries during winter dry and summer wet seasons (after Alongi, 1988a). Values are means ± 1 S.D.

	Morgan/McIvor Estuary			Lockhart Estuary		
	Low	Mid	High	Low	Mid	High
winter	1.1 ± 0.3	0.9 ± 0.1	0.5 ± 0.3	2.0 ± 1.7	0.2 ± 0.3	0.8 ± 0.2
summer	1.2 ± 0.3	0.3 ± 0.3	0.1 ± 1.7	0.7 ± 0.4	0.9 ± 0.6	0.7 ± 0.2
	Claudie Estuary			**Escape Estuary**		
	Low	**Mid**	**High**	**Low**	**Mid**	**High**
winter	3.4 ± 0.5	3.4 ± 0.1	2.3 ± 0.1	3.0 ± 1.4	4.4 ± 0.9	0.0 ± 0.0
summer	1.6 ± 0.7	3.3 ± 1.0	1.4 ± 0.6	2.5 ± 0.5	2.3 ± 1.9	1.2 ± 0.5

6.3.3 Microalgae

In tropical sediments, including mangroves, standing stocks of microalgae (as chlorophyll *a*) are generally low, usually ≤ 5μg chl a.g^{-1} DW (Alongi, 1989, 1990a). Species composition of microalgae in mangroves is known only for a few forests in the Caribbean, where diversity is high (Hagelstein, 1938; Reyes-Vasquez, 1975) and dominated by diatoms. In Indian mangroves, chlorophyll *a* concentrations varied little, ranging from 2.6-6.1 μg.g^{-1} DW (Krishnamurthy *et al.*, 1984). Measurable levels were found to a sediment depth of 6 cm. In a mangrove-lined tidal pool in the Gulf of Elat (Red Sea), Sournia (1977) noted low levels of chl *a* with significantly higher concentrations with distance seaward from the trees.

The few studies available indicate little variation in sedimentary chlorophyll *a* with season or among intertidal zones. In Australian mangroves, chlorophyll *a* levels were generally low (Table 2) and inconsistent among estuaries and seasons (Alongi, 1988a). In the dry Australian tropics, microalgal standing stocks peaked during winter (no rain) and were consistently low (< 3-4μg.g^{-1} DW) in summer due to rain and cloud cover (Alongi, 1988b).

Alongi (1988a, b) has attributed the low chlorophyll *a* levels in mangrove sediments to low light intensity under the dense mangrove canopy (and see Chapter 8, this volume). However, the studies of Cooksey *et al.* (1975) and Cooksey and Cooksey (1978) indicate that dissolved organic carbon in the porewaters is inhibitory to benthic diatom growth in mangroves. The composition of the DOC was not examined but it is probable that the inhibitory agent was soluble phenolic compounds such as tannins, which constitute a major fraction of the porewater DOC in some mangroves (Boto *et al.*, 1989). Other factors such as frequency of tidal wetting and grazing by herbivores are probably also partly responsible for the low microalgal standing stocks.

6.3.4 Macroalgae

There is a rich macroalgal flora attached to pneumatophores and bases of mangrove trees throughout the tropics. This epiphytic flora is dominated by red algae, particularly in the genera *Bostrychia, Caloglossa* and *Catenella* (King, 1981; Wee, 1986; Chihara and Tanaka, 1988). Chihara and Tanaka found a distinct vertical zonation of the macroalgae in the brackish water mangroves of East Indonesia. *Rhizoclonium* occurred in the upper intertidal zone, *Bostrychia* in the upper to the mid-intertidal zone, *Caloglossa* in the mid-intertidal zone while *Catenella, Cladophora* and *Gelidium* occurred in the lower intertidal zone. The productivity of the above has not been studied, however, Dor and Levy (1984) estimated that the productivity of the macroalgae in the hardbottom mangroves of Sinai was 200 times that of the phytoplankton. Rodriguez and Stoner (1990) investigated the epiphytic algae associated with *R. mangle* roots in Puerto Rico. Species richness (8 spp) was low compared with other Caribbean forests, but total algal biomass for the lagoon was high (~ 7.4×10^4 kg DW), equivalent to the annual leaf litterfall. Species composition varied within the lagoon in relation to degree of shelter, distance from shore and from the lagoonal inlet and proximity to freshwater.

6.3.5 Protozoa (excluding Foraminifera)

Studies of freeliving protists, comprised mainly of ciliates, flagellates, foraminiferans and amoebae, are comparatively few for marine sedimentary habitats worldwide. Our ignorance of this highly diverse and abundant group can be attributed to several factors: taxonomic problems, lack of expertise and difficulties in extracting these microbes from sediments. Alongi (1986) found that most extraction methods lead to severe underestimates of protist densities, and thus the number of reliable density estimates is low.

Most of the pre-1975 studies of tropical benthic protozoa were taxonomic; at least three such studies have examined protists in and near mangrove habitats (Ganapati and Narasimha Rao, 1958; Margalef, 1962; Aladro and Lopez Ochoterena, 1967; Odum and Heald, 1972). More recent investigations have examined smaller ciliated protists and microflagellates due to improvements in microscopy and an increased awareness of their energetic importance (Snyder, 1985; Larsen and Patterson, 1990). These studies indicate high diversity of species and morphotypes. For instance, in sediments bordering mangroves in Belize, Snyder (1985) found 67 distinct ciliate morphotypes whereas Nouzacede (1976) found a diverse assemblage of ciliates, predominantly of the Geleiidae.

Seasonality and intertidal zonation of protozoan densities have been examined mainly in the mangroves of northeastern Australia (Alongi, 1988a). Densities of ciliates and large flagellates (microflagellates were not studied) were not significantly different among intertidal zones and estuaries (Table 3), but were more abundant in summer than in winter. Flagellates, mainly dinoflagellates, cryptomonads, and some phytoflagellates, were generally more numerous than ciliates (colorless euglenids, spirotrichs and hypotrichs). Small foraminiferans, naked amoebae and amoeboflagellates were noted rarely; their virtual absence may have been due to methodological limitations.

Table 3. Densities of ciliates and large (>15-20 μm) flagellates in surface (0-2 cm) mangrove sediments of three intertidal heights in four northeastern Australian estuaries (after Alongi, 1988a). Values are means ± 1 S.D. W = winter; S = Summer.

Estuaries	**Ciliates ($cells.cm^{-2}$)**		**Flagellates ($cells.cm^{-2}$)**	
	W	**S**	**W**	**S**
Morgan/McIvor				
Low	70 ± 1	200 ± 70	43 ± 12	94 ± 22
Mid	110 ± 60	99 ± 46	100 ± 70	120 ± 80
High	39 ± 18	250 ± 130	20 ± 28	110 ± 30
Lockhart				
Low	17 ± 8	110 ± 10	6 ± 8	90 ± 48
Mid	6 ± 8	210 ± 40	17 ± 8	120 ± 110
High	17 ± 24	180 ± 90	11 ± 16	180 ± 170
Claudie				
Low	26 ± 12	110 ± 30	43 ± 12	99 ± 44
Mid	9 ± 12	81 ± 82	43 ± 12	81 ± 44
High	63 ± 88	230 ± 96	26 ± 12	160 ± 90
Escape				
Low	15 ± 21	28 ± 32	37 ± 10	93 ± 20
Mid	7 ± 10	42 ± 21	52 ± 10	160 ± 80
High	75 ± 42	260 ± 80	7 ± 10	64 ± 27

Seasonality of protozoans has also been examined in the Gangetic estuary of India (Bhattacharya *et al.*, 1987). Gymnamoebae were found to be the dominant benthic group, especially species of the genus *Acanthamoeba*. Bhattacharya *et al.*, followed the seasonal variations in abundance of the dominant species, *A. rhysodes*. They found that *A. rhysodes* populations attained peak abundances during the monsoon and declined throughout the postmonsoon and premonsoon periods. This pattern was attributed to very dry soil conditions during the non-monsoonal periods. In the dry tropics, seasonality is not clear, with wide fluctuations in abundances of ciliates and large flagellates found in mangrove habitats in northeastern Australia (Alongi, 1988b). Neither group correlated with other biota nor with changes in edaphic characteristics.

Vertical distribution of benthic protozoa in tropical mangroves has been examined from only one site in northeastern Australia (Figure 2). Alongi (unpublished data) found that vertical distribution is dependent upon the presence of below ground roots and rhizomes. In mangrove sediments without root matter (to 5 cm depth), ciliate and flagellate densities decline markedly below surface layers (Figure 2a). Within an *Avicennia marina* stand, densities decline below the surface but exhibit a subsurface maxima in the 4-5 cm layer where roots and rhizome material are present. This layer is more oxidized and has more

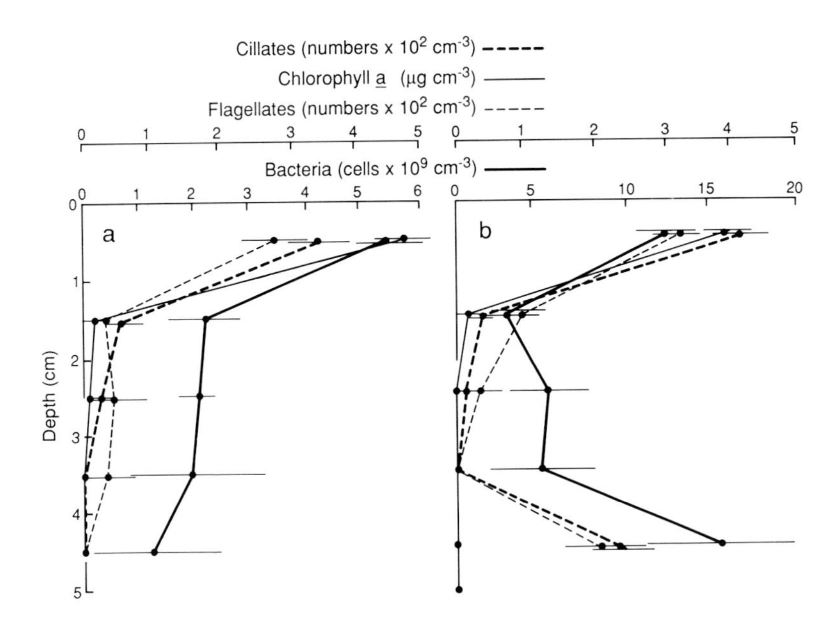

Figure 2. Vertical distribution of protists, microalgae and bacterial densities in (a) unvegetated muds and (b) in mud within the pneumatophores of *Avicennia marina*, Hinchinbrook Island, July 1986 (Alongi, unpublished data).

bacteria, suggesting that the protozoans in subsurface layers are attracted to more oxidized roots and to root-associated bacteria.

Population growth of some mangrove associated protists have been examined in laboratory culture by Ghosh and Choudhury (1987) and Alongi (1990b). Strains of *Acanthamoeba astrohyxis* were cultured in saline agar with *Escherichia coli* over a range of salinities (5-30 ‰) encountered in its natural habitat. Ghosh and Choudhury found that growth varied little with salinity, but strains isolated from mangrove and unvegetated creek bank habitats had different nutritional requirements. Mangrove litter strains grew best with soil extract whereas the strains isolated from unvegetated sediments grew best on distilled agar and bay water agar. Their results indicate the ability of mangrove amoebae to withstand salt stress and to utilize unknown dissolved components in mangrove soils for growth.

Alongi (1990b) found that mangrove litter is a poor food for benthic protists. Populations of mixed species of *Euplotes*, smaller ciliates and zooflagellates (Figure 3), isolated from subtidal sediments receiving outwelled mangrove litter, grew significantly better on mixed cereal (C:N ratio = 17) than on a diet of mangrove litter (C:N ratio = 52). These results suggest that protozoan densities are low in the mangroves of northeastern Australia, in part, because of low food quality. Other factors may be partly responsible for the low abundances, such as low water content and low microalgal concentrations.

The ecological energetics (respiration, secondary production) of protozoa in mangroves, as in other sedimentary environments, are unknown.

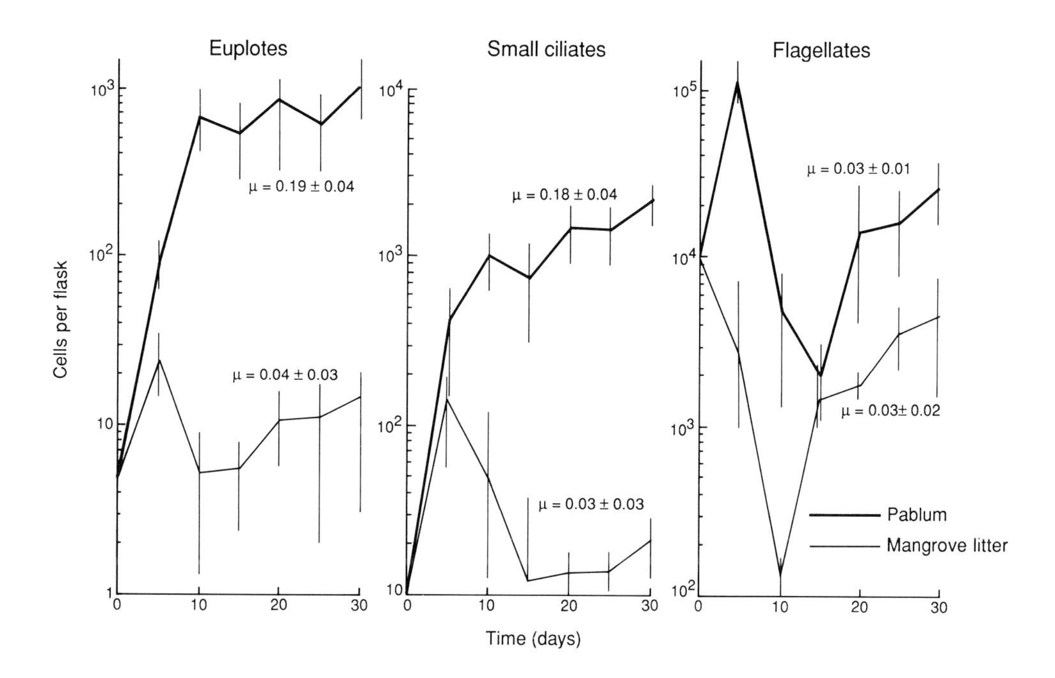

Figure 3. Growth of mixed populations of the ciliate *Euplotes* spp., hymenostomid and hypostomatid ciliates and zooflagellates in tissue cultures with mixed cereal (Pablum) and with mangrove litter. (From Alongi, 1990. With permission from Inter-Research.)

6.3.6 Foraminifera

Foraminifera occur in mangrove sediments, but all studies to date have been descriptive, providing lists of species found and descriptions of new genera and species (Cushman and Bronnimann, 1948a, b; Bronnimann and Zaninetti, 1965; Boltovskoy, 1984). Based on a preliminary study in three mangrove forest zones in Malaysia, Sasekumar (1981) found foraminifera densities varied from 318.10 cm^{-2} in a *Avicennia* forest to 2573.10 cm^{-2} in a *Rhizophora* forest and 338.10 cm^{-2} in a landward *Bruguiera* forest. The common species were *Arenoparella malaysiana, Miliammina polita, Textularia* sp., *Miliammina fusca* and *Haplophragmoides labukensis*. Most species found in mangrove muds are small (< 1 mm), but large species are occasionally found, particularly in fringing forests in close proximity to carbonate deposits (i.e. reef cays) and seagrass beds (Boltovskoy, 1984).

6.3.7 Fungi (including yeasts)

Mangrove fungi are almost exclusively saprobic and belong primarily to the Ascomycetes, Deuteromycetes and Basidiomycetes (see reviews of Hyde and Jones, 1988; Hyde, 1989). Most of the literature of manglicolous fungi deals with their vertical distribution and the taxonomy of species which live on prop roots, trunks and branches (Lee and Baker, 1973; Matondkar *et al.*, 1980b; Araujo *et al.*, 1981; Garg, 1983; Misra, 1986; Hyde, 1990). A less extensive literature is available concerning their role in decompositional processes (Fell and Master, 1973; Fell *et al.*, 1975; D'Souza and D'Souza, 1979a, b; Velho and D'Souza, 1982; Mouzouras, 1989).

Estimates of fungal and yeast colony standing stocks indicate wide variations with geographical location, intertidal zone and sediment depth. Abundances of fungi range from 10^3-10^7 colonies.g^{-1} sediment DW and yeasts range in abundance from 10^2-10^6.g^{-1} DW (Table 4). These estimates are probably low as the methodology to estimate fungal and yeast abundance is poor.

Vertical and horizontal zonation occurs in most forests on submerged tree parts (Hyde, 1990). Terrestrial species develop on trunks above the high tide mark and niche overlap may occur at or below this level between terrestrial and marine species (Lee and Baker, 1973; Hyde, 1990).

In sediments, highest fungal counts are recorded at the surface with decreasing abundance with increasing soil depth (Garg, 1983). Only Matondkar *et al.,* (1980b) have provided evidence of temporal changes in mangroves in India. They found that fungal

Table 4. Dilution plate estimates of fungal and yeast populations in various tropical mangrove habitats.

Location	Habitat	Counts.g^{-1} DW sediment	Reference
Heeia, Hawaii	Rhizosphere soil		Lee and Baker, 1973
	Low intertidal	$^{a}5.0 \times 10^6$, 5.1×10^{6b}	
	Mid intertidal	$^{a}6.6 \times 10^6$, 7.3×10^{6b}	
	High intertidal	$^{a}8.0 \times 10^6$, 7.9×10^{6b}	
	Non-rhizosphere soil		
	Low intertidal	$^{a}3.2 \times 10^4$, 3.6×10^{5b}	
	Mid intertidal	$^{a}9.6 \times 10^4$, 6.8×10^{5b}	
	High intertidal	$^{a}1.4 \times 10^4$, 7.2×10^{5b}	
Goa, India	Chapora, Mandori, Zuari mangrove muds	0.02×10^3 1.2×10^{5d}	D'Souza and D'Souza, 1979
Goa, India	Mandori-Zuari mangroves	2.2×10^{4a} (pre-monsoon) 5.2×10^{4b} (monsoon) 3.8×10^{4b} (post-monsoon)	Matondkar *et al.*, 1980a,b, 1981
Bombay, India	*Avicennia* roots	$0.31.6 \times 10^{4b}$	Araujo *et al.*, 1981
	Avicennia leaf litter	$0.68.6 \times 10^{5b}$	
	mangrove muds	0.012×10^{6c}	

[a] Actinomycete only, [b] Fungi, [c] Yeasts and Fungi, [d] Yeasts

densities were highest during the monsoon season, ascribing this result to increased litterfall during periods of heavy rainfall.

Fungi are known to be important participants in the decomposition of mangrove seedlings and leaves (Fell and Master, 1973; Fell *et al.*, 1975) and wood components (Mouzouras, 1989). Yeasts participate in decomposition processes, but to a lesser extent (Velho and D'Souza, 1982), and appear to exhibit some nitrogen fixing ability (D'Souza and D'Souza, 1979a). Most fungal species attack wood of submerged tree parts, including roots, whereas bark-covered wood is not readily invaded. Fungal hyphae grow mainly in the less-lignified secondary cell layers ('soft rot') of wood, facilitating later attack by woodborers. Fungi are, however, early colonizers of leaves and seedlings, exhibiting species succession and facilitating colonization of bacteria, protozoa, meiofauna and other invertebrates (Fell *et al.*, 1975). The role of fungi in nitrogen fixation and immobilisation and carbon flow will be discussed in chapters 9 and 10, this volume).

6.4 Meiofauna

Densities of mangrove meiofauna were recorded by Kondalarao (1983) in his investigations of the fauna of the Godavari River in India. He found very high abundances (mean = 2130 per 10 cm^{-2}) in surface sediments and noted the dominance of nematodes. Three mangrove forest zones in the west coast of peninsular Malaysia were studied at monthly intervals for a year by Sasekumar (1981). He found mean total numbers of meiofauna ranged from 1129 to 407 per 10 cm^2 with the highest values in the seaward *Avicennia* forest and the lowest values in the landward *Bruguiera* forest. The mid-tide level *Rhizophora* forest harboured 584 individuals per 10 cm^2. Nematodes dominated the meiofauna followed by harpacticoid copepods and oligochaetes except in the *Avicennia* forest where a kinorhynch constituted 5% of the total and was the third most common taxa. Earlier studies provided estimates of densities, but most of this data is not readily comparable to modern investigations, being mainly descriptive (and usually taxonomic), having used inadequate and/or unspecified sampling procedures, or providing densities in incomparable units (e.g. Gerlach, 1957).

Subsequent investigations, mainly in northeastern Australia and India, have indicated generally low (< 500 individuals per 10 cm^2) densities of meiobenthos. Krishnamurthy *et al.*, (1984) examined the meiofauna of mangroves in the Bay of Bengal and recorded total densities in the range of 35-280 individuals per 10 cm^2. Similarly low densities (36-245 individuals per 10 cm^2) have been recorded in mangroves of southern Cuba (Lalana Rueda and Gosselck, 1986) and in estuaries along the eastern coast of Cape York peninsula in Australia (Alongi, 1987b). Highest numbers have been recorded on a mangrove creek bank in the Kakinada Bay system on the east coast of India (Kondalarao and Ramana Murty, 1988). Total mean numbers of animals ranged from 491 to 5924 individuals per 10 cm^2, with nematodes being the dominant group. Other Indian sites have very abundant populations (Govindan *et al.*, 1983; Varshney, 1985), but the organic matter concentrations measured indicate that most or all of these locations are polluted, probably from untreated or only primary treated sewage. High densities have also been found on many other Indian mudbanks adjoining mangroves (Krishnamurthy *et al.*, 1984).

The limited studies available indicate that seasonality depends upon proximity to the equator, with seasonal variation increasing away from this demarcation. Great differences, however, can be found even within a biogeographical zone. For instance, in India, there is little distinct seasonality of meiobenthos in mangroves within the Bay of Bengal, but further north, monsoonal rains greatly influence seasonality (Kondalarao and Ramana Murty, 1988) where lowest densities are recorded in the post flood period of December-January. In the wet Australian tropics, Alongi (1987b) found that meiofaunal densities in mangroves were highest during the summer monsoon. In the dry tropics, lowest densities are recorded in spring and summer when sediment temperatures can exceed 40°C (Alongi, 1988b). In subtropical areas, seasonal patterns follow those well-known for temperate areas where maximum abundances are recorded in late summer (Dye, 1983b, c; Hodda and Nicholas, 1985).

Vertically, most meiofauna are recorded in the top 10 cm, as in most temperate habitats, but some workers (Dye, 1983c) have recorded densities down to at least 50 cm. High subsurface densities usually coincide with more positive redox levels and sandier substrates.

Intertidal zonation of meiobenthos in mangroves has been studied in subtropical Africa (Dye, 1983b, c), in temperate southern Australia (Hodda and Nicholas, 1985) and in tropical northern Australia (Alongi, 1987b). Dye (1983a, b) found high densities of meiobenthos in the mid-intertidal in Africa, whereas densities generally decreased with topographic height in Australia. All of the Indian studies cited earlier were conducted in low intertidal mangroves or adjoining creek banks, where maximum concentrations are most likely.

Meiofauna living on leaf litter, prop roots and wood have been examined by several workers (Fell *et al.*, 1975; Krishnamurthy *et al.*, 1984; Alongi, 1987c), with densities generally lower than in the sediments. In the Caribbean, Fell *et al.* (1975) noted very rapid rates of colonization on submerged leaves with densities up to > 1 per cm^2 of leaf surface. Most organisms were nematodes and harpacticoid copepods. Densities of meiofauna on leaf litter, prop roots and pieces of wood in tropical Australian mangroves were similarly low (< 5 per leaf), but a specific litter nematode fauna was not evident (Alongi, 1987a).

The distribution and abundance of mangrove meiobenthos is evidently affected by a variety of factors acting either antagonistically or synergistically, as found in temperate intertidal habitats. Krishnamurthy *et al.*, (1984) attributed seasonal and zonational patterns to changes in salinity, as influenced by monsoonal rains, and availability of microbial food. Dye (1983b) found a good correlation of meiofaunal densities with redox potential, but poor relationships with temperature and pH. In the Kakinada Bay system, Kondalarao and Ramana Murty (1988) related faunal density patterns to the monsoons and sediment type. In Australia, the inverse relationship between tidal height and abundance is probably due to frequency of tidal inundation, changes in temperature and desiccation (Hodda and Nicholas, 1985; Alongi, 1987b). Within intertidal zones, density levels correlate with different factors (Alongi, 1987b). In the low and mid-intertidal forests, correlations were found with organic content and Eh, whereas the low abundances in the high intertidal related best to grain size and water content. It is apparent that monsoonal rains play an important role in regulating meiofauna numbers in wet tropical forests.

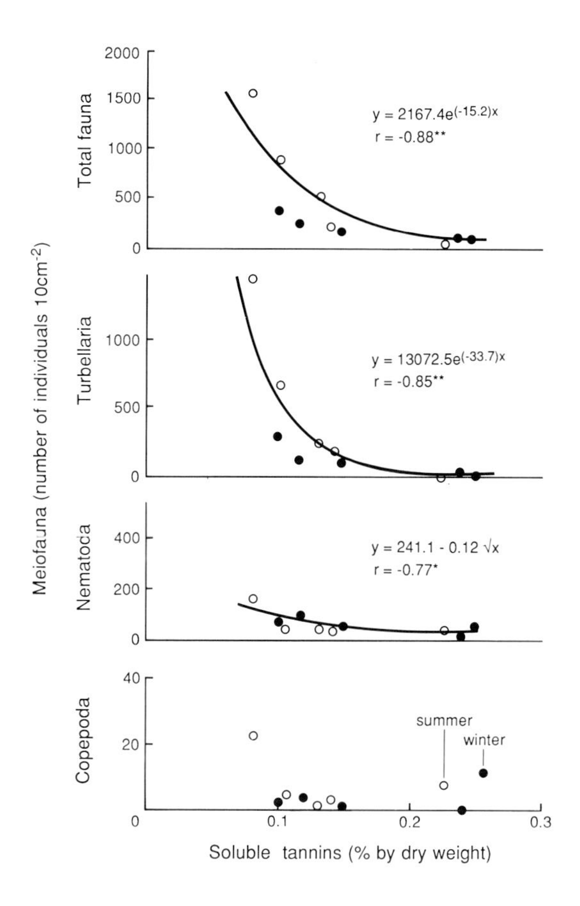

Figure 4. The relationship between soluble tannin concentrations and densities of various meiofaunal taxa in four mid-intertidal forests in tropical Australia. (From Table 1, Alongi, 1987a.)

It has been recently shown that soluble tannins derived from mangroves have a significant negative impact on meiofaunal densities in tropical Australia (Alongi, 1987b). Nearly all of the dominant taxa correlated negatively with concentrations of sediment tannins, particularly in the low and mid (Figure 4) intertidal forests. The critical concentration appears to be within the 0.10-0.15 (% by dry weight) range. Tannins also have a negative impact on population growth. Laboratory populations of the deposit-feeding nematode *Terschellingia longicaudata* were reared on separate diets (equivalent N rations) of mixed cereal, *Avicennia marina* leaves (0.8% tannins) and on leaves of *Rhizophora stylosa* (7.2% tannins). The worms did not grow on fresh, tannin-rich leaves of *R. stylosa,* but growth was significantly better on the *Avicennia* diet and best on a diet of tannin-free, mixed cereal (Alongi, 1987b).

These results suggest that plant-derived tannins may account for the generally low meiofaunal numbers recorded within most tropical mangrove forests but other factors also play a role. This statement is best supported by laboratory experiments conducted by Tietjen and Alongi (1990) with the nematode *Monhystera* sp., isolated from the same mangrove forests as *T. longicaudata*. *Monhystera* sp. was fed a diet of fresh *Avicennia marina* detritus, mixed cereal, and a separate diet of fresh *R. stylosa* litter, but including mixed cereal. *Monhystera* sp. populations died in an earlier experiment when fed *R. stylosa* detritus (identical to the result for *T. longicaudata*), but with a mixture of *R. stylosa* detritus and mixed cereal, population growth of the worms exceeded those on a diet of mixed cereal alone. These results suggest that the high N content of the cereal (5% by DW) more than compensated for the high tannin content of the *R. stylosa* detritus.

Meiofaunal organisms may have special adaptations to survive the harsh conditions found in mangrove sediments. For instance, Nicholas *et al.*, (1987) found that some of the dominant mangrove-dwelling nematodes of southern Australia contain densely packed intracellular inclusions, particularly in intestinal cells, containing concentrated Si, P, S, K, Ca, Fe, Na, Zn and Al. These inclusions are thought to be a mechanism of detoxification. Detoxification of HS in these worms appears to occur by deposition of insoluble metal sulphides rather than by oxidation to elemental sulphur.

Reproductive adaptations may also play a role in their survival in tropical mangroves. Hopper *et al.*, (1973) found that nematodes associated with *R. mangle* litter in Florida exhibited shorter generation times up to 33-35°C and some species survived up to temperatures of 39°C. Such adaptations, and the lack of them, may be partly responsible for the low to moderate levels of species diversity found for mangrove nematodes (Decraemer and Coomans, 1978; Krishnamurthy *et al.*, 1984; Alongi, 1987b, 1990c) and harpacticoid copepods (Kandalarao, 1984; Kondalarao and Ramana Murty, 1988).

A variety of feeding types occur among meiofaunal assemblages in mangroves. In India, Krishnamurthy *et al.* (1984) observed that the sediment fauna was comprised primarily of deposit feeders whereas the decaying leaf and root litter harbored mainly predatory and omnivorous nematodes. Leaf and prop root epiphytes were inhabited mostly by bacterial and algal feeding species. In Australia, Alongi (1987c) examined variation in the trophic and species composition of nematodes among intertidal zones and estuaries. He found low to moderate species diversity with generally few species at each site. Multivariate analyses, however, revealed a fair degree of separation among intertidal zones and estuaries in species composition. Alongi speculated that such compositional differences are due to factors peculiar to each estuary and zone, fostering the establishment of fairly distinct communities over time. Deposit feeders were the dominant trophic group (50%) at all sites, with nearly equal relative abundance of epistrate feeders (28%) and omnivore/predators (22%). In temperate Australian mangroves, species richness appears to be greater than in the tropics (Hodda and Nicholas, 1985), although species more typical of terrestrial or freshwater habitats were also found, particularly a specific stylet-bearing fauna.

Composition of harpacticoid copepod assemblages in tropical mangroves and adjacent habitats have been examined only in the Kakinada Bay system in India by Kondalarao (1984) and Kondalarao and Ramana Murty (1988). Species composition was found to change from the estuarine headwaters to the lower marine reaches of the estuary, with decreasing diversity seaward. Numbers of species per site ranged from 6 to 17 with *Pseudostenhalia secunda* being ubiquitous. Variation in diversity was also found to be a function of sediment type, with lowest richness in muds and highest richness in fine muddy sands. Seasonally, diversity was lowest during monsoons. Two main communities were identified: (1) a mangrove mud assemblage dominated by *P. secunda* and *Stenhelia longifurca* and (2) a coastal fine sand community characterized by *Amphiascoides* sp. and *Hastigerella* sp.

Trophic interactions (competition, predation, amensalism) undoubtedly must play an important role in structuring these meiofaunal communities, but studies of benthic trophic processes in tropical mangroves are very limited. It is generally well established that bacteria, microalgae, fungi and some protists are fed upon by meiobenthic organisms, but little empirical evidence from mangrove habitats is available.

Evidence from tropical mangroves in Australia (Alongi, 1988a) suggests that the dynamics of benthic bacterial and microalgal assemblages may not be tightly coupled to the dynamics of protozoan and meiofaunal consumers (see Chapter 10, this volume for details).

At the opposite end of the food web, negative interactions of meiofauna with macrofauna may also partly explain the low meiofaunal densities. Predation, disturbance and/or competition for food and space are likely modes of interaction. The study of Dye and Lasiak (1986) on fiddler crab-meiofauna interactions suggests such an effect. In caging experiments, they found that meiofaunal numbers increased two to fivefold in cages excluding crabs. Meiofauna were not ingested, indicating that disturbances or competition was the cause. Dye and Lasiak (1986) suggested that meiofauna may also migrate downwards to avoid such interactions, but no evidence was provided.

Meiofauna may be an important food source for larger organisms, but the evidence from mangroves is scant. Krishnamurthy *et al.*, (1984) stated that in Indian mangroves, nematodes are consumed by prawns and carnivorous fishes such as *Leiognathus* spp. and *Oxyurichthys* sp., but data on their contribution to gut volume was not provided. It is likely that they are an important source of nutrition for some animals, but not for others. In Malaysia, Chong and Sasekumar (1981) and Sasekumar (1981) found that meiofauna were a large component in the gut contents of some organisms such as prawns, but not in other animals, such as gastropods and large infaunal crustaceans. For temperate systems, Gee (1989) has shown conclusively that in intertidal and subtidal habitats, meiofauna (harpacticoid copepods) are an important food for the small juveniles of flatfish and salmonids.

6.5 Intertidal Macrobenthos

6.5.1.Infauna

Most of the literature concerning infaunal communities in tropical mangroves is descriptive, with most information detailing variations in species composition with tidal height but

Table 5. Densities of macrobenthos (number of individuals.m^{-2}) in some tropical mangrove and adjacent sand and mudflat sediments. Values are means (±1 S.D.).

Location	**Habitat**	**Total densities**	**Species per site**	**Reference**
Klang, Malaysia	Mangroves, low intertidal	70 ± 102 (epifauna) 137 ± 91 (infauna)	0-11 7-24	Sasekumar, 1974
Morrumbere estuary, S. Africa	Mangroves, low intertidal	170 ± 152	31-74	Day, 1975
	Sandflat	242 ± 235	42-103	
Surin Island, Thailand	Mangroves, low intertidal	4	5	Frith *et al.*, 1976
	Mangroves, mid intertidal	10	8	
	Mangroves, high intertidal	28	34	
	Mudflat	26	11	
	Sandflat	43	11	
Phuket Island, Thailand	Mangroves, low intertidal	80 ± 28	26	Frith, 1977
	Mangroves, mid intertidal	218 ± 34	92	
	Mangroves, high intertidal	129 ± 65	60	
	Mudflat	52	36	
	Sandflat	147 ± 20	47	
Kuala Selangor, Malaysia	Mudflat near mangroves	304 ± 247	22	Broom, 1982
Northwest Cape, N.W. Australia	*Avicennia* forest	992 ± 722	122	Wells, 1983
	Rhizophora forest	257 ± 390	59	
	Mudflat	473 ± 319	31	
	High intertidal flat	1 ± 3		
Cochin estuary, India	Mangroves, low intertidal	5872 (pre-monsoon) 420 (monsoon)	N.A.	Kurian, 1984
	Subtidal	16,000 (pre-monsoon) 1036 (monsoon)		
Ka Yao Yai, Thailand	Mangroves, low intertidal	49	43	Nateewathana and Tantichodok, 1984
	Mangroves, mid intertidal	107	55	
	Mangroves, high intertidal	142-178	70	
	Mudflat	247	109	
	Sandflat	190	48	
Ranong, Thailand	Mangroves, mid intertidal	1190	49	Wada *et al.*, 1987
Irimote Island, The Ryukyus	Mangrove, low-mid intertidal	59-652 (mostly epifauna)	15-32	Shokita *et al.*, 1989
Missionary Bay, N.E. Australia	Mangrove, mid intertidal	2.1 + 1.9 (epifauna) 89 + 62 (infauna)* 62.5 (cryptofauna)	1-7 1-3 19-31	Cragg, Robertson and Sasekumar unpublished data.

* excluding crabs.

providing little quantitative data on variations with season (e.g. Macnae and Kalk, 1962; Berry, 1972; Frith and Frith, 1978; Hutchings and Recher, 1982). Many workers have used different sampling techniques and sieve sizes as well as sampling to different sediment depths, usually in only one season or at one site. Until recently (Warren, 1990), most techniques used to estimate crab numbers, for instance, were subjective and underestimated true abundances.

A few studies have, however, provided comparable estimates of infaunal abundances (Table 5). The data show that densities are generally low compared with other benthic habitats, and vary seasonally and with sediment type. Highest densities generally are found on adjacent unvegetated mudbanks (Kurian, 1984) or in organically polluted mangroves (Diwivedi and Varshney, 1986). For instance, in mangroves polluted by sewage near Bombay, Diwivedi and Varshney recorded a maximum population density of 39,685 individuals.m^{-2}, composed mainly of large foraminifera (98% of total abundance).

Vertical distribution of infauna may be complex depending upon forest type and the distribution and biomass of underground roots and rhizomes (Wada *et al.*, 1987; Shokita, 1989). In mangroves of southern Thailand, Wada *et al.*, (1987) investigated the vertical distribution of underground vegetation and infauna to a depth of one meter. They found that most animals occurred in the upper 20 cm, although some polychaetes and decapods were found deeper. A similar pattern was found by Shokita (1989) in a different forest system near Ranong in Thailand, with grapsid crabs being the dominant subsurface dwelling group.

Several reasons can be offered for the low infaunal densities: (1) control by physical forces (monsoons, high temperatures, desiccation), (2) competition with an abundant epifauna (see section 6.5.2), (3) predation by epifauna and nekton, (4) poor quality of food and (5) chemical defences by the mangroves. Clearly, the forest floor of most mangroves is a physically controlled environment being subjected to monsoon rains, heat, and desiccation (Hanley, 1985). These forces vary with intertidal zone as there appears to be no consistent pattern in zonation of infaunal densities. In Thailand, Frith (1977) and Frith *et al.*, (1976) found highest densities in the mid and high intertidal zones with the lowest numbers at the seaward fringe. In Australia, Wells (1983) recorded the most abundant fauna on adjacent mudflats.

Monsoonal rains also exert a strong seasonal force on the macrobenthos. For example, Kurian (1984) noted a drastic decrease in densities during the monsoons in India, which was attributed to erosion and reduced salinity. On Sagar Island further north, the monsoons affected some phyla but not others (Nandi and Choudhury, 1983). Polychaetes appeared to be seriously affected, whereas actinarians were not.

As with the meiofauna, tannins may also play a role in regulating infaunal densities. Neilson *et al.*, (1986) and Giddins *et al.*, (1986) conducted studies on the litter-feeding habits of the sesarmid crab, *Neosarmatium smithi*, a common decapod in northern Australian mangroves. They found that consumption rates of *Ceriops tagal* litter by crabs correlated negatively with flavolan concentration in leaves with consumption rates increasing when tannin content was reduced by aging.

Tannins may also tie-in with the proposition that food quality for infauna is low. Bacteria are abundant in mangrove sediments, but protozoans, microalgae and meiofauna are usually not (see earlier sections). Assuming infauna require a balanced diet to sustain growth and reproduction, a variety of foods need to be consumed in order to obtain all of the essential nutrients, such as amino acids (Leh and Sasekumar, 1986). Mangrove litter, a major food for most decapods in the mangroves, is of poor nutritional quality, with a wide but generally high C:N ratio (Giddins *et al.*, 1986; Robertson, 1988). In food preference studies with the mangrove amphipod *Parhyale hawaiensis*, Poovachiranon *et al.*, (1986) found that there was a highly significant increase in food preference with increasing degrees of *R. stylosa* leaf decomposition. Feeding rate related positively to leaf nitrogen and starch, but negatively with toughness (wax), crude fiber and tannins.

Bacteria and microalgae are consumed and assimilated by some macrobenthos, but the proportion of carbon and nitrogen requirements of these invertebrates obtained from microbes is not known. Robertson (1986) suggested that bacteria (and perhaps fungi) may fulfill a significant, but unknown, fraction of crab nitrogen requirements. Dye and Lasiak (1987) found that the fiddler crabs *Uca vocans* and *U. polita* and the gastropods *Clithon oualaniensis* and *Cerithidea cingulata* in Australian mangroves ingested and assimilated labelled bacteria and microalgae. The crabs preferred bacteria over microalgae, exhibiting a very high (> 98%) assimilation efficiency for the bacteria. Bacterial assimilation was slightly lower (66-81%).in the gastropods, which showed no food preferences. Dietary requirements and the degree of microbial ingestion (and preferences) varies with phyla and even among species of the same genus (Table 6).

Although many phyla are constituents of mangrove infaunal assemblages, decapods are nearly always the dominant group (see references in Table 6). Decapods are represented by the anomuran families Callianassidae and Coenobitidae and the brachyuran families Leucosidae, Portunidae, Xanthidae, Ocypodidae and Grapsidae, but only the last two contribute large numbers (see review of Jones 1984). As with nearly all other dominant fauna in the mangroves, crabs exhibit clear patterns in vertical zonation and with intertidal position (Table 7). They are important in reducing leaf export (see chapter 10, this volume), influencing forest structure via seed predation (see chapter 5, this volume) and are undoubtedly significant competitors with other fauna and are predators/consumers of other benthic groups, especially microbes. It is probable that their burrowing activities greatly influence sedimentary facies and nutrient chemistry (Robertson, 1991; Smith *et al.*, 1991). Jones (1984) details their physiological adaptations and biogeographical patterns which will not be discussed here.

Recent studies conducted in Southeast Asia have determined some aspects of the reproductive habits, population dynamics and secondary productivity of mangrove crabs (Macintosh, 1984; Pinto, 1984). Pinto investigated the population fluctuations of grapsids of the dominant genus *Chiromantes* and found two population peaks, one during the first inter-monsoon period and the other towards the end of the southwest monsoon. Ovigerous females peaked during the inter-monsoon period. Pinto (1984) suggested that rainfall was a primary factor in initiating the breeding cycle and moulting period of the grapsids.

Table 6. Trophic types and food preferences in some mangrove crabs (adapted from Jones, 1984).

Species	Common habitat	Trophic Type	Preferred Foods
Ocypode ceratophthalmus	landward edge of mangroves	omnivore	insects, molluscs, carrion, leaves
Cardisoma carniflex	high intertidal fringe	omnivore	coconuts, carrion, leaves
Sesarma ricordi	high intertidal fringe	omnivore	vegetation, carrion, crabs
Sesarma spp (general)	mid, high mangroves	omnivore/ detritivore	leaves, organic matter, carrion , plants
Goniopsis cruentata	low intertidal mangroves	omnivore	mud, leaves, seedlings
Aratus pisoni	mangrove trees	herbivore	leaves
Ucides cordatus	high intertidal mangroves	herbivore	leaves, seedlings
Scylla serrata	lower fringe	carnivore	crabs, fish, carrion
Callinectes sp.	channels, waterways	carnivore	bivalves, fish, crabs
Thalamita sp.	channels, waterways	carnivore	bivalves, fish, crabs
Metopograpsus messor	mid intertidal	carnivore	bivalves, fish, crabs
Heloecius sp.	low, mid high mangroves	detritivore	microbes, detritus
Paracleistostoma	low, mid, high mangroves	detritivore	microbes, detritus
Cleistostoma	mid, high mangroves	detritivore	detritus
Uca spp. (general)	low mangroves	detritivore	microbes, detritus
Scopimera sp.	fringe sands	detritivore	microalgae, meiofauna
Dotilla sp., *Ilyoplax* sp., *Dotillopsis* sp., *Mictyris* sp	fringe sand and low mangroves	detritivore	surface detritus

Ocypodid and grapsid crabs also dominate the infauna of Malaysian mangroves with densities routinely occurring within the range of 10 to 70 crabs.m^{-2} (Macintosh, 1984). These crabs are productive, with *Metaplax*, *Uca* and *Sesarma* populations accounting for a total annual production of 0.9 to 17 g.m^{-2}. These estimates are modest compared with production figures for temperate intertidal crabs, but the crabs are an integral part of the mangrove food web, being fed upon by a variety of animals, including birds, snakes and fish (Macintosh, 1984).

6.5.2 Epifauna (including inhabitants of various tree components)

In some mangrove forests, the abundances and biomass of the above ground and tree-dwelling assemblages can exceed those of the infauna (Sasekumar, 1974; Frith *et al*., 1976; Shokita *et al*., 1989). On the sediment surface, gastropods and crustaceans are the major groups and exhibit clear intertidal zonation which has been well documented (see references in Macnae, 1968; Berry, 1975a; Wells, 1980a; Cantera *et al.*, 1983; Plaziat, 1984; Cook and Garbett, 1989). Zonation patterns have been attributed to physico-chemical factors such as frequency of tidal inundation, etc., as well as food sources, competition and predation. In their review of gastropod biogeography, Cantera *et al.*, (1983) attributed zonation to wave energy, wetness and salinity noting that the assemblages inhabiting the sediment surface had a higher species richness than those gastropods living on hard substrates. Vermeij (1974) found high shore littorinids and potamidids were larger and more slender, and neritids larger and more globose

Table 7. Intertidal zonation of mangrove crab genera for different regions of the world. Numbers in parantheses are number of species found (adapted from Jones, 1984; Australian data from Davie, 1982, George and Jones, 1982 and S. Frusher, unpub. data).

Zone	Tropical America	Australia	East Africa	Indo-Malaysia	Red Sea
A. High	*Cardisoma* (1)	*Cardisoma* (1)	*Cardisoma* (1)	*Cardisoma* (1)	*Ocypode* (1)
intertidal	*Sesarma* (3)	*Neosarmatium* (1)	*Neosarmatium* (2)	*Neosarmatium* (2)	*Dotilla* (1)
(EHWS to	*Pachygrapsus* (1)	*Macrophthalmus* (1)	*Sesarma* (2)	*Neoepisarma* (6)	*Uca* (1)
MHWS)	*Cyclograpsus* (1)	*Sesarma* (5)	*Uca* (2)	*Parasesarma* (2)	
	Ocypode (1)	*Cleistocoeloma* (1)		*Cleistocoeloma*	
	Ucides (1)	*Cleistostoma* (1)		*Tiomanum* (1)	
	Uca (6)	*Uca* (3)		*Uca* (2)	
	Ocypode (3)				
B. Mid	*Sesarma* (3)	*Sesarmoides* (1)	*Helice* (2)	*Sesarmoides* (1)	*Macrophthalmus* (2)
intertidal	*Pachygrapsus* (2)	*Sesarma* (1)	*Sesarma* (1)	*Sesarma* (8)	*Uca* (2)
(MHWS to	*Goniopsis* (1)	*Metopograpsus* (2)	*Neosarmatium* (2)	*Metopograpsus* (2)	*Metopograpsus* (1)
MSL)	*Uca* (7)	*Helice* (1)	*Eurycarcinus* (1)	*Helice* (1)	*Paraeleistostoma* (1)
		Cleistocoeloma (1)			*Paracleistostoma* (1)
		Myomenippe (1)			
		Epixanthus (1)			
		Australoplax (1)			
	Eurytium (1)	*Uca* (1)	*Metopograpsus* (1)	*Uca* (6)	
	Menippe (1)	*Ilyoplax* (2)	*Uca* (4)	*Metaplax* (1)	
	Ucides (1)	*Macrophthalmus* (1-9)	*Ilyograpsus* (1)	*Ilyoplax* (3)	
	Panopeus (2)	*Samartium* (2)	*Macrophthalmus* (1)	*Pachygrapsus* (1)	
		Leipocten (1)		*Sarmatium* (1)	
		Heteropanope (1)		*Nanosesarma* (1)	
C:Low	*Panopeus* (1)	*Scylla* (1)	*Neosarmatium* (2)	*Ilyoplax* (1)	
intertidal	*Eurytium* (1)	*Thalamita* (1)	*Uca* (1)	*Macrophthalmus* (1)	
(MSL to	*Pachygrapsus* (2)	*Macrophthalmus* (1-9)	*Scylla* (1)	*Eurycarcinus* (1)	
ELWS)	*Uca* (1)	*Uca* (11)		*Uca* (2)	
	Callinectes (1)			*Metaplax* (1)	
				Scylla (1)	

EHWS = extreme high water springs; MHWS = mean high water springs; MSL = mean sea level; ELWS = extreme low water springs than their lower-shore counterparts, and attributed these as adaptations for extremes of temperature and dessication. The oxygen consumption of three common mangrove gastropods in air and seawater has been studied (Houlihan, 1979). *Nerita birmanica, Cerithidea obtusa* and *Cassidula aurisfelis* are essentially air breathing animals with relatively high metabolic rates.

Above the sediment surface, gastropods, particularly those of the genera *Littorina, Cerithidea, Nerita, Cassidula* and *Ellobium,* predominate on the trunks, prop roots and lower branches of trees, (Cook and Garbett, 1989). These animals feed on the organic debris and algae on the tree bark and move vertically in synchrony with the tides (Berry, 1975). Further down on the tree trunks, lives an encrusting fauna dominated by bivalves, with occasionally heavy infestations of barnacles, which may in turn provide a rich and mobile cryptofauna of isopods, amphipods and errant polychaetes with a safe refuge (Shokita *et al*., 1989). Figure 5 depicts a typical aboveground biocoenosis on *Rhizophora* trees.

The studies of Sutherland (1980), Lalana Rueda and Gosselck (1986) and Ellison and Farnsworth (1992) have provided standing stock and species composition estimates of tree-dwelling fauna. On prop roots of *Rhizophora mangle* in Venezuela, Sutherland (1980) found a rich fauna consisting of sponges, bryozoans, polychaetes, tunicates, hydroids, molluscs and arthropods. Large differences in species composition among roots were attributed to chance resulting from low rates of recruitment and low availability of new roots. Persistence of colonizing species was high (several years) because of resistance to invading larvae and low rates of predation. There was little change in species composition over an 18 month period. In Cuba, lowest densities and biomass of macrofauna occurred on prop roots of *R. mangle* during the dry season, whereas peak biomass (8-17 g $DW.m^{-2}$) was obtained in the wet season. Lalana Rueda and Gosselck (1986) pointed to the fact that zonation in mangrove trees is three dimensional, with uptree, downtree and intertidal facets. Vertical patterns are governed by mobility, frequency of tidal inundation and competition, whereas horizontal patterns, are caused by changes in environmental conditions expected from land to sea. Lateral patterns are presumably a result of inherent patchiness/variability in the distribution of tree species. A high biomass of tree epifauna has been recorded in the estuarine mangroves of Sg.Buloh in Selangor, Malaysia. The tree trunks and aerial roots are densely populated by barnacles, oysters, littorinids, a mytilid bivalve and a thaid gastropod. Biomass values ranged from a high of 1014 kg dry $wt.ha^{-1}$ in the foreshore *Avicennia* forest to a low of 29.7 kg dry $wt.ha^{-1}$ in the high shore *Bruguiera* forest (Tee, 1982).

The epifauna on prop roots can influence the growth of mangrove trees. The original debate about the role of boring isopods in prop root growth (e.g. Rehm and Humm, 1973; Simberloff, *et al*., 1978) has been clarified by the recent work of Perry (1988) and Ellison and Farnsworth (1992). Perry showed experimentally that while isopods and barnacles may cause a 52-62% decrease in net aerial root production for trees in Costa Rica, these effects are mediated by predators of the epifauna. In Belize, Ellison and Farnsworth found that sponges and ascidians growing on the aerial prop roots inhibit their colonization by boring isopods, and thus indirectly facilitate root growth.

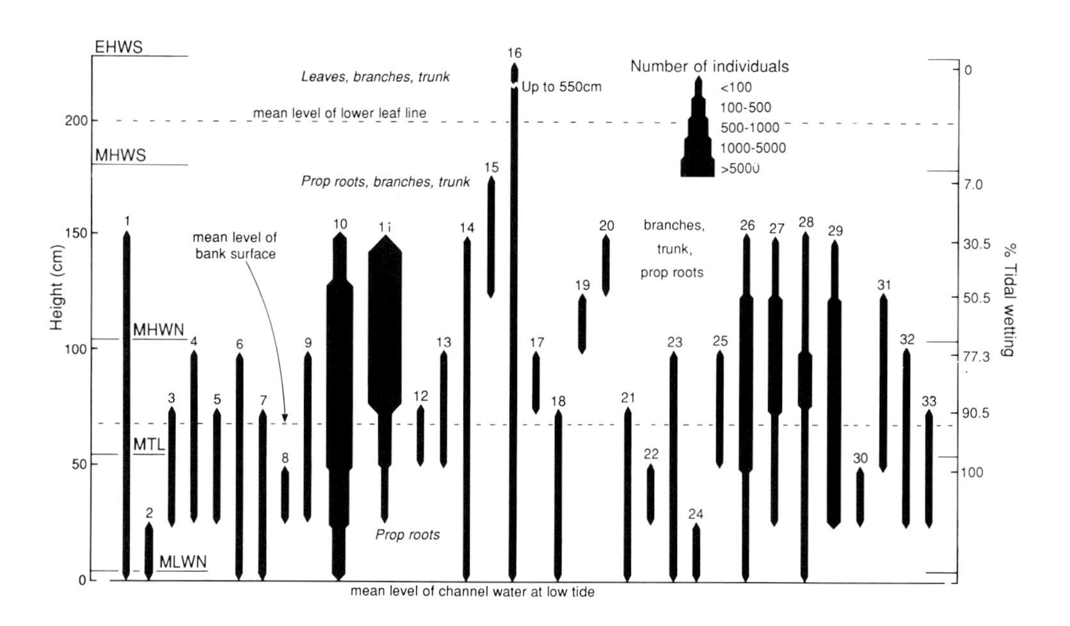

Figure 5. Vertical zonation and abundance of epibenthos on *Rhizophora apiculata* trees, Phuket Island (modified from Frith *et al.*, 1976). Species codes: 1 = Sea anemone spA; 2 = Sea anemone sp.B; 3 = Nemertine sp.A.4 = *Lepidonotus kumari*, 5 = *Petrolisthes* sp.; 6 = *Clibanarius padavensis*, 7 = *Diogenes avarus*; 8 = *Leipocten sordidulum;* 9 = *Tylodiplax tetratylophora*; 10 = *Balanus amphitrite;* 11 = *Chthamalus withersii*; 12 = *Ligia* sp.; 13 = *Sphaeroma walkeri;* 14 = *Nerita birmanica;* 15 = *Littorina carinifera*; 16 = *Littorina scabra*; 17 = *Assiminea brevicula;* 18 = *Cerithidea cingulata;* 19 = *Cerithidea obtusa;* 20 = *C. breve;* 21 = *C. patulum;* 22 = *Capulus* sp.; 23 = *Murex capucinus*; 24 = *Nassarius jacksonianus;* 25 = *Onchidium* sp.; 26 = *Brachidontes rostratus*; 27 = *Isognomon ephippium;* 28 = *Enigmonia aenigmatica*; 29 = *Saccostrea cucullata*; 30 = *Diplodonta globosa*; 31 = *Teredo* sp.; 32 = *Xylophaga* sp.; 33 = *Trapezium sublaevigatum.*

Microhabitats also play a role in fostering niche avoidance and survival within the forest. The mangrove whelk *Telescopium telescopium*, for instance, uses refuges to avoid harsh environmental conditions (Lasiak and Dye, 1986), clustering under shade to avoid heat stress. Activity is synchronised to onset of tides. In Western Australian mangroves, the mudwhelks *Terebralia sulcata* and *T. palustris* co occur, but where the two species overlap, *T. palustris* is found in finer deposits higher on the shore (Wells, 1980b). Niche separation has also been found to occur in littorinids on mangrove trees in Thailand where five species of *Littoraria* live on trees on Phuket Island and are separated spatially: three species are restricted to bark, one lives on leaves and the fifth species migrates between leaves and bark (Cook and Garbett, 1989). Intraspecific competition for food occurs, as demonstrated by the gastropod, *Bembicium auratum* in a temperate mangrove forest (Branch and Branch, 1980). The biology (Berry, 1975a; Morton, 1976a and 1976b) and breeding biology (Berry, 1975b) of several species of mangrove molluscs has been the subject of investigations in Southeast Asia.

A rich wood-boring fauna exists in trapped logs and woody components in mangrove forests (Berry, 1975; Day, 1975; Rao, 1986; Santhakumaran, 1986). Teredinid molluscs, especially of the genera *Bactronophorus, Dicyathifer, Lyrodus* and *Teredo*, are the dominant group. Rao (1986) noted that in India the species composition of woodborers differed between locations. He suggested that mangroves support resident populations of teredinids that persist because of eurytolerance to annual monsoons. Woodborers are ecologically important because they stimulate the decomposition of wood and the activity of the nitrogen fixing bacterial flora. In Australian forests, Cragg and Robertson (in prep.) found a diverse wood-dwelling fauna (55 species) consisting of teredinids, brachyurans, gastropods, amphipods, polychaetes, fishes and insects. Total numbers per site ranged from 42-400 individuals with mean numbers per log ranging from 5 to 29.

Nearly all faunal (and presumably floral) groups living on or above the forest floor are fed upon by fishes during high tide, and, in the case of some crabs, prawns and fishes, also caught and fed upon by man. The diets of mangrove-associated fish vary among forests depending upon the availability of prey items and differences in species composition. Sasekumar *et al.*, (1984) found that in a Malaysian mangrove forest, grapsid crabs, sipunculids and encrusting fauna were the major prey items, whereas in an adjacent estuary, fish feeding was more diverse with most categorised as carnivores (89%) and fewer fish species as detritivores (9%) and omnivores (2%). The trophic role of microbes, meiofauna and macrofauna is less clear concerning actual amounts of biomass transferred up the mangrove food chain.

6.6 Concluding remarks

Sufficient information is available concerning species composition and abundance of infauna and epifauna in tropical mangroves in relation to intertidal position and environmental conditions. However, the number of such studies is few compared with data for temperate intertidal habitats.

It is clear that more process-oriented information is needed for benthic communities in mangroves. Adequate knowledge of trophic interactions, rates of detrital decomposition by benthos, physiological adaptations and estimates of secondary production is lacking, despite the fact that such information is necessary to properly assess the energetic role of benthos in tropical mangrove ecosystems.

6.7 References

Aladro, L., and Lopez Ochoterena, E., 1967. Protozoarios ciliados de Mexico. 14. Algunas aspectes biologicas de quince especies colectodas en la Laguha de Manginda, Veracruz. *Revista Sociedad Mexicana de Historia Natural* **28**:5571.

Alongi, D.M., 1986. Quantitative estimates of benthic protozoa in tropical marine systems using silica gel: a comparison of methods. *Estuarine, Coastal and Shelf Science* **23**:443-450.

Alongi, D.M., 1987a. Intertidal zonation and seasonality of meiobenthos in tropical mangrove estuaries. *Marine Biology* **95**:447-458.

Alongi, D.M., 1987b. The influence of mangrove derived tannins on intertidal meiobenthos in tropical estuaries. *Oecologia* **71**:537-540.

Alongi, D.M., 1987c. Interestuary variation and intertidal zonation of freeliving nematode communities in tropical mangrove systems. *Marine Ecology Progress Series* **40**:103-114.

Alongi, D.M., 1988a. Bacterial productivity and microbial biomass in tropical mangrove sediments. *Microbial Ecology* **15**:59-79.

Alongi, D.M., 1988b. Microbial meiofaunal interrelationships in some tropical intertidal sediments. *Journal of Marine Research* **46**:349-365.

Alongi, D.M., 1989. The role of softbottom benthic communities in tropical mangrove and coral reef ecosystems. *Reviews in Aquatic Sciences* **1**:243-280.

Alongi, D.M., 1990a. The ecology of tropical softbottom benthic ecosystems. *Oceanography and Marine Biology Annual Reviews* **28**:381-496.

Alongi, D.M., 1990b. Abundances of benthic microfauna in relation to outwelling of mangrove detritus in a tropical coastal region. *Marine Ecology Progress Series* **63**:53-63.

Alongi, D.M., 1990c. Community dynamics of freeliving nematodes in some tropical mangrove and sandflat habitats. *Bulletin of Marine Science* **46**:358-373.

Araujo, A., D'Souza, J., and Karande, A., 1981. Studies on fungi and yeasts from the west coast of India. *Indian Journal of Marine Science* **10**:341-348.

Ayyakkannu, K., and Chandramohan, D., 1971. Occurrence and distribution of phosphate solubilizing bacteria and phosphatase in marine sediments at Porto Novo. *Marine Biology* **11**:201-208.

Berry, A.J., 1972. The natural history of West Malaysian mangrove faunas. *Malayan Nature Journal* **25**:135-162.

Berry, A.J., 1975a. Molluscs colonizing mangrove trees with observations on *Enigmonia rosea* (Anomiidae). *Proceedings of the Malacological Society of London* **41**:589-600.

Berry, A.J., 1975b. Patterns of breeding activity in West Malaysian gastropod molluscs. *Malaysian Journal of Science* **3A**:49-59.

Bhattacharya, A., Ghosh, M., and Choudhury, A., 1987. Seasonal abundance of *Acanthamoeba rhysodes* (Singh, 1952) (Protozoa: Gymnamoebia) in a mangrove litter-soil ecosystem of Gangetic Estuary, India. *Journal of Protozoology* **34**:403-405.

Boltovskoy, E., 1984. Foraminifera of mangrove swamps. *Physis* **42**:1-9.

Boonruang, P., 1984. The rate of degradation of mangrove leaves, *Rhizophora apiculata* BL. and *Avicennia marina* (Forsh) Vierh. at Phuket Island, western peninsular Thailand. In: Soepadmo, E., Rao, A.N., Macintosh, D.J. (Eds), *Proceedings of the Asian Symposium on Mangrove Environments: Research and Management*, pp. 200-208, University of Malaya, Kuala Lumpur.

Boto, K.G., 1984. Waterlogged saline soils. In: Snedakar, S.C. and Snedakar, J.G. (Eds), *The mangrove ecosystem: research methods*, pp. 114-130, UNESCO, Paris.

Boto, K.G., and Wellington, J.T., 1983. Phosphorus and nitrogen nutritional status of a Northern Australian mangrove forest. *Marine Ecology Progress Series* **11**:63-69.

Boto, K.G., and Wellington, J.T., 1984. Soil characteristics and nutrient status in a Northern Australian mangrove forest. *Estuaries* **7**:61-69.

Boto, K.G., Alongi, D.M., and Nott, A.L., 1989. Dissolved organic carbon bacteria interactions at sedimentwater interface in a tropical mangrove system. *Marine Ecology Progress Series* **51**:243-251.

Branch, G.M., and Branch, M.L., 1980. Competition in *Bembicum auratum* (Gastropoda) and its effect on microalgal standing stock in mangrove muds. *Oecologia* **46:**106-114.

Broom, M.J., 1982. Structure and seasonality in a Malaysian mudflat community. *Estuarine, Coastal and Shelf Science* **15**:135-150.

Bronnimann, P., and Zaninetti, L., 1965. Notes sur *Litola salsa* (Cushman et Bronnimann) 1948, un Foramenifere de la mangrove d l'ile de la Tinite, W.I. *Archives de Sciences (Geneva)* **18**:608-615.
Cantera, J., Arnaud, P.M., and Thomassin, B.A., 1983. Biogeographic and ecological remarks on molluscan distribution in mangrove biotopes. 1. Gastropods. *Journal of Molluscan Studies* **12A**:10-26.

Chihara, M., and Tanaka, T., 1988. Species composition and ecology of macroalgae in mangrove brackish areas of East Indonesia. In: Ogino, K., and Chihara, M., (Eds), *Biological System of Mangroves.* A report of East Indonesian Mangrove Expedition 1986, pp. 7-20, Ehime University, Japan.

Chong, V.C., and Sasekumar, A., 1981. Food and feeding habits of the white prawn, *Penaeus merguiensis. Marine Ecology Progress Series* **5:**185-191.

Cook, L.M., and Garbett, S.B., 1989. Patterns of variation in mangrove littorinid molluscs on Phuket Island. *Phuket Marine Biological Centre Research Bulletin, No.* **53**, 14 pp.

Cooksey, K.E., Cooksey B., Evans, P.M., and Hildebrand, E.L., 1975. Benthic diatoms as contributors to the carbon cycle in a mangrove community. In: Persoone, G., and Jaspero, E., (Eds). *10th European Marine Biology Symposium,* pp. 165-178, Universa Press, Welteren.

Cooksey, K.E., and Cooksey, B., 1978. Growth influencing substances in sediment extracts from a subtropical wetland: investigation using a diatom assay. *Journal of Phycology* **14**:347-352.

Cragg, S.M., and Robertson, A.I., (in prep). Cryptofauna of decomposing wood in tropical Australian mangrove forests.

Cundell, A.M., Brown, M.S., Stanford, R., and Mitchell, R., 1979. Microbial degradation of *Rhizophora mangle* leaves immersed in the sea. *Estuarine, Coastal Marine Science* **9**:281-286.

Cushman, J.A., and Bronnimann, P., 1948a. Some new genera and species of Foraminifera from brackish water of Trinidad. *Contributions from the Cushman Foundation for Foraminiferal Research* **24**:15-21.

Cushman, J.A., and Bronnimann, P., 1948b. Additional new species of arenaceous Foraminifera from shallow waters of Trinidad. *Contributions from the Cushman Foundation for Foraminiferal Research* **24**:37-42.

Davie, P., 1982. A preliminary checklist of brachyura (Crustacea: Decapoda) associated with Australian mangrove forests. *Operculum* **5**:204-207.

Day, J.H., 1975. The mangrove fauna of the Mozambique estuary, Mozambique. In: Walsh, G.E., Snedaker, S.C., and Teas, H.J., (Eds), *Proceedings of the International Symposium on the Biology and Management of Mangroves*, Honolulu, pp. 415-430, University of Florida, Gainesville.

Decraemer, W., and Coomans, A., 1978. Scientific report on the Belgian expedition to the Great Barrier Reef in 1967. Nematodes. XII. Ecological notes on the nematode fauna in and around mangroves on Lizard Island. *Australian Journal of Marine and Freshwater Research* **29**:497-508.

Dor, I., and Levy, I., 1984. Primary productivity of the benthic algae in the hard-bottom mangal of Sinai. In: Por, F.D. and Dor, I., (Eds)., *Hydrobiology of the mangal*, pp. 179-191, Dr W. Junk Publishers, The Hague.

D'Souza, G., and D'Souza, J., 1979a. Studies on estuarine yeasts. I. Nitrogen fixation in relation to ecological features. *Mahasagar* **12**:93-101.

D'Souza, N.A., and D'Souza, J., 1979b. Studies on estuarine yeasts. III. Pectinolytic yeasts in mangroves. *Mahasagar* **12**:163-168.

Dwivedi, S.N., and Varshney, P.K., 1986. Intertidal macrobenthic fauna of deteriorated mangrove ecosystem off Seven Banglows, Bombay. *Geobios* **13**:112-117.

Dye, A.H., 1983a. A method for the quantitative estimation of bacteria from mangrove sediments. *Estuarine, Coastal and Shelf Science* **17**:207-215.

Dye, A.H., 1983b. Composition and seasonal fluctuations of meiofauna in a Southern Africa mangrove estuary. *Marine Biology* **73**:165-170.

Dye, A.H., 1983c. Vertical and horizontal distribution of meiofauna in mangrove sediments in Transkei, Southern Africa. *Estuarine, Coastal and Shelf Science* **16**:591-598.

Dye, A.H., and Lasiak, T.A., 1986. Microbenthos, meiobenthos and fiddler crabs: trophic interactions in a tropical mangrove sediment. *Marine Ecology Progress Series* **32**:259-267.

Dye, A.H., and Lasiak, T.A., 1987. Assimilation efficiencies of fiddler crabs and deposit feeding gastropods from tropical mangrove sediments. *Comparative Biochemistry and Physiology* **87A**:341-344.

Ellison, A.M., and Farnsworth, E.J., 1992. Belizean mangrove root epibionts: patterns of distribution and abundance, and effects on root growth. *Hydrobiologia/Developments in Hydrobiologia.*, in press.

Fell, J.W., and Master, I.M., 1973. Fungi associated with the degradation of mangrove (*Rhizophora mangle* L.) leaves in South Florida. In: Stevenson, L.H., and Colwell, R.R., (Eds), *Estuarine Microbial Ecology*, pp. 455-473, University of South Carolina Press, Columbia.

Fell, J.W., Cefalu, R.D., Master, I.M., and Tallman, A.S., 1975. Microbial activities in the mangrove (*Rhizophora mangle*) leaf detrital system. In: Walsh, G.E., Snedaker, S.C., and Tes, H.J., (Eds), *Proceedings of the International Symposium on the Biology and Management of Mangroves*, pp. 661-679, University of Florida, Gainesville.

Frith, D.W., 1977. A preliminary list of macrofauna from a mangrove forest and adjacent biotopes at Surin Island, Western Peninsular Thailand. Phuket Marine Biological Centre Research Bulletin, No. 17, 14 pp.

Frith, D.W., Tantanasiriwong, R., and Bhatia, O., 1976. Zonation of macrofauna on a mangrove shore, Phuket Island. *Phuket Marine Biological Center Research Bulletin, No.* **10**, 37 pp.

Frith, D.W., and Frith, C.B., 1978. Notes on the ecology of fiddler crab populations (Ocypodidae: Genus *Uca*) on Phuket, Surin Nua and Yao Yai Islands, Western Peninsular Thailand. *Phuket Marine Biological Center Research Bulletin, No.* **25**, 13 pp.

Ganapati, P.N., and Narasimha Rao, M.V., 1958. Systematic survey of marine ciliates from Visakhapatam. *Andhra University Memoirs in Oceanography* **2**:75-90.

Garg, K.L., 1983. Vertical distribution of fungi in Sunderban mangrove mud. *Indian Journal of Marine Science* **12**:48-51.

Gee, J.M., 1989. An ecological and economic review of meiofauna as food for fish. *Zoological Journal of the Linnean Society* **96**:243-261.

George, R.W., and Jones, D.S., 1982. A revision of the fiddler crabs of Australia (Ocypodidae:*Uca*). *Records of the Western Australian Museum, supplement No.* **14**, 99pp.

Gerlach, S.A., 1957. Marine nematoden aus dem Mangrove gebiet von Cananeia. III. Brasilianische Meeresnematoden. *Abhandlungen der Mathematisch Naturivessenschaftlichen Klasse in Mainz* **5**:3-48.

Ghosh, M., and Cloudhury, A., 1987. Aspects of culture: *Acanthamoeba astronyxis* (Ray and Hayes 1954) from Bay of Bengal coasts, India. *Proceedings of the Indian Academy of Sciences* **96**:63-69.

Giddins, R.L., Lucas, J.S., Neilson, M.J., and Richards, G.N., 1986. Feeding ecology of the mangrove crab Neosarmatium smithi (Crustacea:Decapoda:Sesarmidae). *Marine Ecology Progress Series* **33**:147-155.

Gillan, F.T., and Hogg, R.W., 1984. A method for the estimation of bacterial biomass and community structure in mangrove associated sediments. *Journal of Microbiological Methods* **2**:275-293.

Gomes, H.R., and Mavinkurve, S., 1982. Studies on mangrove swamps of Goa. II. Microorganisms degrading phenolic compounds. *Mahasagar* **15**:111-115.

Govindan, K., Varshney, P.K., and Desai, B.N., 1983. Benthic studies in South Gujarat estuaries. *Mahasagar* **16**:349-356.

Hagelstein, R., 1938. The Diatomaceae of Puerto Rico and the Virgin Islands. *Scientific Surveys of Puerto Rico and Virgin Islands* **8**:313-412.

Hanley, J.R., 1985. Why are there so few polychaetes (Annelida) in Northern Australian mangroves? In: Bardsley, K.N., Davie, J.D.S., and Woodroffe, C.D., (Eds), *Coastal and Tidal Wetlands of the Australian Monsoon Region*, ANU North Australia Research Unit, Mangrove Monograph, No. 1, pp. 239-250.

Hesse, P.R., 1963. Phosphorus relationships in a mangrove swamp mud with particular reference to aluminium toxicity. *Plant and Soil* **19**:205-219.

Hodda, M., and Nicholas, W.L., 1985. Meiofauna associated with mangroves in the Hunter River estuary and Fullerton Cove, Southeastern Australia. *Australian Journal of Marine and Freshwater Research* **36**:41-50.

Hoppe, H.G., Gocke, K., Zamorans, D., and Zimmerman, R., 1983. Degradation of macromolecular organic compounds in a tropical lagoon (Cienago Grande, Colombia) and its ecological significance. *Internationale Revue der Gesamten Hydrobiologie* **68**:811-824.

Hopper, B.E., Fell, J.W., and Cefalu, R.C., 1973. Effect of temperature on life cycles of nematodes associated with the mangrove (*Rhizophora mangle*) detrital system. *Marine Biology* **23**:293-296.

Houlihan, D.F., 1979. Respiration in air and seawater of three mangrove snails. *Journal of Experimental Marine Biology and Ecology* **41**:143-161.

Hutchings, P.A., and Recher, H.F., 1982. The fauna of Australian mangroves. *Proceedings of the Linnean Society of New South Wales* **106**:83-121.

Hyde, K.D., 1989. Ecology of tropical marine fungi. *Hydrobiologia* **178**:199-208.

Hyde, K.D., 1990. A study of the vertical zonation of intertidal fungi on *Rhizophora apiculata* at Kampong Kapok mangroves in Brunei. *Aquatic Botany* **36**:255-262.

Hyde, K.D., and Jones, E.B.G., 1988. Marine mangrove fungi. *P.S.Z.N.I. Marine Ecology* **9**:15-33.
Jones, D.A., 1984. Crabs of the mangal ecosystem. In: Por, F.D. and Dor, I. (Eds). *Hydrobiology of the Mangal*, pp. 89-109, Dr W. Junk Publishers, The Hague.

King, R.J., 1981. The macroalgae of mangrove communities in eastern Australia. *Phycologia* **20**:107-108 (abstract).

Kondalarao, B., 1983. Distribution of meiofauna in the Gautami-Godavari estuarine system. *Mahasagar* **16**:453-457.

Kondalarao, B., 1984. Distribution of meiobenthic harpacticoid copepods in Gautami-Godavari estuarine system. *Indian Journal of Marine Science* **13**:80-91.

Kondalarao, B., and Ramana Munty, K.V., 1988. Ecology of intertidal meiofauna of the Kakinada Bay (Gautami-Godavari estuarine system), east coast of India. *Indian Journal of Marine Science* **17**:40-47.

Krishnamurthy K., Sultan Ali, M.A., and Prince Jeyaseelan, M.J., 1984. Structure and dynamics of the aquatic food web community with special reference to nematodes in mangrove ecosystems. In: Soepadmo, E., Rao, A.N., and Macintosh, D.J., (Eds), *Proceedings of the Asian Symposium on Mangrove Environments: Research and Management*, pp. 429-452, University of Malaya and UNESCO, Kuala Lumpur.

Kurian, C.V., 1984. Fauna of the mangrove swamps in Cochin estuary. In: Soepadmo, E., Rao, A.N., and Macintosh, D.J., (Eds), *Proceedings of the Asian Symposium on Mangrove Environments: Research and Management*, pp. 226-230, University of Malaya and UNESCO, Kuala Lumpur.

Lalana Rueda, R., and Gosselck, F., 1986. Investigations of the benthos of mangrove coastal lagoons in Southern Cuba. *Internationale Revue der Gesamten Hydrobiologie* **71**:779-794.

Larsen, J., and Patterson, D.J., 1990. Some flagellates (Protista) from tropical marine sediments. *Journal of Natural History* **24**:801-937.

Lasiak, T., and Dye, A.H., 1986. Behavioural adaptations of the mangrove whelk, *Telescopium telescopium* (L.) to life in a semiterrestrial environment. *Journal of Molluscan Studies* **52**:174-179.

Lee, B.K.H., and Baker, G.F., 1973. Fungi associated with the roots of red mangrove, *Rhizophora mangle*. *Mycologia* **65**:894-906.

Leh, C.M.U., and Sasekumar, A., 1986. The food of sesarmid crabs in Malaysian mangrove forests. *Malayan Nature Journal* **39**:135-145.

Limpsaichol, P., 1978. Reduction and oxidation properties of the mangrove sediment, Phuket Island, Southern Thailand. *Phuket Marine Biological Center Research Bulletin, No.* **23**, 13 pp.

Macintosh, D.J., 1984. Ecology and productivity of Malaysian mangrove crab populations (Decapoda: Brachyura). In: Soepadmo, E., Rao, A.N., and MacIntosh, D.J., (Eds). *Proceedings of the Asian Symposium on Mangrove Environments: Research and Management*, pp. 354-377, University of Malaya and UNESCO, Kuala Lumpur.

Macnae, W., 1968. A general account of the fauna and flora of mangrove swamps and forests in the Indo-West-Pacific region. *Advances in Marine Biology* **6**:73-270.

Macnae, W., and Kalk, M., 1962. The ecology of mangrove swamps at Inhaca Island, Mozambique. *Journal of Ecology* **50**:19-34.

Margalef, R.I., 1962. *Comunidades naturales*. Instituto de Biologie Marina Universidade Puerto Rico, Rio Predras Special Publicacao, 469 pp.

Matondkar, S.G.P., Mahtani, S., and Mavinkurve, S., 1980a. Seasonal variations in the microflora from mangrove swamps of Goa. *Indian Journal of Marine Science* **9**:119-120.

Matondkar, S.G.P., Mahtani, S., and Mavinkurve, S., 1980b. The fungal flora of the mangrove swamps of Goa. *Mahasagar* **13**:281-283.

Matondkar, S.G.P., Maktani, S., and Mavinkurve, S., 1981. Studies on mangrove swamps of Goa. I. Hetertrophic bacterial flora from mangrove swamps. *Mahasagar* **15**:111-115.

Milward, N.E., 1982. Mangrove dependent biota. In: Clough, B.F., (Ed.), *Mangrove ecosystems in Australia*, pp. 121-139, Australian National University Press, Canberra.

Morton, B., 1976a. The biology and functional morphology of the southeast Asian mangrove bivalve, *Polymesoda (Geloina) erosa* (Solander, 1786) (Bivalve: Corbiculidae). *Canadian Journal of Zoology* **54**:482-500.

Morton, B., 1976b. The biology, ecology and functional aspects of the organs of feeding and digestion of the S.E. Asian mangrove bivalve, Enigmonia aenigmatica (Mollusca: Anomiacea). *Journal of Zoology, London* **179**:437-466.

Misra, J.K., 1986. Fungi from mangrove muds of Andaman Nicobar Islands. *Indian Journal of Marine Science* **15**:185-189.

Mouzouras, R., 1989. Decay of mangrove wood by marine fungi. *Botanica Marina* **32**:65-69.

Nandi, S., and Choudhury, A., 1983. Quantitative studies on the benthic macrofauna of Sagar Island, intertidal zones, Sunderbans, India. *Mahasagar* **16**:409-414.

Nateewathana, A., and Tantichodok, P., 1984. Species composition, density and biomass of macrofauna of a mangrove forest at Ko Yao Yai, Southern Thailand. In: Soepadmo, E., Rao, A.N., and MacIntosh, D.J., (Eds), *Proceedings of the Asian Symposium on Mangrove Environments: Research and Management*, pp. 258-285, University of Malaya and UNESCO, Kuala Lumpur.

Neilson, M.J., Giddins, R.L., and Richards, G.N., 1986. Effects of tannins on the palatability of mangrove leaves to the tropical sesarmid crab *Neosarmatium smithi*. *Marine Ecology Progress Series* **34**:185-187.

Nicholas, W.L., Goodchild, D.J., and Stewart, A., 1987. The mineral composition of intracellular inclusions in nematodes from thiobiotic mangrove mud flats. *Nematologica* **33**:167-179.

Nouzacede, M., 1976. Cytologie fonctionelle et morphologie experimentale de quelques protozoans cilies mesopsammiques geants de la famille des Geleridae (Kahl). *Bulletin Station Biologie D'Archachon* **28**:1-397.

Odum, W.E., and Heald, E.J., 1972. Trophic analyses of an estuarine mangrove community. *Bulletin of Marine Science* **22**:679-738.

Perry, D.M., 1988. Effects of associated fauna on growth and productivity in the red mangrove. *Ecology* **69**:1064-1075.

Pinto, M.L., 1984. Some ecological aspects of a community of mangrove crabs occurring within the Islets of Negombo Lagoon (Sri Lanka). In: Soepadmo, E., Rao, A.N., and MacIntosh, D.J., (Eds), *Proceedings of the Asian Symposium on Mangrove Environments: Research and Management*, pp. 311-330, University of Malaya and UNESCO, Kuala Lumpur.

Plaziat, J.C., 1984. Mollusk distribution in the mangal. In: Por, F.D., and Dor, I., (Eds), *Hydrobiology of the Mangal*, pp. 113-143, Dr W. Junk Publishers, The Hague.

Poovachiranon, S., Boto, K.G., and Duke, N.C., 1986. Food preference studies and ingestion rate measurements of the mangrove amphipod *Parhyale hawaiensis* (Dana). *Journal of Experimental Marine Biology and Ecology* **98**:129-140.

Potts, M., 1979. Nitrogen fixation (acetylene reduction) associated with communities of heterocystous and nonheterocystous blue green algae on mangrove forests of Sinai. *Oecologia* **39**:359-373.

Potts, M., 1980. Blue green algae (Cyanophyta) in marine coastal environments of the Sinai peninsula: distribution, zonation, stratification and taxonomic diversity. *Phycologia* **19**:60-98.

Potts, M., and Whitton, B.A., 1980. Vegetation of the intertidal zone of the lagoon of Aldabra, with particular reference to the photosynthetic prokaryotic communities. *Proceedings of the Royal Society of London, Series B* **208**:13-63.

Rao, M.V.L., 1986. Notes on woodborers from the mangrove of the Godavari estuary, India. In: Thompon, M.F., Sarojini, R., and Nagabhushanam, R., (Eds), *Biology of Benthic Marine Organisms Indian Ocean*, pp. 579-584, Balkcma, Rottcrdam.

Reyes Vasquez, G., 1975. Diatomeas litorales de la famelia Naviculaceae, de la Laguna La Restinga, Isla De Margarita, Venezuela. *Boletim Instituto Oceanografico Universidad de Oriente* **14**:199-249.

Robertson, A.I., 1986. Leafburying crabs: their influence on energy flow and export from mixed mangrove forests (*Rhizophora* spp) in northeastern Australia. *Journal of Experimental Marine Biology and Ecology* **102**:237-248.

Robertson, A.I., 1988. Decomposition of mangrove leaf litter in tropical Australia. *Journal of Experimental Marine Biology and Ecology* **116**:235-247.

Robertson, A.I., 1991. Plant-animal interactions and the structure and function of tropical mangrove forest ecosystems. *Australian Journal of Ecology* **16**:433-443.

Rodriguez, C., and Stoner, A.W., 1990. The epiphyte community of mangrove roots in a tropical estuary: distribution and biomass. *Aquatic Botany* **36**:117-126.

Sahoo, A.K., Sah, K.D., and Gupta, S.K., 1985. Studies on nutrient status of some mangrove mud of the Sunderbans. In: Bhosale, L.J. (Eds), *The Mangroves*, pp. 375-377, Shivaji University, Kolhapur.

Santhakumaran, L.N., 1986. Marine woodboring of mangrove forests. UNESCO Technical Report RAS/79/002, pp. 109-124.

Sasekumar, A., 1974. Distribution of macrofauna on a Malayan mangrove shore. *Journal of Animal Ecology* **43**:51-69.

Sasekumar, A., 1981. The ecology of meiofauna on a Malayan mangrove shore. Ph.D Thesis, University of Malaya, Kuala Lumpur, 183pp.

Sasekumar, A., Ong, T.L., and Thong, K.L., 1984. Predation of mangrove fauna by marine fishes. In: Soepadmo, E., Rao, A.N. and Macintosh, D.J. (Eds), *Proceedings of the Asian Symposium on Mangrove Environments: Research and Management,* pp. 378-384, University of Malaya and UNESCO, Kuala Lumpur.

Shokita, S., 1989. Macrofauna community structure and food chains in the mangal. UNESCO Report on Life History of Selected Species of Flora and Fauna in Mangrove Ecosystems, RAS/79/002, 46 pp.

Shokita, S., Sanguansin, J., Nishijima, S., Soemodihardjo, S., Abdullah, A., Hai He, M., Kasinathan, R., and Okamoto, K., 1989. Distribution and abundance of benthic macrofauna in the Funaura mangal of Iriomote Island, The Ryukyus. *Galaxea* **8**:17-30.

Smith, T.J. III, Boto, K.G., Frusher, S.D., and Giddins, R.L., 1991. Keystone species and mangrove forest dynamics: the influence of burrowing by crabs on soil nutrient status and forest productivity. *Estuarine, Coastal and Shelf Science* **33**:419-432.

Snyder, R.A., 1985. Ciliated protists from the western Atlantic Barrier Reef system, Carrier Bow Cay (Belize). *Proceedings of the Vth International Coral Reef Congress*, **5**:215-220.

Sournia, A., 1977. Notes on primary productivity of coastal waters in the Gulf of Eilat (Red Sea). *Internationale Revue der Gesamten Hydrobiologie* **62**:813-824.

Stanley, S.O., Boto, K.G., Alongi, D.M., and Gillan, F.I., 1987. Composition and bacterial utilisation of free amino acids in tropical mangrove sediments. *Marine Chemistry* **22**:13-30.

Sutherland, J.P., 1980. Dynamics of the epibenthic community on roots of the mangrove *Rhizophora mangle*, at Balia de Buche, Venezuela. *Marine Biology* **58**:7584.

Tee, A.C.G., 1982. Some aspects of the ecology of the mangrove forest at Sungai Buloh. II. Distribution pattern and population dynamics of tree dwelling fauna. *Malayan Nature Journal* **35**:267-277.

Tietjen, J.H., and Alongi, D.M., 1990. Population growth and effects of nematodes on nutrient regeneration and bacteria associated with mangrove detritus from northeastern Queensland (Australia). *Marine Ecology Progress Series* **68**:169-180.

Uchina, F., Hambali, G.G., and Yarazawa, M., 1984. Nitrogen fixing bacteria from warty lenticellate bark of a mangrove tree, *Bruguiera gymnorhiza* (L.) Lamk. *Applied and Environmental Microbiology* **47**:44-48.

Varshney, P.K., 1985. Meiobenthic study off Mahim (Bombay) in relation to prevailing organic pollution. *Mahasagar* **28**:27-36.

Velho, S., and D'Souza, J., 1982. Studies on pectinolytic fungi from the mangrove sediments. *Mahasagar* **15**:167-173.

Vermeij, G.J., 1974. Molluscs in mangrove swamps: Physiognomy, diversity and regional differences. *Systematic Zoology* **22**:609-624.

Wada, K., Komiyama, A., and Ogino, K., 1987. Underground vertical distribution of macrofauna and root in a mangrove forest of Southern Thailand. *Publication of the Seto Marine Biological Laboratory* **32**:329-333.

Warren, J.H., 1990. The use of open burrows to estimate abundances of intertidal estuarine crabs. *Australian Journal of Ecology* **15**:277-280.

Wee, Y.C., 1986. Mangrove algae. *Wallaceana* **46**:13-16.

Wells, F.E., 1980a. Comparative distribution of macromolluscs and macrocrustaceans in a Northwestern Australian mangrove system. *Australian Journal of Marine and Freshwater Research* **35**:591-596.

Wells, F.E., 1980b. A comparative study of distributions in the mudwhelks *Terebralia sulcata* and *T. palustris* in a mangrove swamp in Northwestern Australia. *Malacological Reviews* **13**:1-5.

Wells, F.E., 1983. An analysis of marine invertebrate distributions in a mangrove swamp in Northwestern Australia. *Bulletin of Marine Science* **33**:736-744.

7

Plankton, Epibenthos and Fish Communities

A.I. Robertson and S.J.M. Blaber

7.1 Introduction

Indigenous people have exploited the biota of mangrove waters for centuries, and fish and shrimp are still one of the major products harvested from this habitat (Saenger *et al.,* 1983). Indeed, the supposed connection between mangroves and juvenile nekton is often advanced as one of the key arguments for the conservation of mangrove forests (e.g. Odum and Heald, 1972, 1975; Hatcher *et al.,* 1989).

In this chapter we review what is known about the biotic communities inhabiting mangrove waterways. Because relatively little work has been done on the plankton (microbes, phyto- and zooplankton) of mangrove habitats, we have attempted to give a brief summary of worldwide literature. In contrast, there is a growing literature on penaeid prawn and fish communities in mangroves. Because it is not possible to provide a complete review of tropical mangrove fish communities within the constraints of part of one chapter, we have reviewed recent Australian studies of the relationship between habitat variation and fish community composition, the functional roles of mangroves for fish and the question of how important mangroves are as nursery or feeding grounds relative to non-mangrove habitats. We compare the findings from Australian studies with those from other parts of the world (e.g. Odum *et al.,* 1982; Yanez-Arancibia, 1985; Thayer and Sheridan, in press).

Before discussing the biota of mangrove waterways, we consider briefly the variation in some physical properties of water in mangrove systems which have major influences on the biological communities.

7.2 Physical Attributes of Mangrove Waterways

7.2.1 Hydrology

Wolanski *et al.,* (Chapter 3, this volume) have provided an exhaustive review of the physics of water motion in different mangrove-lined estuaries and embayments. However, one special feature of mangrove forests, their influence on the lateral trapping of water within estuaries (Wolanski *et al.*, 1980), has several important implications for the biota of tropical mangrove waterways, and should be emphasised. Lateral trapping results in greatly increased residence times of water in mangrove waterways. For instance, in the tropical dry season water may be trapped within mangrove estuaries for 2-8 weeks, depending on the configuration of the mangrove swamp (Wolanski and Ridd, 1986; Wolanski *et al.*, 1990; Ridd *et al.*, 1990). Trapping is more effective in the upper reaches of such estuaries than near the mouth, where exchange with nearshore water is rapid (eg. Ridd *et al.*, 1990). Even during the wet season, freshwater from short-term flood events may be trapped for weeks within mangrove forests and side creeks of estuaries (Wolanski and Ridd, 1986). However, in the estuaries of large tropical rivers wet season trapping of water by mangrove forests is likely to be minimal (Wolanski, 1989).

Long residence times for water in mangrove waterways is extremely important in controlling the chemistry of estuarine water (section 7.2.2) and has obvious implications for the residence of water-column biota and their dispersal (eg. section 7.5). Lateral trapping of water in mangroves affects the flow rates in small mangrove creeks, such that ebb tide currents greatly exceed flood tide currents, thus causing scouring of sediment, particularly near the mouths of creeks (Wolanski *et al.*, Chapter 3 this volume). Again, this has obvious implications for epibenthic biota in mangrove waterways (section 7.6.1).

7.2.2 Water Chemistry

Catchment size, estuarine geomorphology, tidal range and rainfall patterns interact to control salinity patterns in mangrove waterways (Wolanski, 1989). There is a wide range in the degree of mixing of fresh and salt water in tropical estuaries, from completely flushed systems during floods, where saltwater is found only outside the estuary, through to mangrove creeks in the hot dry season, where there is trapping of water in the upper reaches, which together with evaporation in the shallow waters at creek mouth gives rise to an inverse estuary. The biota of the waterways in any tropical mangrove-lined estuary can thus be subjected to salinities from 0-> 35%, while there may be similar variances in salinities between estuaries. These differences need to be born in mind when comparing studies of biota.

The degree of horizontal and vertical mixing of water in mangrove waterways also influences other major aspects of mangrove water columns, in particular dissolved oxygen and inorganic nutrient concentrations. Boto and Bunt (1981a) found that both the pH and oxygen concentrations in mangrove creeks dropped significantly as one moved upstream, and dissolved oxygen concentrations < 2ppm were often recorded in blind-ending mangrove creeks. High concentrations of dissolved polyphenolic compounds leached from mangrove detritus, can cause drops in the pH as they are oxidized, thus lowering oxygen concentrations. The larger residence time of water in the upstream sections of mangrove creeks presumably leads to higher concentrations of polyphenolics and thus lower oxygen concentrations (Boto and Bunt, 1981a). Although Boto and Bunt (1981a) found no differences in concentrations of ATP along the same upstream-downstream gradient, presumably increased bacterial populations or growth rates in the upstream sections of creeks may also contribute to the lowered oxygen levels.

7.3 Phytoplankton and Primary Production

There have been relatively few studies of phytoplankton species diversity in mangrove habitats, but there is an indication of a relatively low diversity of phytoplankton taxa (Kutner, 1975; Ricard, 1984). It has been suggested that in some areas the low diversity of phytoplankton in *Rhizophora* mangrove habitats is related to the release of tannins by roots and decomposing wood and leaves (e.g.Tundisi *et al.*, 1973). The densities of net phytoplankton cells varies over four orders of magnitude within and between different mangrove waterways (Table 1). For instance, in Guadeloupe, Ricard (1984) recorded densities of net phytoplankton of 2×10^4 to 5×10^8 $cells.l^{-1}$ over an annual cycle. In a detailed study of net phytoplankton cell densities in Brazil, Kutner (1975) recorded an annual variation in phytoplankton cell densities of 1.2×10^5 to 2.0×10^7 $cells.l^{-1}$. The net phytoplankton community was dominated by the diatom species *Skeletonema costatum, Thalassionema nitzschioides, Asterionella japonica, Chaetoceros abnormis* and *Coscinodiscus* spp.. Kutner (1975) also recorded densities of dinoflagellates up to 7.8×10^4 $cells.l^{-1}$ and microflagellates up to 1.85×10^6 $cells.l^{-1}$. However, net phytoplankton usually comprised a relatively small component of the total number of phytoplankton cells in the Brazilian study. It was estimated that throughout most of the year nannophytoplankton constituted greater than 80% of the total phytoplankton cell densities. Only during periods of diatom blooms were net phytoplankton cells a dominant component of the total phytoplankton cell densities (Kutner, 1975). Nannophytoplankton was also a major constituent of the phytoplankton in Indian estuaries and in Guadeloupe (Qasim *et al.*, 1972; Ricard, 1984).

The concentrations of chlorophyll *a* in mangrove waterways is highly variable (Table 1). For instance, in pristine mangrove systems such as in Missionary Bay in tropical Australia and in the Fly River in New Guinea there is a relatively narrow range of chlorophyll concentrations from 0.15 to 5.07 $\mu g.l^{-1}$. By contrast, in areas close to large human populations, or in regions where large monsoonal rainfall delivers high concentrations of nutrients to enclosed mangroves lagoons, chlorophyll *a* concentrations may reach 60 $\mu g.l^{-1}$.

Phytoplankton productivity in mangrove waterways may be quite high (Table 1). In the coastal lagoons of the Ivory Coast production may be up to 5 $gC.m^{-3}.d^{-1}$. Such high production rates occur in lagoons which receive significant quantities of nitrogen and phosphorus from adjacent human populations. However, even in systems which do not have high eutrophication rates, phytoplankton primary production can still be substantial. For instance in the coastal lagoons of Mexico, daily production can be up to 2.4 $gC.m^{-3}$. Phytoplankton productivity appears to be significantly lower in estuarine mangrove areas than in lagoons fringed by mangroves, or in open embayments fringed by mangroves. For instance, in the large delta of the Fly River in Papua New Guinea, Robertson *et al.*, (1992) measured daily production rates ranging from 22 to 693 $mgC.m^{-2}$, similar to values measured in Malaysian estuarine mangrove systems (Table 1).

Recently Teixeira and Gaeta (1991) have shown that the picoplankton component (cells $<1\mu m$) of mangrove phytoplankton in southeastern Brazil is responsible for 3 to 29% of ^{14}C uptake. In offshore coastal waters the picoplankton was responsible for 19 to 40% of net primary production.

Table 1. Phytoplankton cell counts, chlorophyll *a* concentrations and net primary production in mangrove waterways; * = ^{14}C method, ** = light/dark bottle O_2 method.

Location	Habitat	Net-phytoplankton Cell Counts ($n°.l^{-1}$)	Chlorophyll *a* ($\mu g.l^{-1}$)	Production ($mgC.m^{-3}.d^{-1}$) (*$mgC.m^{-2}.d^{-1}$*)
Africa				
Gambia[1]	Estuarine mangroves		0.3-8.2	1-445
Ghana[2]	Coastal lagoon, surface			626-1992**
	Coastal lagoon, bottom			222-600**
Ghana[3]	Coastal lagoon, surface			385-1420**
	Coastal lagoon, bottom			120-320**
Ivory Coast[4]	Coastal lagoons		10.3-16.4	*200-5000***
Mauritania[5]	Mangrove embayment		0.46-3.60	580*
	Mangrove creek		0.20-1.07	215*
Americas				
Guadeloupe[6]	Mangrove channel	$2x10^4$-$5x10^8$	10-60	8-1700*
Mexico[7]	Coastal lagoon	10^5-10^8	0.3-8.2	*1200**
Brazil[8]	Estuarine mangroves	$1.2x10^5 2.0x10^7$	1.08-19.26	110-500* *100-800*
Mexico[9]	Coastal lagoon, Pacific			*2450***
Asia				
India[10]	Estuarine waters	$1.7x10^5$-$4.9x10^5$	2.5-14.0	*232-1211*** *266-833**
India[11]	Coastal lagoon	$7x10^3$-$7.4x10^6$	4.36-39.8	60-662**
India[12]	Estaurine mangroves		2.1	*190-1540**
Thailand[13]	Embayment			*560-2410**
Malaysia[14]	Estuarine mangroves			*274-959***
Malaysia[15]	Estuarine mangroves		0.53-21.20	10-1068**
Australia and New Guinea				
New Guinea[16]	Estuarine mangroves		0.25-5.07	22-693*
Australia[17]	Mangrove creek		1.3	

Sources: 1. Healey *et al.*, 1988; 2. Kwei, 1977; 3. Pauly, 1975; 4. Pages *et al.*, 1981; Iltis, 1984; 5. Sevrin-Reyssac, 1980; 6. Ricard, 1984; 7. Gomez-Aguirre, 1977 in Ricard, 1984; Day *et al.*, 1982; 8. Teixeira *et al.*, 1965; Teixeira *et al.*, 1969; Tundisi *et al.*, 1973; Kutner, 1975; 9. Edwards, 1978; 10. Gopinathan, 1972; Qasim *et al.*, 1969; Qasim, 1973; 11. Sunderaraj and Krishnamurthy, 1973; 12. Pant *et al.*, 1980; 13. Wium-Andersen, 1979; 14. Ong *et al.*, 1984; 15. Lee *et al.*, 1984; 16. Robertson *et al.*, 1992; 17. Boto and Bunt, 1981b.

Table 2. Comparison of the contribution of phytoplankton and mangrove primary productivity to the total *in situ* production in lagoonal and estuarine mangrove ecosystems.

Site and source	Area (Km^2)	Productivity ($gC.m^{-2}.d^{-1}$)	Production (tonnes $C.d^{-1}$)	%
Terminos Lagoon (Mexico)[1]				
Phytoplankton	1,600	1.2	1,920	50
Mangrove	1,300	1.3	1,690	43
Seagrass	100	2.6	260	7
Fly River Estuary (Papua New Guinea)[2]				
Phytoplankton	3,100	0.18	558	20
Mangrove	874	1.9-2.7	2214	80

Sources: 1. Day *et al.*, (1982); 2. Robertson *et al.*, (1991 and 1992)

The relative importance of phytoplankton to total mangrove system primary production varies with the geomorphology of the study site, the rates of flow of water and consequently the turbidity and rates of delivery of nutrients. For instance, Day *et al.*, (1982) have estimated that in the Terminos Lagoon in Mexico 50% of the total carbon fixed per day comes from phytoplankton and 43% from the surrounding mangrove swamps (Table 2). In this lagoon there are large areas of relatively clear, shallow water conducive to phytoplankton production. A similar situation is also observed in lagoonal systems in the Ivory Coast and the Cochin backwaters in India (Pages *et al.*, 1981; Qasim *et al.*, 1969). In the estuarine mangrove systems typical of most regions of southeast Asia, central America and tropical South America, high turbidity, large fluctations in salinity and the relatively small ratio of open waterway to mangrove forest area (0.02 - 0.32, Chapter 3, this volume) all ensure that the contribution of phytoplankton to total estuarine primary production is likely to be relatively small. For instance in the Fly River delta in Papua New Guinea the daily production by mangrove forests has been estimated at approximately 2,214 tonnes carbon. This represents 80% of the estimated total estuarine productivity, the remaining 20% being contributed by water column phytoplankton production (Table 2). However, considering the relatively poor nutritional quality of mangrove detrital material (see Chapter 10, this volume) it is quite possible that phytoplankton production may play a more important role in supporting higher trophic levels in estuarine mangrove systems than has previously been acknowledged.

7.4 Microbial Communities

Despite the central role of microbial communities in remineralization processes and food chains in nearshore habitats (eg. Cole *et al.*, 1986; Sherr and Sherr, 1988) there have been very few studies of microbial communities in tropical mangrove waterways. Several studies have attempted to enumerate and categorize the functional groups of bacteria using plating techniques (e.g. Pant *et al.*, 1980; Carmouze and Caumette, 1985), but there have been only four quantitative studies of the densities of bacteria in mangrove waterways using reliable, direct-count techniques. Healey *et al.*, (1988) estimated that bacterial densities in the mangrove dominated estuary of the Gambia River in West Africa varied between 1 and 2 x 10^6 cells.ml^{-1}, similar to values obtained from temperate estuarine waters (Cole *et al.*, 1988). Eighty percent of bacterial cells in the water column of the Gambia River estuary were free cells, the remainder attached to detrital particles (Healey *et al.*, 1988). Seasonal changes in free bacteria densities were small, but concentrations of attached bacteria were higher in the Gambia estuary during rising and flood tide periods when the suspended solid concentrations were at their maximum. Free bacteria were also found in greater densities in regions of the Gambia estuary with high chlorophyll *a* concentrations, which were usually adjacent to luxuriant mangrove forests. Healey *et al.*, (1988) also measured the uptake of ^{14}C-labelled glucose by bacterial communities and found that incorporation of this substrate occurred at a greater rate than in temperate estuaries. They also found positive correlations between glucose uptake rates and densities of attached bacteria and the concentrations of suspended solids in the water column. These results suggest that particle - associated bacteria were the important and metabolically active members of the plankton assemblage. Given that phytoplankton biomass (chlorophyll *a*) and production were low in the Gambia estuary (Table 1) and that bacterial biomass and glucose uptake rates were high, Healey *et al.*, (1988) pointed to the dominance of heterotrophy over autotrophy in the water column of this estuary.

In tropical Australia, Revelante and Gilmartin (University of Maine unpub. data) estimated that during the dry season bacterial cell densities in a mangrove creek ranged from 0.9 to 3.3 x 10^6 cells.ml^{-1} (mean 1.85 x 10^6). In an adjacent open bay, ~ 3 km from the creek, densities were slightly lower with a range of 0.8 to 1.2 x 10^6 (mean 1.05 x 10^6). Bacterial carbon estimates for the two habitats were 26-93 (mean 52) µgC.l^{-1} in the creeks and 23-34 (mean 30) µgC.l^{-1} for the embayment.

Robertson *et al.*, (1992) measured bacterial standing stocks and production in the water column of the Fly River estuary in Papua New Guinea. They recorded bacterial densities of between 10^4 and 10^5 cells per ml, significantly lower then densities recorded in a variety of temperate and tropical water columns (Cole *et al.*, 1988; Healey *et al.*, 1988) but argued that there were problems with direct counts in situations where suspended solid concentrations are ≥ 1g.l^{-1}. However, bacterial production rates in the Fly estuary ranged between 20.4 and 498.1 mgC.m^{-3}.d^{-1}, similar to the average of bacterial production estimates in temperate estuaries (Cole *et al.*, 1988). By combining estimates of water column respiration rates and bacterial production rates Robertson *et al.*, (1992) indicated that bacteria were responsible for up to 48% of the total water column metabolism in the estuary of the Fly River. As in the Gambia River, total water column metabolism in the estuary of the Fly greatly exceeded *in situ* water column primary production.

In Ebrie Lagoon, a mangrove fringed lagoon in the Ivory Coast, Caumette *et al.*, (1983) studied the large populations of phototrophic bacteria which develop in a brown surface layer on the lagoon during some periods of the year. The phototrophic bacterial community was dominated by *Rhodopseudomonas* spp., *Chromatium gracile, Chlorobium vibrioforne,*

C. phaeopacteroids and *Pelodictyon* spp.. It was estimated that 41% of the total algal production in the water column of the lagoon of 2.2 $gC.m^{-3}.d^{-1}$ was due to phototrophic bacteria. Gut content analysis of the dominant copepod in the lagoon, *Acartia clausi,* showed that phototrophic bacteria, especially *Rhodospirillaceae* and *Chromatium,* formed a major proportion of its diet.

It is surprising, given the emphasis on the importance of mangrove detritus decomposition in tropical estuaries, and the hypotherized role of mangrove detritus in supporting coastal food chains, that there have been so few studies of microbial community stocks and dynamics. Whether microbial communities are important food resources for water column animal communities in mangrove forests, needs to be addressed if we wish to test hypotheses about the links between mangrove primary productivity and nearshore secondary production.

7.5 Zooplankton Communities

The degree of freshwater flushing in estuaries, and the seasonal variation in salinities in other mangrove systems such as coastal lagoons, are the main factor controlling the species composition of mangrove zooplankton communities (Grindley, 1984). Within estuaries there is often a well defined shift in zooplankton species composition with decreasing salinities; a stenohaline marine component penetrating only to the mouths of estuaries (eg., species such as *Corycaeus*), a euryhaline marine component penetrating further up the estuary (e.g., species of *Paracalanus, Parvocalanus* and *Oithona*), a true estuarine component (eg., species of *Pseudodiaptomus*) and a freshwater component comprising species normally found only in fresh water (eg., species of *Diaptomus*) (Grindley, 1984). The species characteristic of these components are of course different in various parts of the world (e.g., see Table 3).

Most of the zooplankton surveys in Table 3 were conducted in the mainstreams of tropical mangrove estuaries or close to mangrove forests growing on the edge of lagoons or embayments, rather than directly within the forest itself. An obvious feature of most studies is the abundance of species of the cyclopoid copepod genus *Oithona* (Table 3; and see Ambler *et al.*, 1991). Other important copepod taxa are the harpacticoids *Pseudodiaptomus* spp. and the calanoids *Acartia* spp. *Paracalanus* spp. and *Parvocalanus* spp. Because of the difference in mesh sizes of nets used to capture zooplankton, caution is required when comparing the results of different studies. In most studies that were performed in estuaries or coastal lagoons, total zooplankton densities range from 10^4 to 10^5 individuals per m^3. Biomass figures for mangrove zooplankton are few and variable, ranging from <1 to 623 $mg.m^{-3}$ (Table 3). These values are substantially higher than those recorded offshore from mangrove habitats. For instance, Robertson *et al.*, (1988) showed that there was an order of magnitude higher density of zooplankton in mangrove habitats in tropical Australia then in the waters of an adjacent embayment, some 10 kilometers from their mangrove sampling sites.

Although holoplankton dominate zooplankton communities in the mainstream of mangrove estuaries and in mangrove-lined embayments, a variety of meroplankton are

Table 3. Zooplankton community standing stocks (densities and dry mass) and dominant taxa (> 80% of total numbers) in tropical mangrove waterways. A = converted from wet weight, assuming dry weight = 0.019 wet weight (Omori, 1969); B = only adult copepods.

Location	Habitat	Salinity Range (‰)	Net Mesh Size (μm)	Density (no.m-3)	Biomass (mg.m-3)	Dominant Taxa
Africa						
Ivory Coast[1,2]	Coastal lagoon	0-30	64	up to 2.5 x105	up to 48.2	*Acartia clausi*, *Pseudodiaptomus hessei*, *Oithona brevicornis*, *Paracalanus* sp., rotifers, *Penilia* sp., *Evadine* sp.
Americas						
Venezuela[3]	Coastal lagoon	6-35	70	$9.0\text{-}1.28 \times 10^5$	2.3-55.7	*Oithona hebes*, *Brachionus plicatilis*, *Favella panamensis*.
Brazil[4]	Estuary	3-24	50	$6 \times 10^2\text{-}1.5 \times 10^5$		*Oithona* sp., *Pseudodiaptomus acutus*.
Brazil[5]	Estuary	14-32	50	$6 \times 10^4\text{-}2.2 \times 10^5$		*Pseudodiaptomus acutus*, *Euterpina acutifrons* *Acartia lilljeborgii*, *Oithona orals*, *Paracalanus* sp.
Puerto Rico[6]	Embayment	30-37	202	$1.3 \times 10^2\text{-}4.2 \times 10^3$		*Acartia tonsa*, *Pseudodiaptomus cokeri*.
Asia						
India[7,8]	Estuary	5-20	75	$1.0 \times 10^3\text{-}4.9 \times 10^4$	40-176[A]	-
India[9]	Estuary	3-33	119	$1.2 \times 10^2\text{-}7.4 \times 10^4$		*Oithona* spp.
Thailand[10]	Estuary	NA	100	$2.0 \times 10^4\text{-}1.7 \times 10^5$		*Oithona brevicornis*, *Acartia* spp., *Corycaeus* spp., *Microsetella norvegica*
Singapore[11]	Estuary	14.5-29.4	>10	$3.7 \times 10^3\text{-}1.1 \times 10^5$[B]		*Acartia* spp.
Australia, New Guinea						
New Guinea[12]	Estuary	0-25.4	105	$1.5 \times 10^2\text{-}1.7 \times 10^4$	<1-623	*Oithona aurensis*, *Parvocalanus crassirostris*, *Oithona simplex*, *Oncea media*, *Oithona attenuata*, Bivalve larvae.
Australia[13]	Estuary	7-38	105	$2.0 \times 10^2\text{-}6.1 \times 10^4$		*Parvocalanus crassirostris*, *Oithona simplex*, *Paracalanus* spp.,

Table 3 continued

Australia[14]	Estuary	0-35	290	_	_	*Oithona australis*, *Euterpina acutifrons*. *Gladioferens pectinatus*, *Calamoccia trifida*, *Boeckella fluvialis*, *Sulcanus conflictus*, *Pseudodiaptomus colefaxi*, *Oithona brevicornis*.

Sources: 1. Arfi *et al.*, 1987; 2. Caumette *et al.*, 1983; 3. Zoppi de Roa, 1974; 4. Teixeira *et al.*, 1969; 5. Tundisi *et al.*, 1973; 6. Youngbluth, 1980; 7. Bhunia and Choudury, 1982; 8. Sakar *et al.*, 1984; 9. Shanmugan *et al.*, 1986; 10. Marumo *et al.*, 1985; 11. Chua, 1973; 12. Robertson *et al.*, 1990; 13. Robertson *et al.*, 1988; 14. Kennedy, 1978.

seasonally abundant within the side creeks and forests of mangrove habitats. In tropical Australia, Robertson *et al.*, (1988) sampled in a variety of microhabitats within an estuarine complex. A mangrove forest site was sampled using a specially designed, floating pump system (Dixon and Robertson, 1986), while small mangrove drainage creeks, the mainstream of the mangrove dominated estuary, seagrass flats at the entrance to the mangrove estuary and an offshore (10 km) station were sampled with oblique net tows. Zooplankton community structure often differed among mangrove habitats, particularly in the wet season, but the zooplankton of the mangrove habitats always clustered separately from seagrass and bay samples in classification analyses, mainly due to shifts in the copepod fauna and the abundance of meroplanktan taxa, particularly invertebrate eggs and brachyuran zoea, in mangrove habitats (Robertson *et al.*, 1988). Mangrove and seagrass habitats exhibited marked seasonality in densities of most zooplankton taxa, and during the early wet season (Dec), >70% of the numbers of zooplankton in creeks and forest habitats was meroplankton (Figure 1). There were often also significant tidal variations in zooplankton densities and community structure in mangrove creeks; low tide densities were usually significantly lower than high tide values (Robertson *et al.*, 1988). The finding of marked differences in the proportion of meroplankton among mangrove habitats in eastern Australia (Figure 1) contrasted strongly with previous studies of tropical near shore zooplankton (see Table 3), reflecting the more intensive sampling of all mangrove microhabitats in Robertson *et al.*'s, (1988) study rather than truly depauperate meroplankton in other regions. Recent studies on crab larvae in a Costa Rican mangrove system (Dittel and Epifanio, 1990) have also shown high densities of crab larvae (~ 1000 larvae .m^{-3}) in the zooplankton.

Dittel and Epifanio (1990) investigated temporal patterns in the abundance of larval crabs from several families. They found that spawning by adult crabs followed definite lunar cycles and the larvae of several taxa exhibited tidally rhythmic changes in abundance. Early (zoea 1) larval stages of *Uca* spp., Grapsidae, Xanthidae, *Pinnotheres* spp., and *Petrolisthes* spp. were most abundant during ebb spring tides (Figure 2), suggesting export to the open sea from the mangrove habitat. In contrast advanced zoeal stages and megalopae appeared to take advantage of nocturnal flood tides to be recruited back into the estuary (Figure 2).

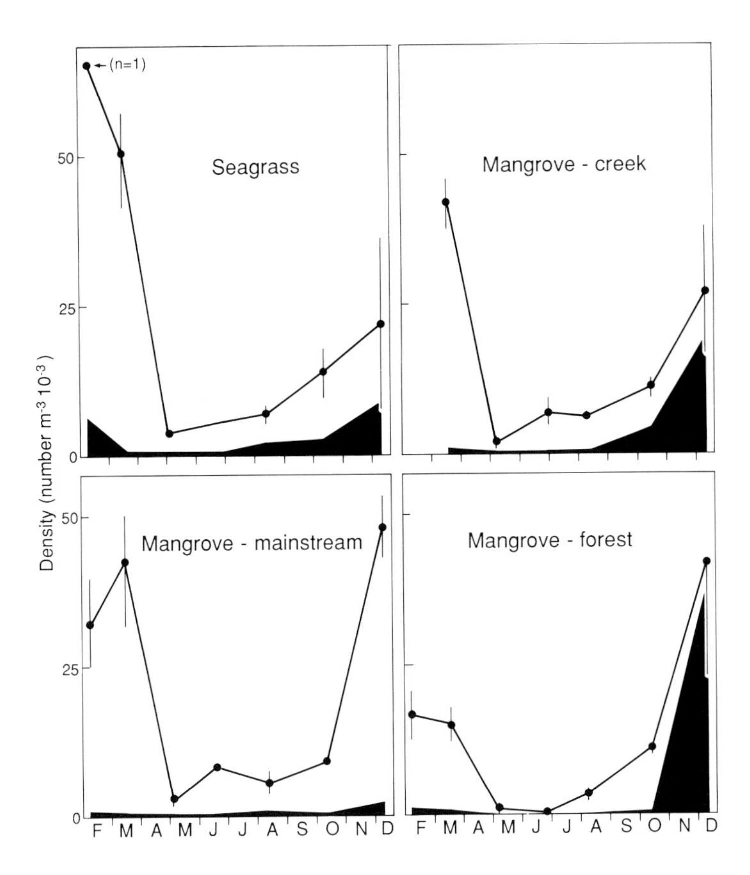

Figure 1. Mean (±1SE) total densities of zooplankton in microhabitats within Alligator Creek, a mangrove lined estuary in tropical northeastern Australia. Dark areas show the proportion of the total zooplankton made up by meroplankton (from Robertson *et al.*, 1988).

7.6 Epibenthos

7.6.1 Community studies

There has been only one detailed study of the epibenthic communities of mangrove waterways. Daniel and Robertson (1990) sampled the epibenthos of tidal mangrove creeks in Missionary Bay and in the Murray River estuary in tropical Queensland, on a seasonal basis using a specially designed beam trawl with high pressure water jets mounted within the trawl (cf. Penn and Stalker, 1975). They investigate the relationship between exported mangrove detritus and faunal communities and the way that community structure of epibenthos changed with distance from mangrove habitats. Missionary Bay has an extensive (50 km^2) mangrove forest dissected by several tidal creeks, only one of which has significant freshwater input. In contrast, the nearby Murray River catchment receives a rainfall of > 2000 $mm.y^{-1}$ and estuarine salinity ranged from 3.2-31.0 ‰ at the sampling sites used in the study.

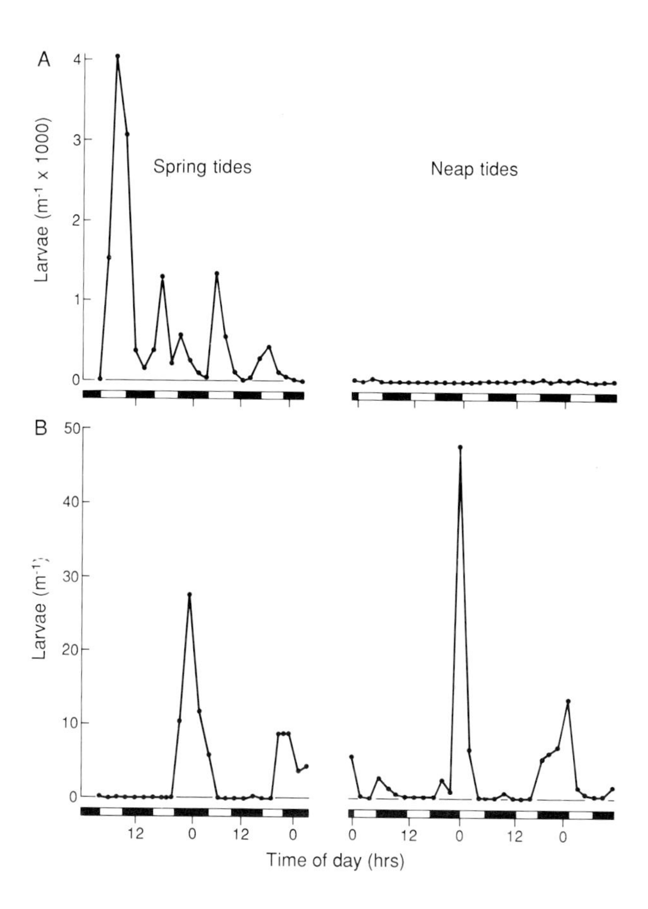

Figure 2. Abundance of (a) *Uca* spp. zoea stage 1 and (b) *Uca* spp. megalopae over five consecutive tidal cycles during spring and neap tide periods in the water column of a mangrove estuary in Costa Rica. Shaded areas are hours of darkness. Light and dark bars on the x-axis refer to ebb and flood tide periods, respectively. From Dittel and Epifanio (1990).

In mangrove creeks in Missionary Bay the epibenthos was dominated by caridean shrimps, mysids, tanaeids, polychaetes, penaeid prawns, small fish and crabs. Small hymenosomatid crabs (90% of Br = brachyura, Figure 3) were often a major component of the catch in the creek in Missionary Bay receiving freshwater input (creek1) and within the estuary of the Murray River (Figure 3). A notable feature of the densities of epibenthos within mangrove creeks in Missionary Bay was the highly variable pattern among creeks at different times of the year. Indeed, only penaeid prawns showed a clear seasonal pattern of abundance across different creeks and estuaries, with densities being significantly higher in the late wet season early dry season (May) than during the rest of the year (Daniel and Robertson, 1990).

Small fish and penaeid prawns dominated the biomass of epibenthos of mangrove habitats in both Missionary Bay and the Murray estuary (Figure 3). There was no clear seasonal pattern in epibenthic biomass at either site and mean biomass ranged from zero to 400 $mg.m^{-2}$ (Figure 3).

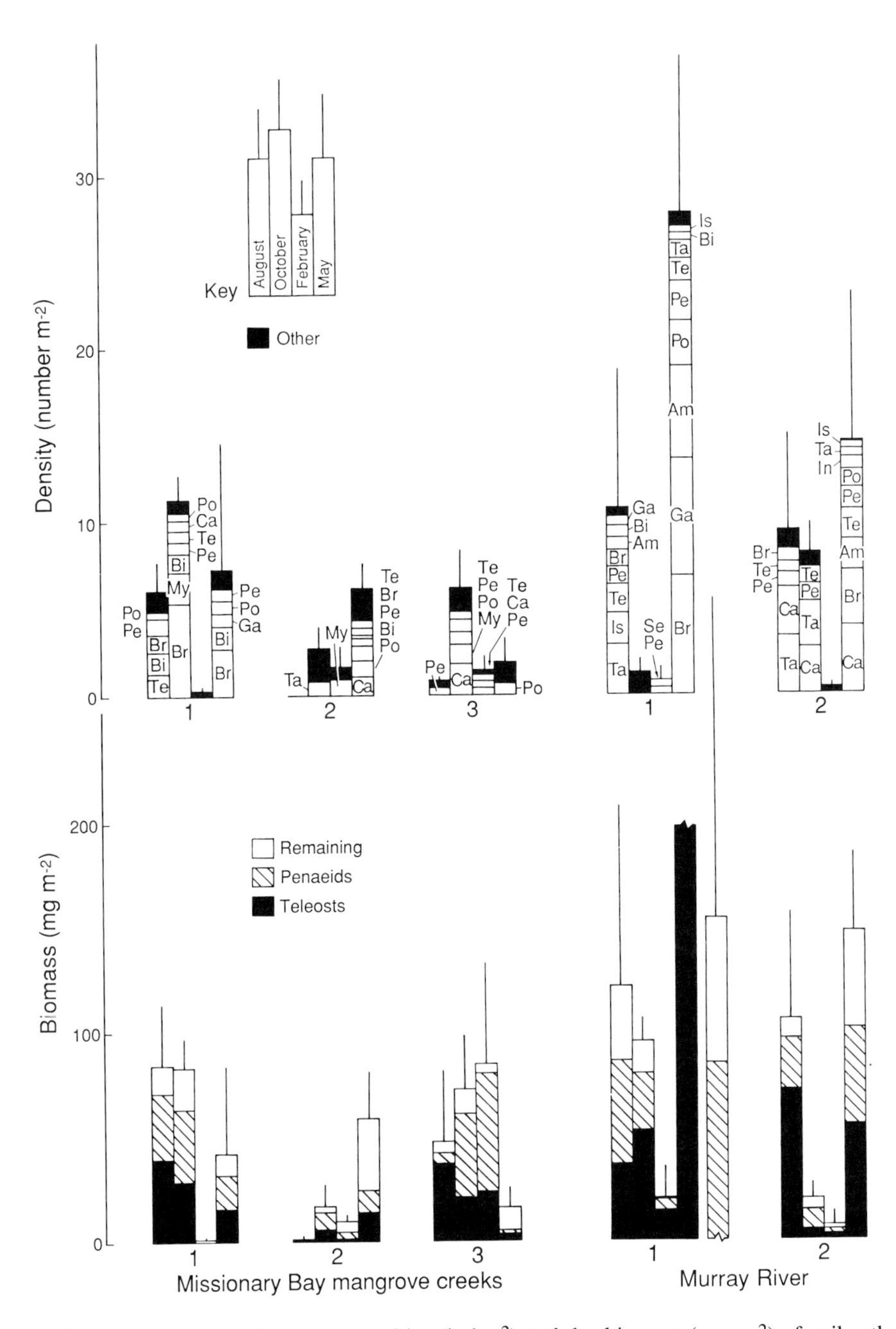

Figure 3. Seasonal changes in the mean densities (ind.m[2]) and dry biomass (mg.m-[2]) of epibenthos in three tidal (high salinity) mangrove creeks in Missionary Bay and two estuarine mangrove sites in the adjacent Murray River in north Queensland. Bars are standard errors. Am: amphipods; Bi: bivalves; Br: brachyurans; Ca: carideans; Cu: cumaceans; Ga: gastropods; Is: isopods; My: mysids; Pe: penaeids, Po: polychaetes; Ta: tanaeids; Te: teleost fishes; Se: sergestid shrimps; O: other taxa (from Daniel and Robertson, 1990).

The standing stock estimates for epibenthos reported by Daniel and Robertson (1990) appear to be much lower than those in other shallow water marine habitats. For instance, densities of macrocrustaceans (greater than 3 mm in size) in Florida seagrass beds averaged approximately 90 per metre squared (Gore *et al.*, 1981); in mangrove waterways, mean total densities were always less than 60 ind.m^{-2} (Figure 3). Total densities of juvenile shrimps and blue crabs (*Callinectes sapidus*) in vegetated and non-vegetated habitats in Texas saltmarshes (Zimmerman and Minello, 1984) were generally much higher than total epibenthos densities recorded in the mangrove waterways in tropical Australia (saltmarsh 1.2 to 80.2 ind.m^{-2}; mangroves 0 to 2.9 ind.m^{-2}).

The mangrove waterways sampled by Daniel and Robertson (1990) represent a physically harsh environment for most epibenthic taxa. The velocity of tidal currents in these mangrove creeks are high (up to 200 cm.sec^{-1}; Wolanski *et al.,* 1980). The scouring effect in mangrove creeks produced by high current speeds is particularly severe during spring tide periods. Epibenthic organisms are thus continually exposed to severe habitat disruption. Some relatively benign microhabitats exist within creeks, where detritus exported from forests accumulates behind snags or in the lee of mud-banks (see below), but in general the bottoms of mangrove waterways are far more disturbed than in many other tropical sedimentary environments.

The long (weeks) retention times of water in mangrove creeks during the dry season in northeastern Australia (see 7.2.1) means that there is reduced tranport of detritus and animals out of creeks by tides. The result is marked differences in the communities of epibenthos between mangrove creeks and adjacent embayments, as observed by Daniel and Robertson (1990). In contrast, during the wet season, Daniel and Robertson (1990) observed greater variations in epibenthic community structure within mangrove habitats than between mangrove habitats and offshore embayments. This shift to greater homogeneity of epibenthic community structure across the inshore-offshore gradient in the wet season was a result of high current speeds during spring tides, which caused whole sections of the epibenthos to be transported out of mangrove creeks along with mangrove detritus (Boto and Bunt, 1981b; Daniel and Robertson, 1990).

Daniel and Robertson (1990) also found that the mass of mangrove detritus present in mangrove waterways was a very good predictor of the standing stocks of epibenthos. Greater than 34% of the total variance in the biomass of small fish and penaeid prawns was explained by the mass of mangrove detritus captured in their trawling operations. It is likely that small fish respond to mangrove detritus as a feeding habitat because of the increased availability of small crustaceans and polychaete prey amongst clumps of decomposing leaves on the bottom of mangrove creeks (Daniel and Robertson, 1990). Exported mangrove detritus on the bottom of mangrove waterways is likely to serve as a useful shelter from predation for juvenile penaeid shrimps. Since juvenile penaeids feed on a combination of organic matter, meiofauna and small macrofauna taken from the surface of the mud (Robertson, 1988) it appears unlikely that penaeids would respond directly to mangrove detritus as a food source. For penaeids it is also unlikely that the exported detritus represents increased living space *per se,* as most shrimps are known to bury themselves in the surface layers of the sediments (e.g. Hindley, 1975; Hill, 1985).

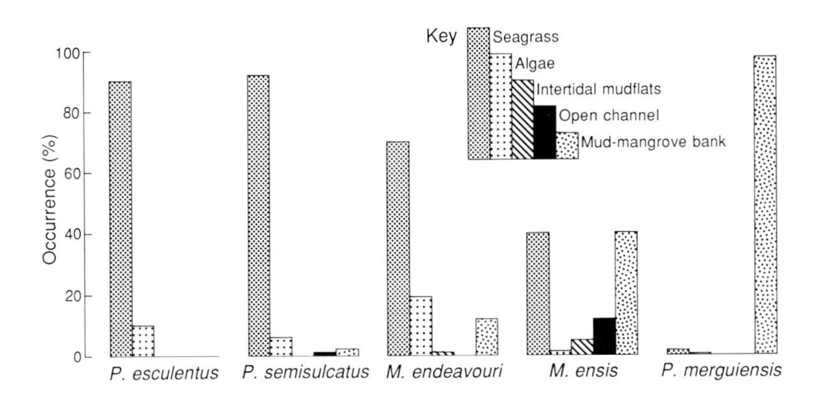

Figure 4. The percentage of the total catch of five species of juvenile penaeid prawns captured in five nearshore habitats in the Embley River estuary, tropical Australia (from Staples *et al.*, 1985).

7.6.2 Penaeid prawns

It has long been known that many of the commercially important penaeid species harvested on the continental shelves of tropical regions have their nursery areas within the coastal zone (e.g. Dakin, 1938; Kutkuhn, 1966). Extensive sampling has also shown that juvenile penaeid prawns are often highly abundant in mangrove habitats (Africa; e.g., Branford, 1981; de Freitas, 1986; Mexico, e.g., Edwards, 1978; Puerto Rico e.g., Stoner, 1988; India, e.g., Achuthankutty and Nair, 1980; Sambasivam, 1985; Thailand, e.g., Boonruang and Janekaru, 1985; Malaysia, e.g., Chong *et al.*, 1990; New Guinea, e.g., Frusher, 1983; and Australia, e.g., Staples *et al.*, 1985; Robertson and Duke, 1987). All of these studies reveal high densities of juvenile penaeids within small mangrove creeks and on the margins of the mainstreams of estuaries fringed by mangrove vegetation. When simultaneous sampling of a variety of inshore habitats has been undertaken it is clear that several species of penaeids occur only in mangrove-associated waters during their juvenile phase. In tropical Australia Staples *et al.*, (1985) compared the abundance of penaeid species in a variety of estuarine habitats (Figure 4). They showed clearly that juveniles of *Penaeus merguiensis* are only found in mangrove associated waterways, while *Penaeus semisulcatus* and *P. escualentus* are much more abundant in seagrass habitats. Other species, e.g. *Metapenaeus ensis* were commonly found in both seagrass and mangrove habitats (Figure 4).

Sampling within the forested areas of mangrove habitats reveals great variation in the number of penaeids associated directly with this microhabitat. In Australia and Malaysia several studies have recorded high numbers of juveniles of the commercially important *Penaeus merguiensis* inhabiting mangrove forests at high tide (Robertson, 1988; Houston, 1978 quoted in Vance *et al.*, 1990; Chong *et al.*, 1990). In contrast, in Florida there is a paucity of penaeids in *Rhizophora* prop root habitats (Thayer and Sheridan, 1992).

The movement into and out of mangrove waterways by penaeids is best illustrated by the detailed studies of *Penaeus merguiensis* in the Gulf of Carpentaria in northern Australia. Offshore egg production by adult *P. merguiensis* has two peaks, one in September-October, the other in March (Figure 5). The September-October peak in egg production originates

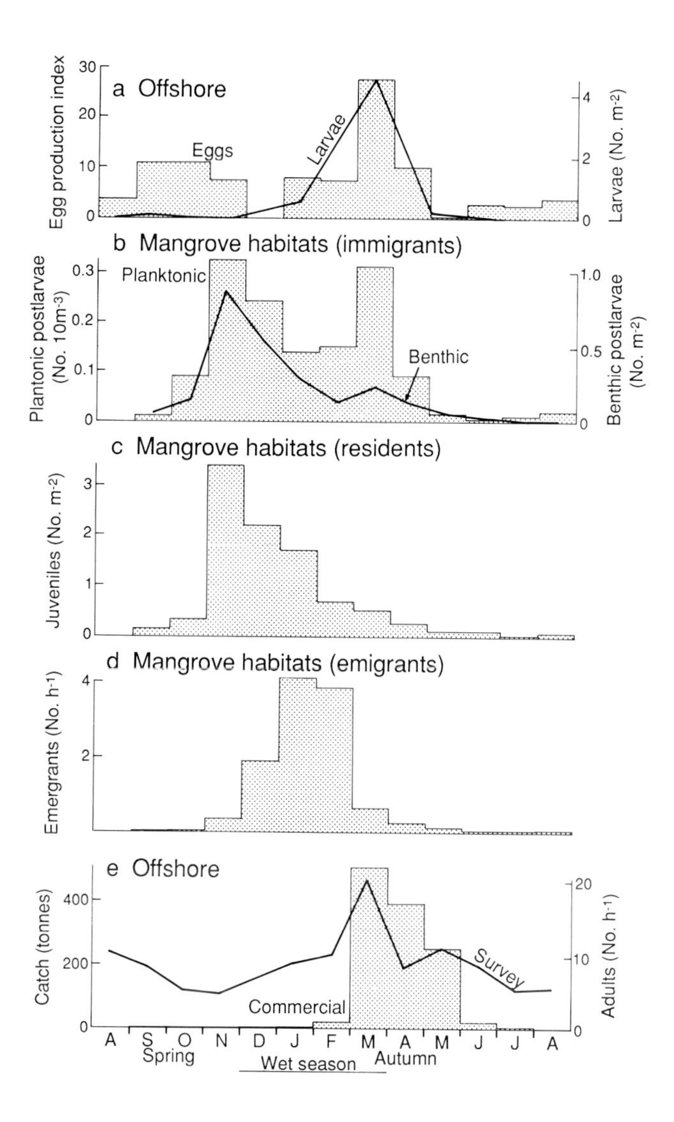

Figure 5. Sequence of life history events for *Penaeus merguiensis* in the southern Gulf of Carpentaria, Australia. Histograms show the abundances of eggs and larvae offshore in the Gulf (a); postlarvae and juveniles in mangrove habitats (b-d); and adults in the offshore commercial catches (e). From Rothlisberg *et al.*, (1985).

from a small number of large adult females, and the March peak comes from a large number of females present offshore during the commercial fishing season. Larval abundance follows a simiiar pattern but the numbers in September-October are much lower than those in March. There are two peaks in the immigration of benthic postlarvae into mangrove-lined estuaries, one occurring in November and the other in March. Usually, the November peak is much more consistant than the March peak. The apparent anomaly between major spawning in March and major recruitment in November is explained by the seasonal changes in the advection of larvae from spawning areas to the nursery grounds, combined with seasonal differences in the settlement success of postlarvae (Rothlisberg *et al.*, 1985). The main period of postlarval immigration into mangrove waterways thus occurs at a time of relatively low numbers of females and larvae offshore. Superimposed on this disproportionate survival of larvae is the differential settlement of post larvae in the mangrove waterways in November (Figure 5). Post larvae that arrive in the estuary after the wet season (March), have a relatively low settlement success compared with the November immigrants; a smaller proportion thus survive through to the juvenile stage. Juveniles from the November generation remain in the estuary for one to four months and begin emigrating from mangrove waterways during the wet season. The main wave of recruits to the commercial fishery appears offshore in March and April (Figure 5) when the majority of prawns are six to seven months old.

Within mangrove waterways very small juvenile *P. merguiensis* (2 to 4 mm carapace length) prefer to inhabit upstream sections of small creeks off the main channel of mangrove habitats (Figure 6). These small juveniles also undergo quite large movements during each daily tidal cycle. Juveniles clearly use the inundated mangrove forests at high tide; on ebb tides they are caught in large numbers in trap nets set in the small gutters which drain mangrove forests (Robertson, 1988), and very few juveniles are caught on banks adjacent to mangrove when the mangrove forests is inundated at high tide (Staples and Vance, 1979). It appears that juvenile *P. merguiensis* resist downstream movement by ebb tides until the water level - and therefore the amount of available habitat within mangrove forests and drainage forests - is quite low. They then enter the main river and concentrate in the turbid shallow waters close to the waters edge near low tide. On the flood tide most of these prawns move back upstream with tidal currents and enter creeks and the mangrove forests (Vance *et al.*, 1990).

The exact mechanism for selection of mangrove habitats by *P. merguiensis* postlarvae, and the subsequently greater numbers of juveniles in mangrove habitats relative to other adjacent nearshore habitats (e.g. Staples *et al.*, 1985) is not certain. It is clear from sampling carried out in the Embley River in the Gulf of Carpentaria in tropical Australia that postlarval

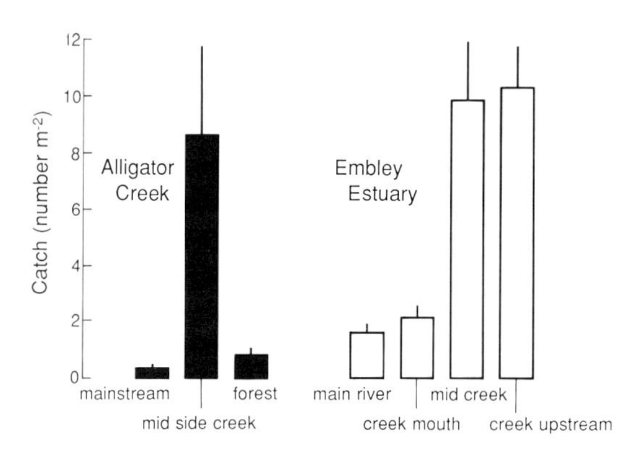

Figure 6. Use of different mangrove microhabitats by postlarval and juvenile *Penaeus merguiensis* in Alligator Creek and the Embley estuary in north Queensland. Data are means + 1SE. See Figure 7 for location of the two sites. From Robertson (1988) and Vance *et al.*, (1990).

P. merguiensis settle, at least in small numbers, in all parts of the estuary, but that the main concentration of postlarvae is along mangrove edges and in particular in the upsteam reaches of small mangrove-lined creeks off the mainstream of the estuary (Vance *et al.*, 1990). One flood tide is not sufficient to transport postlarvae from the mouth of the estuary to the upstream limits of their preferred habitat. Therefore, after initial contact with the substrate they must either move back into the water column to be carried upstream by subsequent flood tides or walk across the substrate to their final settlement point. The alternative is to remain where they first settle and be subjected to increased predation and decreased growth if the habitat is less than optimal. Differential survival among microhabitats is probably important in determining the observed differences in the distribution of settled postlarvae and small juveniles. It is likely that once postlarvae have finally settled in the small creeks, they remain there until they have reached at least five millimetres carapace length (Vance *et al.*, 1990). It is not clear that settling postlarvae of *P. merguiensis* distinguish between mangrove lined banks of the mainstream of estuaries and similar banks in small side creeks. Vance *et al.*, (1990) argue that postlarvae which settle along the main river banks probably remain there but suffer increased mortality from fish predators relative to those that settle in small side creeks of estuaries.

While it is likely that the mangrove habitat provides increased physical structural complexity that will decrease the efficiency of predatory fish in feeding on juvenile *P. merguiensis,* there has not been any conclusive experimental evidence which point to the cause of higher numbers of juvenile *P. merguiensis* in mangroves as compared to other adjacent habitats, such as are available for prawns in some other shallow water systems (eg. Minello and Zimmerman, 1986). Given the rate at which mangrove forests are being removed in southeast Asia (eg., Hatcher *et al.*, 1989) and the importance of the commercial catch of *P. merguiensis* in the region (FAO, 1990), experimental evidence of the causes of the links between this prawn and mangrove waterways is urgently needed.

Further evidence of the close relationships between penaeids and mangrove habitats in the tropics is provided by significant correlations between the estimated maximum sustainable yield (MSY) of penaeids and the area of mangrove habitats (AM) in several regions of the world (Macnae, 1974; Turner, 1977; Martosubroto and Naamin, 1977; Staples *et al.*, 1985). In a recent reworking of most of these data, Pauly and Ingles (1986) showed that the relationship,

$$\log_{10} \mathrm{MSY} = 2.41 + 0.4875 \log_{10} \mathrm{AM} - 0.0212\mathrm{L},$$

where L = degrees of latitude, explained 53% of the variance in the dependent variable (MSY). In their discussion of this relationship the authors went as far as to suggest that since the relationship is logarithmic, the impact of a given reduction of mangrove area on penaeid production will become greater as the remaining area is reduced. This may also be interpreted to mean that destruction of mangrove forests may have the greatest negative impact on penaeid fisheries in regions with only small mangrove areas (Pauly and Ingles, 1986).

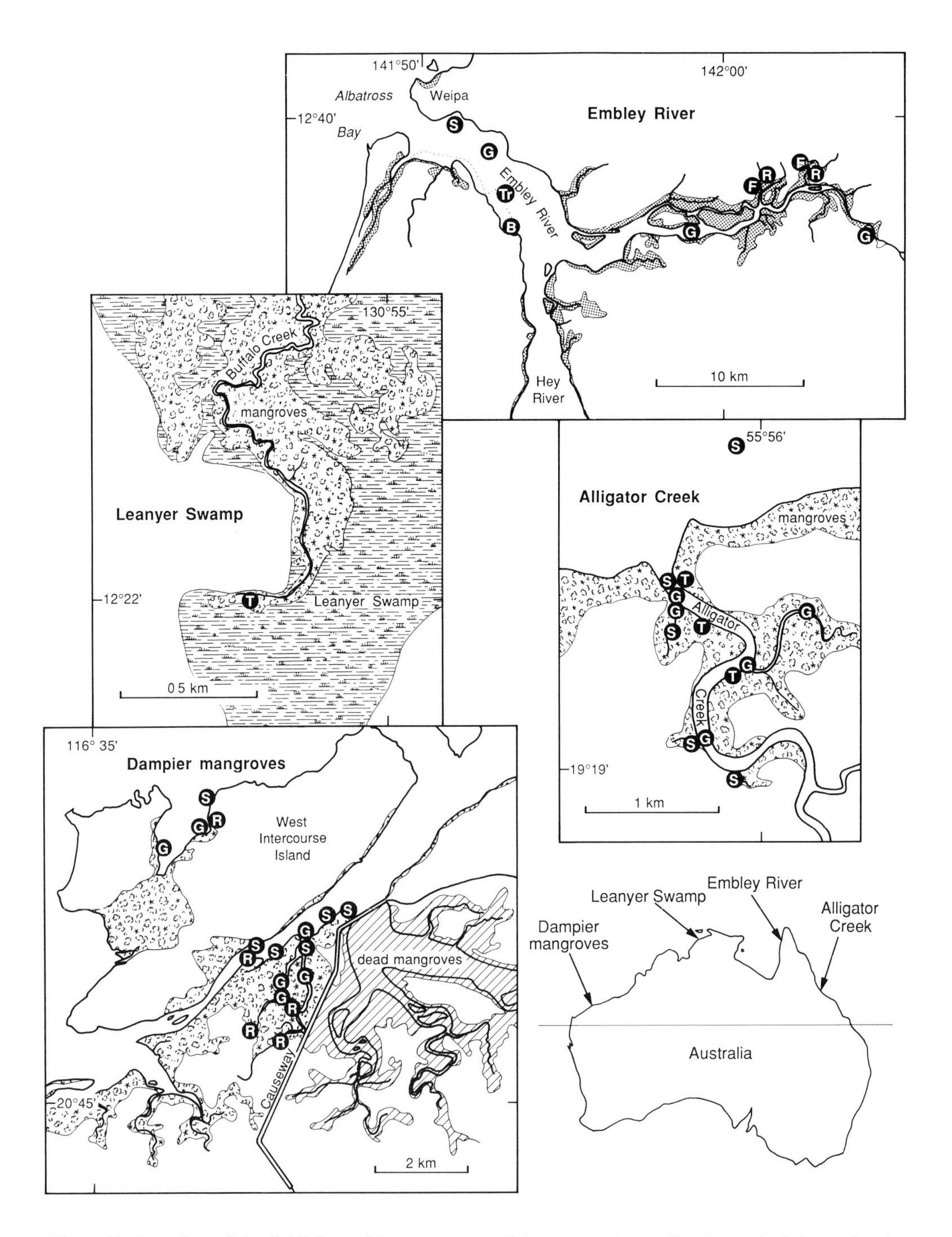

Figure 7. Location of the field sites of four mangrove fish community studies in tropical Australia, the extent of mangrove forests and sampling sites in each of the microhabitats sampled at each location. S = seine net sites, G = gill net sites, T = trap net sites, R = rotenone sites, B = block net sites, Tr = trawl net sites and F = sites where biomass was estimated by fishing to extinction.

While such regression models offer appealing evidence of a link between mangroves and commercial fisheries they have a variety of analytical and theoretical drawbacks. The MSY data is itself questionable and subject to a high degree of error like most fisheries data. Secondly, even if the data are assumed to be correct, nearly half of the variance in MSY is *not* explained by area of mangrove or latitude, and is caused by other factors. In addition, three hypotheses might explain the *cause* of a correlation between mangrove area and commercial penaeid yields. Firstly, mangrove waterways act as nursery grounds for penaeids, which as they mature, move offshore and enter the commercial fishery. Strong evidence in support of this hypothesis comes from the many surveys of postlarval and juvenile penaeids in nearshore habitats (see earlier this section) which show some penaeids to be mangrove-associated as juveniles. However, as pointed out earlier the *causal* link between these juvenile penaeids and mangroves has not been established experimentally. In addition, a large proportion of the penaeids contributing to commercial catches are not mangrove-associated as juveniles, for example *Penaeus semisulcatus* and *P. esculentus* in tropical Australia (Staples *et al.*, 1985; Vance *et al.*, 1990). This is likely to be the reason why catches of penaeids were sustained while there was widespread removal of mangroves on the west coast of India (Macnae, 1974).

The second hypothesis that might explain high penaeid catches adjacent to mangrove areas is that mangroves provide a materials (carbon, nitrogen, phosphorus) subsidy to nearshore waters via outwelling. Evidence for this hypothesis is equivocal (Boto and Wellington, 1988; Robertson *et al.*, 1989; and see Chapters 9 and 10, this volume). However, as mentioned earlier in this section, outwelled macroparticulate detritus can provide important habitat for juvenile penaeids outside mangrove waterways (Daniel and Robertson, 1990).

Finally, as Longhurst and Pauly (1987) have indicated, tropical regions which produce the greatest biomass of penaeids also receive the greatest discharge of terrestrially derived sediment and nutrients, and there are several examples of correlations between river run off and catches of prawns on adjacent continental shelf systems (e.g., Browder, 1985).

Despite the strong inference of a causal link between mangroves and penaeids from the 'nursery-ground' data, no data exists that show unequivocally a significant drop in penaeid prawn catches caused by reduction of mangrove habitat.

7.7 Studies of Mangrove Fishes In Tropical Australia

There have been four recent and relatively long term studies of mangrove fishes by different research groups in tropical Australia (see Figure 7 for location of sites). Alligator Creek in northeastern Queensland was studied by the Australian Institute of Marine Science (Robertson, 1988; Robertson and Duke, 1987, 1990a,b), while the CSIRO Division of Fisheries studied the Embley River estuary (Blaber *et al.*, 1989, 1990a,b; Brewer *et al.*, 1989 and 1991; Salini *et al.*, 1990), Leanyer Swamp in the Northern Territory (Davis, 1988) and mangrove habitats in the Dampier region of Western Australia (Blaber *et al.*, 1985; Blaber, 1986). With the exception of the Leanyer Swamp study, which was restricted to an extreme high intertidal microhabitat, all studies used similar gear to sample fish from most

Table 4. Physical and biological characteristics of the sites used for recent mangrove fish community studies in tropical Australia. For location of each site, see Figure 7. NA = not applicable.

Characteristics	**Alligator Creek**	**Embley estuary**	**Leanyer Swamp**	**Dampier mangroves**
Dominant mangrove genera	*Rhizophora* *Avicennia* *Ceriops*	*Rhizophora* *Bruguiera* *Avicennia* *Ceriops*	*Avicennia*	*Avicennia*
Relative diversity of microhabitats present	Medium	High	Low	Medium
Maximum tidal range in region (m)	3.5	2.6	7*	5.6
Turbidity	Medium	Medium	High	Low
Water temperature (°C)	21-31	25-32	25-39	17.0-31.3
Salinity range (‰)	30-38	5-35	0-51	35.7-39.6
Length of estuary (km)	4	30	NA	NA
Mean annual rainfall (mm)	1215	1722	1659	265

* this swamp only flooded when tidal height >6.8m.

microhabitats present at each location. All sites were in the arid zone tropics, subject to long dry seasons and short west seasons, although the sites have a range of mean annual rainfalls (Table 4). After considering briefly the communities of fishes in each site, we compare the studies in light of differences in the physical characteristics of each site (Table 4), and then compare the findings from the Australian studies with those from other regions.

7.7.1 Alligator Creek

Four main habitat types were identified in this estuary: the intertidal mangrove forests that are flooded at high tide, small tributary creeks (depth at low tide ~ 0.5 m), the main channel with its shallow marginal mudbanks and a seagrass flat at the mouth of the estuary (Figure 7). The emphasis of work in Alligator Creek was on the use of mangrove habitats by juvenile fish.

One hundred and twenty-eight species of 43 families were recorded in Alligator Creek. The fish catch was dominated numerically by the families Engraulidae, Ambassidae, Leiognathidae, Clupeidae and Atherinidae. Two species, the perchlet *Ambassis gymnocephalus* and the anchovy *Stolephorus carpentariae* made up 52% of the total catch and together with 18 other species (Table 5) made up > 96% of total numbers.

Intertidal mangrove forests

A large number of small species used the intertidal forests at high tide. Perchlets, *Ambassis gymnocephalus,* anchovies *Stolephorus carpentariae, S. nelsoni,* and *Encrasicholina devisi,* ponyfish *Leiognathus equulus* and *L. splendens* and archerfish *Toxotes chatareus* were all important components (by numbers and biomass) of the fish community using intertidal forests, although their abundances showed definite seasonal patterns. Greater densities of fish occured in the wet season than the dry season. The wet season is the period of greatest recruitment of juvenile fish to the community (Robertson and Duke, 1990b) and also the time when zooplankton abundance is highest in the mangrove forests (Robertson *et al.*, 1988). *Stolephorus* anchovies were most abundant in the forest in the wet season while other anchovies, *Thryssa brevicauda* and *T. hamiltoni,* and the clupeid *Escualosa thoracata* were more common later in the year. Post-larvae and juveniles of engraulids and clupeids moved into the intertidal forest areas in vast numbers on flood tides and fed on the abundant zooplankton during the wet season but moved out of the estuary altogether in the dry season. In contrast, large numbers of juvenile and sub-adult *Ambassis* and *Leiognathus* use the forest habitat throughout the year. Larger species such as barramundi, *Lates calcarifer*, mangrove snapper, *Lutjanus argentimaculatus* and the sparid *Acanthopagrus berda* also enter the forests at high tide, in search of food. The overall mean (±1se) density and biomass (fresh weight) of fish estimated for the mangrove forest at high tide were 3.5 ± 2.4 individuals $.m^{-2}$ and 10.9 ± 4.5 $g.m^{-2}$, respectively (Robertson and Duke, 1990a).

Small mangrove creeks

Seine netting of this habitat at low tide showed that the two dominant fish in terms of both numbers and biomass in small creeks were the perchlet *Ambassis gymnocephalus* and the ponyfish *Leiognathus equulus* which together form more than 50% of the fish community. Other seasonally abundant species in this habitat are the ponyfish, *Leiognathus splendens,* the anchovy, *Thryssa hamiltoni,* the goby, *Drombus ocyurus* and the sparid *Acanthopagrus berda.* Most of the fish which use the forest at high tide move into shallow creeks on the ebb tide and hence such creeks may support very high standing stocks of fish at low tide. However, some pelagic schooling species such as the *Stolephorus* species and the clupeids *Escualosa thoracata, Sardinella albella* and *Herklotsichthys castelnaui* moved directly into the mainstream of the estuary at low tide and were not abundant in small creeks. Mean (±1se) densities and biomass (fresh weight) for these small creeks were 31.3 ± 12.4 individuals $.m^{-2}$ and 29.0 ± 12.1 $g.m^{-2}$, respectively (Robertson and Duke, 1990a).

Main channel

Seine netting at low tide showed that the main channel of the estuary was dominated by a suite of small species including the benthic clupeid *Nematalosa come,* the pelagic clupeid *Herklotsichthys castelnaui* as well as the perchlet, *Ambassis gymnocephalus.* Larger species

Table 5. The most abundant fish in catches from Alligator Creek, northeastern Australia. Table shows the relative abundance (percentage of the total fish catch), the period of the year when individuals had mature gonads, the range of total lengths in Alligator Creek and the usual total length of mature adults. – = no mature gonads observed; NA = not available; P = permanent resident; LT = long term (~1yr) resident; ST = short term or sporadic user of the estuary. After Robertson and Duke (1990b).

Fish species (Family)	**Percentage of total catch**	**Mature gonads present**	**Range of lengths in Alligator Creek (mm)**	**Maximum Adult length (mm)**	**Residency Status**
Ambassis gymnocephalus (Ambassidae)	29.3	Aug-Feb	7.5-77.5	100	P
Stolephorus carpentariae (Engraulidae)	22.8	_	27.5-67.5	70	ST
Leiognathus equulus (Leiognathidae)	13.8	_	7.5-82.5	250	LT
Encrasicholina devisi (Engraulidae)	6.8	_	37.5-57.5	96	ST
Stolephorus nelsoni (Engraulidae)	6.7	_	27.5-82.5	96	ST
Leiognathus splendens (Leiognathidae)	3.8	_	7.5-57.5	150	LT
Escualosa thoracata (Clupeidae)	3.5	_	17.5-87.5	114	ST
Pranesus eendrachtensis (Atherinidae)	1.9	Oct	17.5-82.5	110	LT
Sardinella spp. (Clupeidae)	1.3	_	17.5-112.5	127	LT
Leiognathus decorus (Leiognathidae)	1.3	_	12.5-62.5	165	LT
Pseudomugil signifer (Melanotaeniidae)	0.8	12 months	12.5-72.5	66	P
Herklotsichthys castelnaui (Clupeidae)	0.7	Aug-Oct	17.5-122.5	135	P
Thryssa brevicauda (Engraulidae)	0.7	_	37.5-52.5	>95	LT
Chelonodon patoca (Tetraodentidae)	0.7	_	7.5-82.5	380	LT
Ambassis buruensis (Ambassidae)	0.6	Dec-Feb	17.5-97.5	125	P
Thryssa hamiltoni (Engraulidae)	0.	_	22.5-137.5	250	LT
Drombus ocyurus (Gobiidae)	0.4	Dec-Feb	12.5-46.5	NA	P
Acanthopagrus berda (Sparidae)	0.3	_	7.5-262.5	380	LT
Nematolosa come (Clupeidae)	0.3	_	37.5-147.5	226	LT
Pomadasys kaakan (Pomadsyidae)	0.2	_	(17.5-157.5)	460	LT

common in the main stream are barramundi, *Lates calcarifer,* queenfish, *Scomberoides* spp., mangrove snapper, *Lutjanus argentimaculatus* and archerfish, *Toxotes chatareus.* Estimated mean (±1se) densities and biomass on mudbanks in the main channel were 0.5 ± 0.1 individuals .m^{-2} and 2.5 ± 0.4 g.m^{-2}, respectively (Robertson and Duke, 1990a).

Seagrass habitat

Ninety one species were recorded from seine nettings in the seagrass habitat at the mouth of Alligator Creek. Catches were dominated by *Stolephorus* anchovies, the ponyfish *Leiognathus splendens* and *L. decorus*, mullet, *Liza subviridis* and the hemiramphid *Arramphus sclerolepis.* Densities of fish in the seagrass habitat were 4 to 10 times less than in mangrove creeks sampled with the same gear (Robertson and Duke, 1987).

Robertson and Duke (1990b) classified the twenty major species in Alligator Creek (those accounting for > 96% of the catch) into three groups on the basis of their residency status in the mangrove habitat; five species were permanent residents, completing their life-cycles in mangrove waterways; eight were `long-term' temporary residents, being present for ~ 1 year as juveniles before moving to other nearshore habitats; and seven were 'short-term' residents or sporadic users of the mangrove habitat (Table 5). Amongst the latter group, four species lived continuously in the mangrove habitat for between 1 and 4 months, while three engraulid species appeared to move rapidly, and often, between mangrove and other nearshore habitats. One of the resident species spawned and recruited throughout the year, but recruitment of juveniles into mangrove waterways was highly seasonal for the remaining species, being concentrated in the late dry season (October) to mid wet season (February) period. Temporary resident species dominated the fish community in the wet season (December-April), but resident species comprised more than 90% of total fish numbers in the mid dry season (August) after temporary residents left the mangroves in the early dry season. Nine of the twenty dominant species in Alligator Creek were strictly dependent on mangrove-lined waterways during all or part of their life-cycle; these were the perchlets, *Ambassis gymnocephalus* and *A. nalua,* the sparid *Acanthopagrus berda,* the tetraodontid, *Chelonodon patoca,* the goby, *Drombus ocyurus,* the clupeids, *Herklotsichthys castelnaui* and *Sardinella* spp., the ponyfish, *Leiognathus equulus,* the grunter, *Pomadasys kaakan,* and the rainbowfish, *Pseudomugil signifer.*

7.7.2 Embley estuary

Open-water mangrove-lined channels and small mangrove creeks occur throughout the Embley estuary, but are the only habitats in the middle and upper reaches of the estuary. Intertidal, sandy-mud beaches occur near the mouth of the river and there are extensive intertidal mudflats adjacent to mangrove forests in the lower reaches of the estuary. Shallow water seagrass flats also occur in the lower reaches of the estuary (Figure 7). The emphasis of work in the Embley was on fish which may be predators of juvenile penaeid prawns, and the study reported only estimates of fish species richness and biomass in habitats within the estuary.

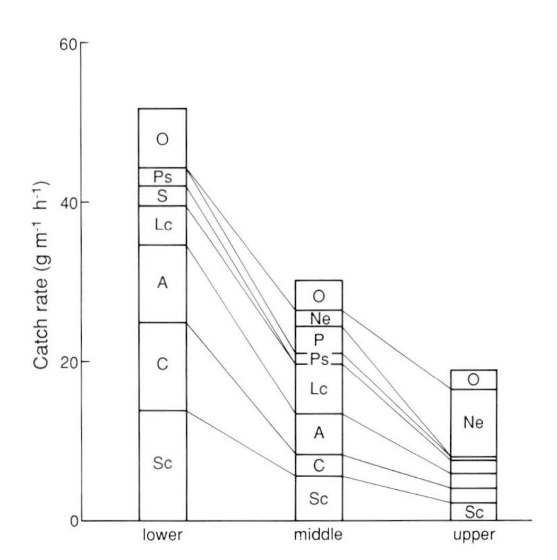

Figure 8. Catch rates for fish in the lower, middle and upper reaches of the main channel of the Embley estuary. A = Ariidae catfish, C = Carcharhinidae sharks, Lc = *Lates calcarifer* (Centropomidae), Ne = *Nematalosa erebi* (Clupeidae), P = Pristidae sharks, Ps = *Polydactylus sheridani* (Polynemidae), S = Sphrynidae sharks, Sc = *Scomberoides commersonianus* (Carangidae) and O = all other taxa. From data in Blaber *et al.*, 1989.

A total of 197 species was recorded from all habitats of the Embley estuary, most of which (92%) were found in the lower reaches of the estuary, whereas only 69 (35%) species and 40 (20%) species were collected in the middle and upper reaches, respectively. The major species recorded in each habitat within the estuary are given below.

Main channel

Regions of this habitat less than 5m deep were sampled with a fleet of 66m long monofilament gill nets of 5 different mesh sizes ranging from 50-150 mm stretch mesh. The sampling thus targeted relatively large and mobile species.

Such nettings captured 127 species during 21 months of sampling (6 dates). The total catch rate of fish, in grams of fish per meter of net per hour, averaged over all seasons decreased from the lower to the upper reaches of the main channel (Figure 8). In the lower reaches of the main channel, the catches were dominated by the carangid *Scomberoides commersonianus,* eight species of carcharinid shark, six species of catfish of the family *Ariidae* and barramundi, *Lates calcarifer.* There were greater catches of *L. calcarifer* in the middle reaches of the main channel and this species made up the greatest percentage of the total catch. Catches of *Scomberoides commersonianus,* ariid catfish and carcharinid sharks decreased markedly in the middle and upper reaches of the main channel, while the reverse was true for the mud herring *Nematalosa erebi,* which dominated the catch in the upper reaches. Blaber *et al.,* (1989) showed that for many taxa there was a temporal component to the downstream-upstream distribution of fish in the main channel. For instance, carcharinid sharks became more abundance in the upper reaches of the mainstream in the late dry season, when salinities were similar along the length of the estuary. However, for some abundant taxa, for example *Lates calcarifer,* temporal changes in biomass did not correspond with seasons over the 21 months of sampling and may have reflected interannual variation in recruitment or other factors.

In the middle and upper reaches of the mainstream, where it was possible to block off a section of the channel, fishing to extinction using gill nets, produced estimates of fish biomass of 16.1 and 7.1 g (fresh wt).m^{-2} (Blaber *et al.*, 1989). These catches were dominated by the same species captured in regular gill net surveys.

Sandy-mud beaches (lower estuary)

Seine netting in this habitat yielded 72 fish species of which the dominant species (i.e. 75% of total biomass) were the sparid *Acanthopagrus berda,* the hemiramphid *Arrhamphus sclerolepis,* the stingray *Himantura uarnek,* the barramundi *Lates calcarifer,* the mullet *Liza vaigiensis* and the carangid *Scomberoides commersonianus.* The mean biomass of fish (over 21 months of sampling) in this habitat was estimated to be 5.03 g.m^{-2} (Blaber *et al.*, 1989).

Seagrass habitat (lower estuary)

Beam trawling at high tide captured 53 fish species in the beds of *Enhales acoroides.* The catch was dominated by small individuals of the cod *Epinephelus suillus,* the teraponid *Pelates quadrilineatus,* the rabbit fish *Siganus canaliculatus,* the apogonid *Apogon fupelli,* the snapper *Lutjanus russeli* and the leatherjacket *Monocanthus chinensis.* The mean (±1se) biomass estimate based on all beam trawl samples was 0.48 ± 0.12 g.m^{-2}. However, rotenone samples in the same habitat, while they produced only 14 species, gave a mean (±1se) biomass estimate of 1.84 ± 0.39 g.m^{-2} (Blaber *et al.*, 1989).

Intertidal mudflats adjacent to mangrove forests (lower estuary).

A single stake netting in July 1 (early dry season) which isolated 9167 m^2, trapped 647.3 kg of fish (i.e. 70.6 g.m^{-2}) of 39 species. The catch was dominated by four teleosts and two species of Dasyatididae (stingrays), which made up of 76% of overall weight. Numerically the most abundant species were the catfish *Arius proximus,* the ephippid, *Drepane punctata* and the gerrid, *Gerres abbreviatus.*

Stake nettings covering such large areas have not been attempted in other mangrove fish studies (Thayer *et al.*, 1987; Morton, 1990; Ley and Montague, 1991). In southeastern Queensland, Morton (1990) used a stake net technique to sample a 3000m^{-2} area of mangrove forest. The maximum catch recorded by Morton (1990) during 13 months of monthly samples was 61.5 g.m^{-2}.

Small mangrove creeks and inlets (all reaches of estuary)

This habitat was sampled by blocking the creeks (2mm mesh net) and poisoning all fish with rotenone. Sixty six species dominated by the puffer, *Tetraodon erythrotaenia,* the mullet, *Liza subviridis,* the mud herring, *Anodontostoma chacunda,* the hemiramphid, *Zenarchopterus buffonis,* the perchlet, *Ambassis nalua,* the archerfish, *Toxotes chatareus* and the sparid, *Acanthopagrus berda* were captured in creeks. Estimated mean biomass of all fish was 8.2 $g.m^{-2}$ (Blaber *et al.* 1989).

The Embley system is unique among the fish studies discussed here in that there was simultaneous sampling of fish in the estuary and in the deeper parts of the adjacent Albatross Bay (Blaber *et al.*, 1990a,b). Of the 197 fish species in the Embley, 106 were not captured in the adjacent bay (59 of these 106 species are however, known to spawn or occur in such embayments elsewhere in Australia; Blaber *et al.*, 1990a,b).

Truly estuarine - dependent species in the Embley belong to three groups (Blaber *et al.*, 1989):

1, Permanent residents (27 species). Most of these species are small fish, living mainly among mangroves and in adjacent creeks, such as the Gobiidae (15 species), Eleotridae (3 species) and Hemiramphidae (2 species).

2, Juveniles of estuarine and shallow marine species (24 species). Juveniles of *Acanthopagrus berda, Ambassis dussumeri, A. gymnocephalus, A. nalua, Amniataba candavittatus, Arius graeffi, A. proximus, Arrhamphus sclerolepis, Caranx papuensis, C. sexfasciatus, Carcharhinus leucas, Epinephelus suillus, Gerres abbreviatus, G. poeti, Lates calcarifer, Liza subviridis, L. vaigiensis, Megalops cyprinoides, Mugil georgii, Sillago analis, S. ingennua, Sphyraena barracuda, S. genie* and *Valamugil buchanani* occurred solely in the Embley estuary.

3, Juveniles of offshore species (14 species). *Anodontostoma chacunda, Chelonodon patoca, Gerres filamentosus, G. oyena, Leiognathus equulus, Lutjanus argentimaculatus, Monocanthus chinensis, Pelates quadrilineatus, Sardinella albella, Sillago lutea, Stolephorus indicus, Terapon jarbua, T. puta* and *Thryssa hamiltoni* also occurred solely in the estuary.

7.7.3 Leanyer Swamp

This tidal swamp is situated at the headwaters of Buffalo Creek to the northeast of Darwin in the Northern Territory (Figure 7), and was studied by Davis (1988) over a 6 month period spanning dry- and wet- seasons. The swamp is flooded periodically by high spring tides, and has numerous shallow (generally <1 m) pools at low tide. Except during the wet season, when they are augmented by freshwater, these pool areas dry out between spring tides. Fish were captured as they moved from the mangrove-lined channel of Buffalo Creek into Leanyer Swamp on spring tides using a fish trap with 1mm mesh. Fish captured in the box-like container at the head of the trap were killed with rotenone.

The composition of the fish fauna entering Leanyer Swamp showed major patterns corresponding to the late dry, early wet and late wet seasons. Thirty eight species from 24 families, mainly juveniles, were recorded entering the swamp (Table 6). Overall, the most abundant species was the scatophagid, *Selenotoca multifasciata,* followed by the perchlet, *Ambassis gymnocephalus* and the megalopid, *Megalops cyprinoides.* However, the first of these species was much more abundant in the early wet (Nov-Dec) than the late wet (Jan-Mar) whereas *Megalops cyprinoides* was more common in the mid wet (Jan). Other species also exhibited differences in patterns of abundance. The two most abundant gobies showed different seasonal patterns: *Hemigobius crassa* was most abundant in the late dry (Oct-Nov) whereas *Pseudogobius* sp. only appeared in the wet season (Nov-March) and was absent during the dry season (Oct).

The movement of the fish community into Leanyer Swamp was determined more by season (i.e. month of the year), tidal sequence and tidal height rather than patterns of salinity and temperature, which varied as much over the short-term as they did with season (Davis, 1988). The seasonal succession of species entering this high intertidal swamp was a reflection of the breeding patterns of each species and the dispersal abilities of their juveniles. The majority of species in Leanyer Swamp are juveniles that use it as a nursery: spawning of many of these, including *Megalops cyprinoides, Lates calcarifer, Gerres oyena, Selenotoca multifasciata, Liza subviridis, L. macrolepis* and *Lutjanus argentimaculatus,* occurs in shallow marine waters or the lower reaches of estuaries (Beumer 1980; Pollard, 1980; Day *et al.,* 1981; Cyrus and Blaber, 1984; Davis, 1985; Blaber, 1987). Other common estuarine species of the region, such as *Atherinomorus endrachtensis, Lutjanus russelli, Acanthopagrus berda* and *Polydactylus sheridani,* rarely enter the swamp from the upper reaches of the estuary. Such tidal swamps are probably not favoured nursery habitats for these species. Adults and juveniles of truly estuarine species, those that spend their entire life cycle in estuaries, and occur in the swamp, include the rainbow fish *Pseudomugil tenellus,* the perchlet *Ambassis gymnocephalus* and *Synbranchus bengalensis.* With the possible exception of *Hypseleotris compressus,* no strictly freshwater species has been found in Leanyer Swamp. This is to be expected as no perennial freshwater steams feed the swamp.

The influence of tidal sequence and tidal height on abundance in the fish trap was considerable for eight species (Davis, 1988). For example, the eleotrid *Bostrichus zonatus* and the mullet *Liza macrolepis* entered on early flood tides while *Sillago analis* only appeared towards the end of the spring tide period. The numbers of the perchlet *Ambassis gymnocephalus,* the ponyfish *Leiognathus brevirostris* and the mangrove snapper *Lutjanus argentimaculatus* were directly related to tidal height. Higher tides enabled juvenile fish to penetrate further into the swamp and benefit from rich food resources available in newly inundated feeding areas (Davis, 1988).

7.7.4 Dampier Mangroves

The mangrove areas studied in this arid region of northwestern Australia were Cleaverville Creek, north Withnell Bay and the West Intercourse Island complex (Figure 7). The three habitats sampled were open channels, intertidal mudbanks and small creeks, which drain almost completely at low tide. A total of 113 fish species from 42 families were recorded from the various mangrove habitats.

Table 6. Major species (those that occurred in more than 10% of trappings or of which >100 individuals were captured during the study period) at Leanyer Swamp, in the Northern Territory of Australia. The total catch and the relative abundance during the major seasons are shown for each species (O = absent; + = rare; ++ = common; +++ = abundant). After Davis, (1988).

Species (Family)	**Total**	**Relative abundance**		
		Dry season	**Early wet season**	**Mid-late wet season**
Selanotoca multifasciata (Scatophagidae)	153416	+++	+++	++
Ambasssis gymnocephalus (Ambassidae)	85054	++	+++	++
Megalops cyprinoides (Megalopidae)	43277	+	+++	+
Pseudomugil tenellus (Melanotaeniidae)	14571	+++	++	++
Pseudogobius sp. (Gobiidae)	14268	O	++	++
Hemigobius crassa (Gobiidae)	7963	++	++	+
Sillago analis (Sillaginidae)	4165	++	++	+
Gerres oyena (Gerridae)	2345	++	++	+
Liza macrolepis (Mugilidae)	1705	++	++	+
Liza dussumieri (Mugilidae)	1680	++	+	O
Terapou jarbua (Teraponidae)	1275	++	++	+
Gobiidae sp. 3	985	++	+	++
Scatophagus argus (Scatophagidae)	499	+	+	+
Amniataba caudavitattus (Teraponidae)	419	+	++	+
Bostrichus zonatus (Eleotridae)	166	+	++	+
Lates calcarifer (Centropomidae)	50	+	+	++
Leiognathus brevirostris (Leiognathidae)	44	+	+	+
Lutjanus argentimaculatus (Lutjanidae)	42	O	+	+

Open water channels

Sixty two species were recorded from gill nettings in this habitat and there is considerable seasonal variation in fish abundance and species composition. In winter (July) and early summer (October) catches were dominated by the caringids *Scomberoides commersonianus, Caranx ignobilis* and *Gnathanodon speciosus,* the sparid *Mylio latus,* the mullet *Valamugil buchanani, Liza subviridis, L. vaigiensis* and *Mugil cephalus* and the scombrid *Scomberomorus semifasciatus.* During summer (January) the channels were dominated by juvenile sharks, *Carcharhinus limbatus,* catfish *Arius* sp.2 and *Arius proximus* as well as the aforementioned Carangidae and Mugilidae. There was also an increase in the catch of *Pomadasys argenteus* and *Nematalosa come* in summer, in contrast to the abundance of both *Mylio latus* and *Scomberomorus semifasciatus* which declined.

Intertidal mudbanks adjacent to mangrove forest

Forty seven species were captured in this habitat using seine nets. The predominant species were the mullets *Valamugil buchanani, V. cunnesius, Liza macrolepis, L. subviridis* and *Mugil cephalus,* the sillaginids *Sillago analis* and *S. maculata,* the gerrids *Gerres oyena* and *G. subfasciatus* as well as the perchlet *Ambassis gymnocephalus* and the atherinids *Allanetta mugiloides* and *Craterocephalus pauciradiatus*.

Small mangrove creeks

A total of 38 species were captured in the small mangrove creeks that drain at low tide. The number of species increased from 15 in winter (July) to 25 in summer (January). This increase was primarily due to the influx of juveniles of larger species whose adults occur outside the mangrove habitat. Much of the biomass however, was contributed by permanently resident estuarine species such as *Ambassis gymnocephalus* and 12 species of Gobiidae, particularly *Acentrogobius moloanus*. Temporarily resident juveniles formed a smaller proportion of the fauna than the small resident species in terms of total number and biomass throughout the year (biomass of residents ranged from 55 - 80%). Large individuals of *Mylio latus, M. palmaris* and the flathead *Platycephalus indicus* entered the creeks at high tide.

The large tidal range in the Dampier mangrove area influences the distribution and movements of many species. Large numbers of fish move into channels, over mudbanks and into small creeks on the rising tide. There is a converse movement on the ebbing tide, particularly at spring lows when even small fish are forced to retreat to shallow intertidal areas largely outside the mangrove creek system, or to remain in isolated pools. During neap tides, water remains in most channels and creeks, and because the water in these mangroves is unusually relatively clear (Table 4), divers have been able to observe large numbers of predators, particularly carangids, among the mangrove roots at high tide (Blaber *et al.*, 1985).

7.7.5 Comparisons of Australian Mangrove Fish Studies

Habitats and physical factors

The number of microhabitats within mangrove systems has a major influence on fish community structure. Large mangrove estuaries such as the Embley in the Gulf of Carpentaria contain **open water channels** up to 6 m deep; extensive **intertidal mudflats** adjacent to **intertidal mangrove forests** and smaller tributary **creeks** usually 1 - 4 km long and up to 3 m deep, as well as **seagrass beds** and **sandy beaches** close to the open sea. Most smaller mangrove estuaries comprise two or more of these components depending on size. For example, Alligator Creek on the Queensland east coast is broadly comparable to one of the tributary creeks of the Embley system with its associated intertidal mangrove forests. Leanyer Swamp at the headwaters of a mangrove estuary represents a mangrove habitat that is well developed in the Northern Territory where tidal ranges are more than 7 m and there is seasonally high rainfall. Such backswamps are less well developed in the Embley and Alligator Creek systems, where tidal ranges are smaller, and are non-existent in the arid Dampier area.

Habitat diversity, and hence fish species richness, is also a function of tidal amplitude, water clarity and salinity fluctuations. For instance, in the mangroves of the Dampier region

Table 7. The forty fish species common to Alligator Creek, Embley estuary and the Dampier region mangrove waterways in tropical Australia. Also shown are broad trophic roles, relative numerical abundance in mangrove fish communities and the life history stages present in mangrove waterways. I: iliophagus; Z: zooplanktivorous; BI: benthic invertebrate feeder; H: herbivorous; P: piscivorous; +++: very abundant; ++: common; +: rare; J: juveniles; A: adults.

Family and Species	**Trophic role**	**Abundance**	**Life-history**
Clupeidae			
Nematalosa come	I	+++	J
Engraulidae			
Thryssa hamiltoni	Z	+++	J
Chirocentridae			
Chirocentrus dorab	P	+	J, A
Ariidae			
Arius graeffei	BI	++	J, A
Exocoetidae			
Strongylura strongylura	P	++	J, A
Hemiramphidae			
Arrhamphus sclerolepis	H	++	J, A
Platycephalidae			
Cymbacephalus nematophthalmus	BI	–	J, A(?)
Platycephalus indicus	BI	+	J, A(?)
Ambassidae			
Ambassis gymnocephalus	Z, BI	+++	J, A
Teraponidae			
Amniataba caudavittatus	BI	+	J
Terapon jarbua	BI	+	J
Sillaginidae			
Sillago analis	BI	+	J
Sillago maculata	BI	+	J
Sillago sihama	BI	+	J
Carangidae			
Caranx sexfasciatus	P	+	J, A
Leiognathidae			
Gazza minuta	P, BI	++	J
Leiognathus equulus	Z, BI	+++	J
Leiognathus decorus	Z, BI	+	J
Gerreidae			
Gerres abbreviatus	BI	+	J
Gerres filamentosus	BI	+	J
Lutjanidae			
Lutjanus argentimaculatus	P, BI	+	J, A
Lutjanus russelli	P, BI	+	J
Monodactylidae			
Monodactylus argenteus	Z	+	J, A
Scatophagidae			
Selenotoca multifasciata	Z	++	J

Table 7 continued

Atherinidae			
Atherinomorus endrachtensis	Z	++	J, A
Mugilidae			
Liza subviridis	I	++	J, A
Liza vaigiensis	I	++	J, A
Valamugil buchanani	I	++	J, A
Valamugil cunnesius	I	++	J, A
Polynemidae			
Eleutheronema tetradactylum	BI	+	J, A
Gobiidae			
Acentrogobius caninus	BI(?)	+	J, A
Acentrogobins viridipunctatus	BI(?)	+	J, A
Drombus triangularis	BI(?)	+	J, A
Favonigobius melanobranchus	BI	+	J, A
Eleotridae			
Butis butis	BI, P	+	J, A
Bothidae			
Pseudorhombus arsius	BI	+	J, A
Pseudorhombus elevatus	BI	+	J, A
Monacanthidae			
Monocanthus chinensis	H, BI	+	J
Tetraodontidae			
Arothron immaculatus	BI	+	J, A
Chelonodon patoca	BI	++	J

water clarity, tidal range and salinity fluctuations differ from those in the other Australian sites studied so far. At Dampier, the tidal range is such that most of the creeks drain at low tide; shallow areas at high and low tide are far apart and hence small fish have to move considerable distances to remain in shallow water. There is usually no freshwater inflow at Dampier and the salinity gradients typical of most tropical mangrove areas are absent (Table 4). Hence, fish which prefer lower salinities are never recorded at Dampier as they are in less arid systems like the Embley, which receive a seasonal flush of freshwater during the wet season. The water at Dampier also has a low turbidity (usually ~3 NTU) without gradients, again in contrast to the other sites considered above, where turbidities are usually medium to high (Table 4). Although juveniles of many species occur in the Dampier mangroves, their numbers are low compared both with other areas (Blaber, 1980; Blaber *et al.,* 1989; Robertson and Duke, 1990b), and with the numbers of adults in the same area. The clear water conditions and deep water at high tide appear to favour predation on juveniles by piscivorous fishes at Dampier. These mangroves are characterised by a high proportion of piscivores in water deeper than 2 m and piscivores always form about 40% of species - in marked contrast to the other mangrove studies where they form about 20% of species (Blaber, 1980; Blaber *et al.*, 1985, 1989; Robertson and Duke, 1990a).

The presence of seagrass beds at the entrances of the Embley estuary and Alligator Creek also enhanced the number of fish species recorded at these sites. This is in some ways similar to many of the Caribbean estuaries and embayments, where seagrasses often occur in close proximity to mangroves and increase local fish species richness (e.g. Thayer *et al.*, 1987).

Ubiquitous fish species

Excluding Leanyer Swamp, which is a habitat that was not sampled elsewhere, 40 species of 24 families are common to the Alligator, Embley and Dampier systems (Table 7) and 72 species occur in both the Alligator and Embley systems in Queensland. In terms of biomass, of the top 20 fish species in Embley creeks, 14 occur in Alligator Creek, and of the top 20 in Alligator Creek, 14 occur in Embley creeks; the following are fish species common to both sites: *Acanthopagrus berda, Ambassis gymnocephalus, A. nalua, Atherinomorus endrachtensis, Chelonodon patoca* and *Drombus ocyurus.* (Blaber *et al.*, 1989; Robertson and Duke, 1990a).

Ambassis gymnocephalus is the dominant zooplanktivore (Martin and Blaber, 1983) in Alligator Creek, tributary creeks at Dampier and in Leanyer Swamp. Although present in the Embley, it formed only 0.16% of fish biomass and had comparatively low numbers in creeks. Other zooplanktivores that were relatively abundant in all systems are the atherind *Atherinomorus endrachtensis* and the ponyfish *Leiognathus equulus* (Table 7). The dominant mullet in all mangrove creeks and intertidal areas at most sites is *Liza subviridis,* although it is unusual that Mugilidae generally were an uncommon group in Alligator Creek during Robertson and Duke's (1990a) sampling period. Larger mullet, *Valamugil buchanani* and *Liza vaigiensis* appear to be more abundant in the open main channels of Australian mangrove systems. *Nematalosa come* is a benthic clupeid with an iliophagous habit similar to that of mullet; it is abundant in both the Alligator and Dampier systems (6% of biomass in each) but the closely related *Anodontostoma chacunda* (10% of biomass) was more abundant in the creeks of the Embley. Among small benthic invertebrate feeders, the puffers *Arothron immaculatus* and *Chelonodon patoca,* the gerrids *Gerres abbreviatus* and *G. filamentosus,* the bothid *Pseudorhombus elevatus,* the whitings *Sillago analis, S. maculata* and *S. sihama* and two teraponids, *Terapon jarbua* and *Amniataba caudavittatus* are equally well represented from Alligator Creek to Dampier (Tablc 7). The detritivorous scatophagid, *Selenotoca multifasciata,* is a seasonally important component of the fauna occurring in mangrove forests, and its juveniles were the most abundant species in Leanyer Swamp.

Gobies are a very species rich part of the fauna of mangrove creeks, but form only a small proportion of overall numbers or biomass in tropical Australia. For example the 15 species in Embley creeks together comprise only 2% of biomass. Widespread species include *Acentrogobius caninus, A. viridipunctatus, Drombus triangularis* and *Favonigobius melanobranchus* (Table 7).

There are two ubiquitous lutjanids: juveniles of both *Lutjanus russelli* and *L. argentimaculatus* are common in all mangrove environments (Table 7) but in low numbers. They generally represent less than 1% of biomass captured with nets in creeks and intertidal forests (e.g. Blaber *et al.*, 1989; Robertson and Duke, 1990a). Adult *L. argentimaculatus* occur in open water channels and move into the forests at high tide to feed, however they are probably under-represented in gill net catches, as evidenced by the numbers taken by hook and line fishermen.

Catfishes of the family Ariidae are prominent omnivores of northern Australian mangrove areas. At least eight species occur in the Embley, of which five (*Arius argyropleuron* (=*macrocephalus*), *A. graeffei, A. mastersi, A. proximus* and *Arius* sp.2) are also found in the Dampier mangrove system. They appear to be less abundant on the east coast of Australia and only *Arius graeffei* has been record in Alligator Creek.

Among piscivores, the most widespread species in mangrove creeks are the carangid *Caranx sexfasciatus,* the flatheads *Cymbacephalus nematphthalmus* and *Platycephalus indicus* and threadfin *Eleutheronema tetradactylum.* In the open water channels the dominant species vary with locality but *Carcharhinus leucas* and *C. limbatus* are common and widespread sharks. Although not captured in Alligator Creek they do occur elsewhere on the east coast of Queensland (Robertson and Duke, 1990a). It is noteworthy that many of the piscivores in mangrove areas also consume large quantities of penaeid prawns (Robertson, 1988; Salini *et al.*, 1990; Salini *et al.*, 1992).

Species with restricted geographic distributions

Restricted distributions may result from geographic barriers or micro- habitat availability and physical conditions in particular estuaries. Geographically restricted species include some very common northern and eastern Australian mangrove species such as the sparid *Acanthopagrus berda* and the barramundi *Lates calcarifer. Acanthopagrus berda* occurs in most mangrove habitats and is omnivorous; the adults are common in open water channels and creeks (up to 20% of biomass) and juveniles in creeks and intertidal forests (up to 4% of total fish community biomass). This species occurs throughout suitable areas of the Indo-west Pacific and in tropical Australia from Queensland to the Northern Territory. However, in northern West Australia it is replaced by an endemic congener *Mylio (Acanthopagrus) latus* - hence its absence from the Dampier mangroves. Also absent from the Dampier mangroves is the barramundi *Lates calcarifer.* In the Embley this important predator forms 11 - 15% of biomass in open water channels and is also abundant in Alligator Creek and in the Northern Territory of Australia (Davis, 1985, 1988). Eighty Mile Beach in West Australia apparently forms a barrier to its western distribution despite suitable habitat in the Dampier region.

The species-rich Gobiidae contains many species with apparently discontinuous distributions. The 27 species recorded from the four Australian mangrove systems are shown in Table 8. Almost all of these have been recorded from throughout the Indo-west Pacific (Hoese, 1986) and yet only four are common to all three of the larger systems being considered here. The extensive sampling that has taken place at these localities makes it unlikely that many more species will be found at each. It is possible that the majority of goby species have specific habitat requirements, although elucidation of these must await further ecological work on the family. Most gobies have long larval durations which permit dispersal over large areas but the proximate factors which determine their distribution also remain a mystery.

Table 8. The occurrence of Gobiidae in the four mangrove systems of northern Australia shown on Figure 7 (+ = present, – = not recorded).

Species	Mangrove System			
	Alligator	Embley	Leanyer	Dampier
Acentrogobius caninus	+	+	–	+
Acentrogobius gracilis	+	+	–	–
Acentrogobius janthinopterus	–		+	–
Acentrogobius moloanus	–	–	–	+
Acentrogobius viridipunctatus	+	+	–	+
Amoya sp.	–	+	–	–
Apocryptodon madurensis	+	–	–	+
Brachyamblyopus sp.	+	–	–	–
Callogobius sp.	–	–	–	+
Crytocentrus sp.	–	+	–	–
Drombus globiceps	+	+	–	–
Drombus ocyurus	+	+	–	–
Drombus triangularis	+	+	–	+
Favonigobius melanobranchus	+	+	–	+
Glossogobius biocellatus	–	+	–	–
Glossogobius celebius	–	+	–	–
Glossogobius circumspectus	–	+	–	+
Hemigobius crassa	–	–	+	–
Incara multisquamata	+	–	–	–
Mugilogobius duospilus	–	–	+	–
Pandaka ludwilli	+	–	–	–
Pandaka rouxi	–	+	+	–
Prionobutis wardi	+	–	–	–
Pseudogobius sp.	+	+	+	–
Redigobius balteatus	+	–	–	–
Yongeichthys criniger	+	–	–	+
Yongeichthys nebulosus	–	–	–	+
Totals	15	15	4	10
Number unidentified species	2	1	1	2

7.7.6 Fish Faunas of Tropical Mangroves: General Considerations

Owing to differences in abiotic factors and topography among systems, as well as variations in sampling methods and intensity, care is needed in comparisons of species richness in mangrove systems worldwide. Here we have restricted comparisons to those studies that are based on long-term sampling (>6 months) in single estuaries or embayments (Table 9), rather than compilations of mangrove fish species for whole countries.

Several clear patterns emerge from such a comparison. The species richness of fish in tropical Australian mangroves is comparable with that of similar areas throughout the Indo-west Pacific (Table 9). In addition, medium and large systems usually have more species than

Table 9. Numbers of fish species recorded from tropical mangrove systems (size of estuaries, based on length; small = <3km; medium = <20 km; large = >20 km)

System	Country	Size	n sp	Author
Indo-west Pacific				
Alligator Creek	Australia	medium	128	Robertson & Duke (1990a)
Trinity	Australia	medium	91	Blaber (1980)
Embley	Australia	large	197	Blaber *et al.*, (1990a)
Leanyer	Australia	small	38	Davis (1988)
Dampier	Australia	large	113	Blaber *et al.*, (1985)
Vellar Coleroom	India	large	195	Krishnamurthy & Jeyaseelam (1981)
Chilka	India	large	110	Jones & Sunjansingani (1954)
Ponggol	Singapore	medium	78	Chua (1973)
Klang	W. Malaysia	medium	102	Chong *et al.* (1990)
Kretam Kechil	E. Malaysia	small	44	Inger (1955)
Purari	Papua New Guinea	large	104	Haincs (1979)
Pagbilao	Philippines	medium	128	Pinto (1988)
Solomon Islands	Solomon Islands	small	8-44*	Blaber & Milton (1990)
Morrumbene	Mocambique	large	113	Day (1974)
Tudor Creek	Kenya	medium	83	Little *et al.*, (1988)
Tropical Atlantic				
Itamaraca	Brazil	large	81	Paranagua & Eskinazi-Leca (1985)
Orinoco	Venezuela	large	87**	Cervigon (1985)
Terminos	Mexico	large	72	Yanez-Arancibia *et al.*, (1988)
Cienaga Grande	Colombia	large	114***	Leon & Racedo (1985)
Grande Terre	Guadeloupe	small	26	Lasserre & Toffart (1977)
West Puerto Rico	Puerto Rico	small	41-65****	Austin (1971), Stoner (1986)

* range from 13 small estuaries
** includes 11 primary freshwater species
*** may include some non-mangrove associated species
**** incorporates 8 mangrove sites

smaller ones; deep open water channels in the larger systems favour more of the larger species, particularly carangids and sharks, in addition to a higher number of incidental marine visitors.

Comparisons with tropical Atlantic mangroves indicate that those of the Indo-west Pacific are more species rich (Table 9) despite the large size of some Atlantic systems (e.g. Terminos). The families and genera living in Caribbean mangroves are similar to those of the Indo-west Pacific, including such families as Ariidae, Carcharhinidae, Carangidae, Gerreidae and Lutjanidae. However, the families Sciaenide, Haemulidae and Sparidae show greater dominance and species numbers in the Caribbean (Yanez- Arancibia, 1988), which may partly be related to the proximity of other habitats in the Caribbean (see below, this section).

Sub-tropical mangroves usually harbour fewer fish species than tropical sites. For example, 48-64 species have been recorded in *Rhizophora* dominated systems in southern

Florida (Thayer and Sheridan, 1992), while 65 and 46 species have been taken from mangrove waterways at Brisbane and Sydney in eastern Australia (Stephenson and Dredge, 1976; Quinn, 1980; Bell *et al.*, 1984).

Fish species composition and richness in any one tropical mangrove system will depend primarily upon, (a) its size and diversity of habitats together with its flood and tidal regimes; (b) its proximity to mangrove and other systems, and (c) the nature of the offshore environment, particularly depth and current patterns. The influence of habitat diversity has already been discussed (section 7.7.5).

Proximity to other mangrove systems ensures colonisation by even those species with no or short larval durations, as well as by movements of adults and juveniles. A corollary of this is that proximity to non- mangrove areas, such as coral reefs may influence fish species composition in the mangroves (Parrish, 1987). However, the effects of the proximity of coral reefs on fish faunas of mangrove estuaries has been discussed by Quinn and Kojis (1985). They showed that mangroves in north-eastern Papua New Guinea are nursery areas for relatively few species of coral reef fish, and suggest that this may apply to the south-west Pacific as a whole. Results from the Solomon Islands and New Caledonia support this suggestion: only 8 to 9% of fish numbers in Solomon Island mangroves are juveniles of reef species. Few of the more than 500 species recorded in coral reefs in the area occur in the mangrove estuaries (Blaber and Milton, 1990). Thollot and Kulbicki (1989) showed that of a total of 497 fish species they studied in coral reefs, soft bottom habitats and mangrove waterways in New Caledonia, only 9 species were ubiquitous. Overlap between soft bottom habitats and mangroves was 36 species, mainly Leiognathidae, Lutjanidae and Sphyraenidae, while only 13 species were common to reefs and mangroves.

Links between mangrove and seagrass habitats may be more important than those between mangroves and coral reefs. For instance in many of the Florida and Caribbean mangrove systems, seagrasses grow adjacent to and among the roots of mangrove trees (e.g. Stoner, 1986; Thayer *et al.*, 1987; Ley and Montague, 1991; Thayer and Sheridan, 1992). Increased protection and living space provided by seagrasses is likely to increase local patterns of fish species richness in these mangrove systems.

The manner in which the offshore environment influences fish community structure is well illustrated by the mangroves of Solomon Islands in the south west Pacific (Blaber and Milton, 1990). Mangrove forests in the Solomon Islands are well developed but are small and isolated from each other by extensive fringing coral reef lagoons. The Solomon Islands are at the western edge of the Pacific Plate (Springer 1982) and are separated from Australia and eastern New Guinea by ocean trenches from 5000 to 9000 m deep (Coleman, 1966). They have shallow-water connections to the Bismark group of Islands, but these in turn are isolated from the northern New Guinea coast by depths greater than 1,000 m. The mangrove fish faunas of the Solomon Islands are closely related to those of New Guinea and northern Australia. Almost one quarter of the species in the Solomon Island estuaries are widely distributed in the Indo-Pacific and occur in similar habitats in Asia and Africa. Although the Solomon Islands chain originated in the Mesozoic (Coleman, 1966), the present high islands are a result of vulcanism that began in the Pliocene (11 to 1 million years BP to present) (Thompson and Hackman, 1969). Hence the mangrove estuaries are on emergent shorelines of recent volcanic origin. Colonisation of these habitats by fishes characteristic of mangrove estuarine areas has probably been from the well-developed and older mangrove areas of northern Australia and New Guinea.

Although almost all the species found in the Solomons' estuaries occur throughout the Australo-Papuan region, a number of families and species common in that region are absent. There are some notable absentees: all ariid catfishes, the centropomid *Lates calcarifer,* all Pomadasyidae and Sciaenidae, as well as the sparid *Acanthopagrus berda.* These taxa form a prominent part of northern Australian mangrove faunas (see section 7.7.5). All, with the exception of *L. calcarifer* and *A. berda,* occur on the north coast of New Guinea (Quinn and Kojis, 1985, 1986) and are prominent in the Australo-Papuan fish fauna. The presence of suitable mangrove habitats in the Solomons suggests that it may not be ecological requirements that have excluded these taxa. Given the recent origin of the mangrove communities of the Solomons, the present mangrove fauna was probably derived from the Australo-Papuan region by dispersal of larvae, juveniles or adults. The deep trenches on the western edge of Solomon Islands would be an effective barrier to movements of many juvenile and adult benthic species. Groups without planktonic larvae may be prevented from reaching the Solomons by the absence of a shallow-water connection. For example, ariid catfishes have large eggs and are mouth brooders (Rimmer and Merrick, 1982) and there is no larval dispersal phase. The nearest mainland mangrove areas to New Georgia are 900 km away in New Guinea (except for patches in the Bismark Archipelago, Percival and Womersley, 1975) and at least 1600 km away in Australia.

With the exception of wide-ranging mobile groups such as carangids, most colonisation of the Solomon mangrove habitats has probably been by larval dispersal. Currents only flow towards the Solomons in February and March and drift times from northern Australia or the Gulf of Papua all exceed 30 days (MacFarlane, 1980). Many fish taxa not recorded in the Solomons appear to have larval durations that are too short to complete the journey from the nearest mangrove areas in New Guinea: *Lates calcarifer* has a larval duration of about 14 days (Salini and Shaklee, 1988, T.L.O. Davis pers. comm.) and Pomadasyidae about 15 days (Thresher, 1984). No data are available for the sparid *Acanthopagrus berda,* but the larvae of the closely related American tropical mangrove species *Archosargus rhomboidalis* begin to abandon the pelagic habit only 12 days after hatching (Houde, 1975). Similarly estuarine sciaenids from Florida have a larval duration of only about 20 days (Peters and McMichael, 1987). In contrast, the taxa that are well represented in Solomons mangroves have larval durations in excess of 20 days. For example, most Gobiidae have larval durations of 30 to 40 days (Thresher, 1984), *Eupomacentrus nigricans* 32 days (R.E. Thresher, pers. comm.) and most apogonids at least 25 days (Thresher, 1984).

It is difficult to compare the densities and biomass of fish in mangrove habitats from different regions of the world, owing to differences in gears used and microhabitats sampled. The most reliable estimates available to date are from southern Florida, where catches have been adjusted for capture efficiencies or at least two diffferent techniques (block netting and

Table 10. Mean density and biomass estimates for fish in tropical and subtropical mangrove systems. Biomass data are fresh weights.

Location and Habitat	Density (no.m^{-2})	Biomass (g.m^{-2})	Method
Americas			
Florida - prop root habitat[1]	8.0	15.0	Block net
Florida - prop root habitat[2]	~1-14		Block net, visual estimates
Mexico - mangrove lined canal[3]	3.1-5.7	7.9-12.5	Seine net
Mexico - mangrove lined lagoon[4]	26.3-161.2	1.2-1.3	Seine net
Mexico - mangrove lined lagoon[5]	-	0.4-3.4	Trawl
South-East Asia			
Malaysia - mangrove creeks[6]	1.3	1.7	Bag/Block net
Australia			
Alligator creek - forest, creek and mainstream[7]	0.5-31.3	2.5-29.0	Trap and seine nets
Embley estuary - creeks and mainstream[8]	-	5.0-16.1	Seine nets, rotenone and fishing to extinction
Moreton Bay - forest[9]	0.27	25.3	Block net
Pacific			
Solomon Islands - creeks[10]	-	11.6	Rotenone

1, Thayer *et al.*, 1987; 2, Ley and Montague, 1991; 3, Warburton, 1978; 4, Edwards, 1978; 5, Yanez-Arancibia *et al.*, 1988; 6, Chong *et al.*, 1990; 7, Robertson and Duke, 1990b; 8, Blaber *et al.*, 1989; 9, Morton, 1990; 10, Blaber and Milton, 1990.

visual censuses) have been used to check estimates (Table 10). Estimates of fish densities in mangrove waters range from 0.3 to 161 fish .m^{-2} (Table 10). Very low densities in Moreton Bay, Australia were estimates from sampling targetting mainly adult fish, while those from Malaysia may represent the results of high fishing pressure (A. Sasekumar, pers. comm.). Most biomass estimates are in the range 7-29 g.m^{-2}. The highest estimates were recorded in subtropical Moreton Bay, Australia, using a block net which was very efficient at capturing large species, and in Alligator creek where fish were netted at low tide when concentrated in creeks. Low biomass estimates in Terminos Lagoon (the third Mexican site, Table 10) were based solely on trawl data, while the Malaysian biomass figures again reflect overfishing.

7.7.7 Utilisation and Dependence of Fishes on Mangroves

The degree to which various inshore marine fishes of the Indo-west Pacific and elsewhere may be dependent on mangrove areas as nurseries has received considerable

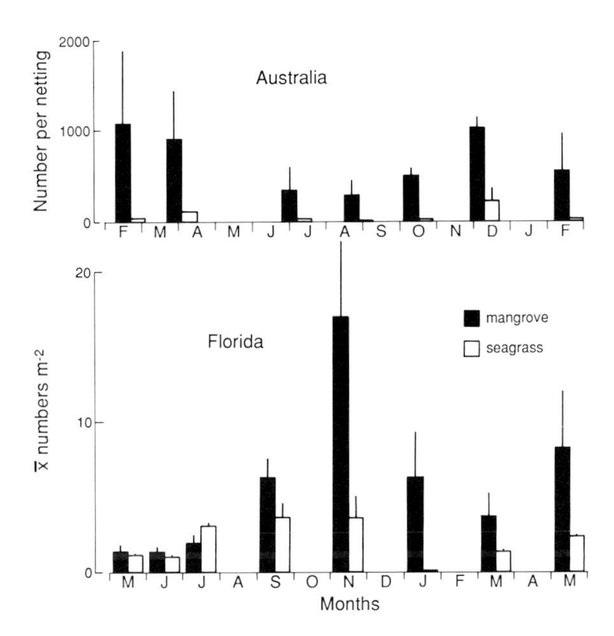

Figure 9. Relative densities of fish in mangrove and adjacent nearshore seagrass habitats in Alligator Creek, north-eastern Australia and Everglades National Park, Florida. The Australian data compares catches taken with a fine-meshed (3mm) seine net in both narrow (3m) mangrove creeks and a nearby seagrass bed (see Figure 7), while the Florida data are based on block net samples within flooded mangrove forests and otter trawling in adjacent seagrass beds. (Data from Robertson and Duke, 1987; Thayer *et al.*, 1987).

attention (e.g. Day *et al.*, 1981; Odum *et al.*, 1982; Thayer *et al.*, 1987; Robertson and Duke, 1987, 1990b, Blaber *et al.*, 1989). The results of these and other studies have shown that mangroves harbor higher densities of juvenile fishes than adjacent habitats (Figure 9) and that most fish species, which use mangroves as juveniles, spawn outside estuaries. In Australia, offshore sampling and estuarine sampling during the study of the Embley showed that of a total of 197 species in the Embley estuary, 15% are permanently estuarine while another 19% live as juveniles only in estuaries. These two groups, which form approximately one third of the species, can be considered 'estuarine-dependent' (Blaber *et al.*, 1989). In the study of Alligator Creek (Robertson and Duke, 1990b) 45% of the 20 numerically dominant species appeared to be dependent on the estuarine habitat.

Estuarine dependence among tropical marine teleosts is not confined to particular families or even genera. In many closely related species, the adults are sympatric offshore but only some of their juveniles are estuarine-dependent. For example, juvenile *Gerres filamentosus* are 'estuarine-dependent' whereas *Gerres subfasciatus* is not (Cyrus and Blaber, 1983; Blaber *et al.*, 1989); juveniles of *Leiognathus equulus* are found only in estuaries, whereas juveniles of other *Leiognathus* species are abundant offshore (Robertson and Duke, 1990b). The reasons for these differences frequently relate to feeding requirements and active preferences for particular physical conditions, such as degree of turbidity (Cyrus and Blaber, 1983; Cyrus and Blaber, 1987).

However, as with penaeid prawns, the role of mangroves in estuarine dependence by fish remains to be clarified. Three broad hypotheses have been advanced to explain the high densities of fish and the dependence of certain species on tropical estuaries which contain large mangrove forests. The first suggests that the turbid waters reduce the effectiveness of large predators on fish (Blaber and Blaber, 1980; Cyrus and Blaber, 1987). Evidence supporting this hypothesis comes from comparisons of fish densities across a variety of coastal habitats which show that the abundances of certain species respond positively to increases in turbidity. Further evidence comes from observations of greater abundances of piscivorous species in mangroves that receive no run-off from the land, and thus have much clearer waters (Blaber *et al.,* 1985 and see Section 7.7.5). Evidence supporting this hypothesis also comes from estuarine systems in which there is little or no mangrove habitat (e.g. the St Lucia system in South Africa) (Cyrus and Blaber, 1987). Thus, if fish are responding solely to turbidity, it is possible that mangrove vegetation has little effect on the dependence of fish on estuaries. However, Robertson and Duke (1987) have shown highly significant differences in the densities and communities of fishes between mangrove and other nearshore habitats that are immediately adjacent to each other and have no differences in turbidity. Thus it appears that factors other than turbidity may also be important in determining the preference of fish for mangrove habitats.

Another hypothesis is that the presence of mangroves in estuaries and embayments are important feeding sites for fish (Chong *et al.,* 1990) or that they increase the supply of food available to juvenile fish. Although fish certainly do feed in mangrove forests and areas immediately adjacent to mangrove forests, there is no evidence to suggest that this is the primary reason for their presence in these habitats. Benthic food resources for fish are not more abundant in mangrove habitats than in adjacent bare sand and mud habitats or seagrass beds (e.g. Nateewathana and Tantichodok, 1984). However, Robertson *et al.* (1988) have shown that mangrove habitats have an order of magnitude greater densities of the preferred zooplankton prey of fish during the late dry - mid wet season recruitment period of fishes than in the middle of the dry season in tropical Australia. There is thus some support for the hypothesis that there is a greater supply of food for recruiting fish in mangroves than in adjacent habitats.

The third hypothesis concerns the structural complexity and increased living space or shelter from predators provided by the prop roots, pneumatophores and fallen logs and branches in mangrove forests and the snags caused by such debris in mangrove creeks. These features are likely to play a role in determining the dependency of some juvenile fish on the mangrove habitat (Thayer *et al.*, 1987; Robertson and Duke, 1990b). The effects of habitat structural complexity on fish assemblages are illustrated by differences in the fish faunas of mangrove estuaries in the Solomon Islands that have different amounts of mangrove debris (Blaber and Milton, 1990). The estuaries with relatively sandy substrates are dominated by mangroves of the genus *Bruguiera* and are clogged with mangrove tree debris. Pomacentridae and some species of Apogonidae and Gobiidae predominate in these estuaries. These species are largely absent from estuaries with soft, muddy substrates, dominated by the mangrove genus *Rhizophora,* which were clear of woody debris. Other species of the family Gobiidae, including burrowing species, are the dominant fish, in these latter estuaries (Blaber and Milton, 1990).

Other data and observations also point to the role of structural complexity in determining fish species richness and standing stocks in mangrove systems. For instance, Daniel and Robertson (1990) observed a highly significant relationship between mangrove detritus (sticks, pieces of wood and leaves) and fish densities and biomass in mangrove creeks. In Sri Lanka local fishermen improve their catches in lagoons, that once were surrounded by mangroves, by creating thickets of dead mangrove sticks on the lagoon floor. Netting around these thickets, once they have been in place for some weeks, produces much higher catches than recorded in adjacent bare mud areas (Pinto, 1986).

Recent work in Florida has provided some of the best evidence for the role of mangrove prop roots in the support of fish populations in estuaries. Thayer *et al.* (1987) and Ley and Montague (1991) have shown that high densities of both juvenile and adult fish occur within the complex of roots within *Rhizophora* forests. Similar data are also available for juvenile and adult fish in Australia (Robertson and Duke, 1990a; Morton, 1990) and Malaysia (Sasekumar *et al.*, 1984). However, as yet there is no experimental evidence to support the hypothesis that such structure decreases the effectiveness of predation in mangrove habitats, relative to other nearshore habitats.

These three hypotheses, which refer basically to turbidity, food and shelter, are probably all important in explaining the importance of mangrove habitats to fish. The relative significance of each hypothesis, in terms of dependency and utilisation of mangroves by fishes, will vary depending upon the fishes in question and the particular nature of each mangrove habitat.

7.8 References

Achuthankutty, C.T., and Nair, S.R.S., 1980. Penaeid prawn population and fry resources in a mangrove swamp of Goa, India. *Symposium Series of the Marine Biological Association of India,* **6**:190-195.

Ambler, J.W., Ferrari, F.D., and Fornshell, J.A., 1991. Population structure and swarm formation of the cyclopoid copepod *Dioithona oculata* near mangrove cays. *Journal of Plankton Research,* **13**:1257-1272.

Arfi, R., Pagano, M., and Saint-Jean, L., 1987. Communautes zooplanctoniques dans une lagune tropicale (la lagune Ebrie, Cote d'Ivoire): variations spation-temporelles. *Revue Hydrobiologie Tropicale,* **20**:21-36.

Austin, H.M., 1971. A study of the ichthyofauna of the mangroves of western Puerto Rico during December, 1967 - August, 1968. *Carribean Journal of Science* **11**:27-39.

Bell, J.D., Pollard, D.A., Burchmore, J.J., Pearce, B.C., and Middleton, M.J., 1984. Structure of a fish community in a temperate tidal mangrove creek in Botany Bay, New South Wales. *Australian Journal of Marine and Freshwater Research* **35**:33-46.

Beumer, J.P., 1978. Feeding ecology of four fishes from a mangrove creek in North Queensland, Australia. *Journal of Fish Biology* **12**:475-490.

Bhunia, A.B. and Choudury, A., 1982. Some ecological considerations for zooplankton production in Chemaguri creek, Sagar Island (South), Sundarbans. *Mahasagar* **15**:247-252.

Blaber, S.J.M., 1980. Fish of the Trinity Inlet system of North Queensland with notes on the ecology of the fish faunas of tropical Indo-Pacific estuaries. *Australian Journal of Marine and Freshwater Research* **31**:137-146.

Blaber, S.J.M., 1986. Feeding selectivity of a guild of piscivorous fishes in mangrove areas of north-west Australia. *Australian Journal of Marine and Freshwater Research* **37**:329-336.

Blaber, S.J.M., 1987. Factors affecting recruitment and survival of Mugilidae in estuaries and coastal waters of the Indo-West Pacific. *American Fisheries Society Symposium* **1**:507-518.

Blaber, S.J.M., and Blaber, T.G., 1980. Factors affecting the distribution of juvenile estuarine and inshore fish. *Journal of Fish Biology* **17**:143-162.

Blaber, S.J.M., Brewer, D.T., and Salini, J.P., 1989. Species composition and biomasses of fishes in different habitats of a tropical northern Australian estuary: their occurrence in the adjoining sea and esturarine dependence. *Estuarine, Coastal and Shelf Science* **29**:509-531.

Blaber, S.J.M., Brewer, D.T., Salini, J.P., and Kerr, J., 1990b. Biomasses, catch rates and patterns of abundance of demersal fishes, with particular reference to penaeid prawn predators, in a tropical bay in the Gulf of Carpentaria, Australia. *Marine Biology* **107**:397-408.

Blaber, S.J.M., and Milton, D.A., 1990. Species composition, community structure and zoogeography of fishes of mangroves in the Solomon Islands. *Marine Biology* **105**:259-268.

Blaber, S.J.M., Salini, J.P., and Brewer, D.T., 1990a. A checklist of the fishes of Albratross Bay and the Embley estuary, north eastern Gulf of Carpentaria. *CSIRO Marine Laboratories Report Series* **210**:1-22.

Blaber, S.J.M., Young, J.W., and Dunning, M.C., 1985. Community structure and zoogeographic affinities of the coastal fishes of the Dampier region of north-western Australia. *Australian Journal of Marine and Freshwater Research* **36**:247-266.

Boonruang, P., and Janekaru, V., 1985. Distribution and abundance of penaeid postlarvae in mangrove areas along the east coast of Phuket Island, Southern Thailand. *Research Bulletin of the Phuket Marine Biological Centre* No. 36, 29 p.

Boto, K.G., and Bunt, J.S., 1981a. Dissolved oxygen and pH relationships in northern Australian mangrove waterways. *Limnology and Oceanography* **26**:1176-1178.

Boto, K.G. and Bunt, J.S., 1981b. Tidal export of particulate organic matter from a northern Australian mangrove system. *Estuarine, Coastal and Shelf Science* **13**:247-255.

Boto, K.G., and Wellington, J.S., 1988. Seasonal variations in concentrations and fluxes of dissolved organic and inorganic materials in a tropical, tidally-dominated mangrove waterways. *Marine Ecology Progress Series* **50**:151-160.

Branford, J.R., 1981. Sediment preferences and morphometric equations for *Penaeus monodon* and *Penaeus indicus* from creeks of the Red Sea. *Estuarine, Coastal and Shelf Science* **13**:473-476.

Brewer, D.T., Blaber, S.J.M., and Salini, J.P., 1989. The feeding biology of *Caranx bucculentus* Alleyne and Macleay (Teleostei:Carangidae) in Albatross Bay, Gulf of Carpentaria; with special reference to predation on penaeid prawns. *Australian Journal of Marine and Freshwater Research* **40**:657-668.

Brewer, D.T., Blaber, S.J.M., and Salini, J.P., 1991. Predation on penaeid prawns by fishes in Albatross Bay, Gulf of Carpentaria. *Marine Biology* **109**:231-240.

Browder, J.A., 1985. Relationship between pink shrimp production on the Tortugas grounds and water flow patterns in the Florida Everglades. *Bulletin of Marine Science* **37**:839-856.

Carmouze, J.P., and Caumette, P., 1985. Les effets de la pollution organic sur les biomasses at activites du phytoplankton et des bacteries heterotrophes dans la lagune Ebrie (Cote d'Ivoire). *Revue Hydrobiologie Tropicale* **18**:183-211.

Caumette, P., Pajano, M., and Saint-Jean, L., 1983. Repartition verticale du phytoplancton, des bacteries et du zooplankton dans un milieu stratifie en Bace de Bietri (Lagune Ebrie, Cote d'Ivoire). Relations trophiques. *Hydrobiologia* **106**:135-148.

Cervigon, F., 1985. The ichthyofauna of the Orinoco estuarine water delta of the West Atlantic coast, Caribbean. In: Yanez-Arancibia, A., (Ed.), Fish community ecology in estuaries and coastal lagoons, pp. 57-78., UNAM Press, Mexico City.

Chong, V.C., Sasekumar, A., Leh, M.U.C., and D'Cruz, R., 1990. The fish and prawn communities of a Malaysian coastal mangrove system, with comparisons to adjacent mudflats and inshore waters. *Estuarine, Coastal and Shelf Science* **31**:703-722.

Chua, T-E., 1973. An ecological study of the Ponggol estuary in Singapore. *Hydrobiologia* **43**:505-533.

Cole, J.J., Findlay, S., and Pace, M.L., 1986. Bacterial production in fresh and saltwater ecosystems: a cross-system overview. *Marine Ecology Progress Series* **43**:1-10.
Coleman, P.J., 1966. The Solomon Islands as an island arc. *Nature* **211**:1249-1251.

Cyrus, D.P., and Blaber, S.J.M., 1983. The food and feeding ecology of Gerreidae Bleeker 1859, in the estuaries of Natal. *Journal of Fish Biology* **22**:373-393.

Cyrus, D.P., and Blaber, S.J.M., 1984. The reproductive biology of *Gerres* (Teleostei) Bleeker 1859, in Natal estuaries. *Journal of Fish Biology* **24**:491-504.

Cyrus, D.P., and Blaber, S.J.M., 1987. The influence of turbidity on juvenile marine fishes in estuaries. Part 1. *Journal of Experimental Marine Biology and Ecology* **109**:53-70.

Dakin, W.S., 1938. The habits and life-history of a penaeid prawn (*Penaeus plebejus*). *Proceedings of the Zoological Society of London* **A108**:163-183.

Daniel, P.A., and Robertson, A.I., 1990. Epibenthos of mangrove waterways and open embayments: community structure and the relationship between exported mangrove detritus and epifaunal standing stocks. *Estuarine, Coastal and Shelf Science* **31**:599-619.

Davis, T.L.O., 1985. Seasonal changes in gonal maturity, and abundance of larvae and early juveniles of barramundi, *Lates calcarifer* (Bloch), in Van Diemen Gulf and the Gulf of Carpentaria. *Australian Journal of Marine and Freshwater Research* **36**:177-190.

Davis, T.L.O., 1988. Temporal changes in the fish fauna entering a tidal swamp system in tropical Australia. *Environmental Biology of Fishes* **21**:161-172.

Day, J.H., 1974. The ecology of Morrumbene estuary, Mocambique. *Transactions of the Royal Society of South Africa* **41**:43-97.

Day, J.H., Blaber, S.J.M., and Wallace, J.H., 1981. Estuarine Fishes. In: Day, J.H., (Ed.), *Estuarine Ecology with particular reference to Southern Africa,* pp. 197-221, Balkema, Cape Town.

Day, J.W., Day, R.H., Barreiro, M.T., Ley-Lon, F., and Madden, C.J., 1982. Primary production in the Laguna de Terminos, a tropical estuary in the southern Gulf of Mexico. *Oceanologica Acta,* **5**(suppl):269-276.

de Freitas, A.J., 1986. Selection of nursery areas by six southeast African Penaeidae. *Estuarine, Coastal and Shelf Science* **23**:901-908.

Dittel, A.I., and Epifanio, C.E., 1990. Seasonal and tidal abundance of crab larvae in a tropical mangrove system, Gulf of Nicoya, Costa Rica. *Marine Ecology Progress Series* **65**:25-34.

Dixon P., and Robertson, A.I., 1986. A compact, self-contained zooplankton pump for use in shallow coastal habitats: design and performance compared to net samples. *Marine Ecology Progress Series* **32**:97-100.

Edwards, R.R.C., 1978. Ecology of a coastal lagoon complex in Mexico. *Estuarine and Coastal Marine Science* **6**:75-92.

FAO, 1990. FAO Yearbook, Fishery Statistics. 1988, Food and Agriculture Organization of the United Nations, Rome.

Frusher, S.D., 1983. The ecology of juvenile penaeid prawns, mangrove crab (*Scylla serrata*) and the giant freshwater prawn (*Macrobrachium rosenbergii*) in the Purari Delta. In: Petr, T., (Ed.), *The Purari-Tropical Environment of a High Rainfall River Basin,* pp. 341-353, Dr W. Junk Publishers, The Hague.

Gopinathan, C.P., 1972. Seasonal abundance of phytoplankton in the Cochin backwaters. *Journal of the Marine Biological Association of India* **14**:568-577.

Gore, R.H., *et al.*, 1981. Studies on decapod crustacea from the Indian River region of Florida XI. Community composition, structure, biomass and species - areal relationships of seagrass and drift algae - associated macrocrustaceans. *Estuarine, Coastal and Shelf Science* **12**:485-508.

Grindley, J.R., 1984. The zooplankton of mangrove estuaries. In: Por, F.D., and Dor, I., (Eds.), *Hydrobiology of the mangal*, pp. 79-88, Dr W. Junk Publishers, The Hague.

Haines, A.K., 1979. An ecological survey of fish of the lower Purari River system, Papua New Guinea. Department of Minerals and Energy, Papua New Guinea. Purari River Hydroelectric Scheme Environmental Studies **6**:1-102.

Hatcher, B.G., Johannes R.E., and Robertson, A.I., 1989. Review of research relevant to conservation of shallow tropical marine ecosystems. *Oceanography and Marine Biology: An Annual Review* **27**:337-414.

Healey, M.J., Moll, R.A., and D'allo, C.O., 1988. Abundance and distribution of bacterioplankton in the Gambia River, West Africa. *Microbial Ecology* **16**:291-310.

Hill, B.J., 1985. Effects of temperature on duration of emergence, speed of movement and catchability of the prawn, *Penaeus esculentus*. In: Rothlisberg, P.C., Hill, B.J., and Staples, D.J., (Eds.). *Second Australian National Prawn Seminar*, pp. 77-83, NPS2, Cleveland, Australia.

Hindley, J.P.R., 1975. Effects of endogenous and some exogenous factors on the activity of the juvenile banana prawn *Penaeus merguiensis. Marine Biology* **29**:1-8.

Hoese, D.F., 1986. Gobiidae. In: Smith, M.M., and Heemstra, P.C., (Eds.), *Smith's Sea Fishes*, pp. 774-807, Macmillan, Johannesburg.

Houde, E.D., 1975. Effects of stocking density and food density on survival, growth and yield of laboratory-reared larvae of sea bream *Archosargus rhomboidalis* (L.) (Sparidae). *Journal of Fish Biology* **7**:115-127.

Iltis, A., 1984. Biomass phytoplanctonique de la lagune Ebrie (Cote d'Ivoire). *Hydrobiologia* **118**:153-175.

Inger, R.E., 1955. Ecological notes on the fish fauna of a coastal drainage of North Borneo. *Fieldiana: Zoology* **37**:47-90.

Jones, S., and Sunjansingani, K.H., 1954. Fish and fisheries of the Chilka Lake with statistics of fish catches for the years 1948-1950. *Indian Journal of Fisheries* **1**:256-344.

Kennedy, G.R., 1978. Plankton of the Fitzroy River Estuary Queensland. *Proceedings of the Royal Society of Queensland* **89**:29-37.

Krishnamurthy, K., and Jeyaseelan, M.J.P., 1981. The early life history of fishes from Pichavaram mangrove ecosystem of India. *Rapports et proces-Verbaux des Reunions, Conseil Permanent International pour l'Exporation de la Mer*, **178**:416-423.

Kutner, M.B., 1975. Seasonal variation and phytoplankton distribution in Cananeia region, Brazil. In: Walsh, G.E., Snedaker, S.C., and Teas, H.J., *Proceeding International Symposium on the Biology and Management of Mangroves,* Honolulu, Vol I, pp. 153-169, University of Florida, Gainesville.

Kutkuhn, J.H., 1966. The role of estuaries in the development and perpetuation of commercial shrimp resources. *American Fisheries Society Special Publication* **3**:16-36.

Kwei, E., 1977. Biological, chemical and hydrological characters of coastal lagoons of Ghana, West Africa. *Hydrobiologia* **56**:157-174.

Lasserre, G., and Toffart, J.L., 1977. Echantillonnage et structure des populations ichthyologiques des mangroves de Guadeloupe on Septembre 1975. *Cybium 3e Serie* **2**:115-127.

Lee, Y.S., Kaur, B., and Broom, M.J., 1984. The effect of palm oil mill effluent on the nutrient status and planktonic primary production of a Malaysian mangrove inlet. In: Soepadmo, E., Rao, A.N., and Macintosh, D.J., *Proceedings of the Asia symposium on mangrove environment: research and mangement,* pp. 575-591, University of Malaysia and UNESCO, Kuala Lumpur.

Leon, R.A., and Racedo, J.B., 1985. Composition of fish communities in the lagoon and estuarine complex of Caragena Bay, Cienaga de Tesca and Cienaga Grande of Santa Marta, Colombian Caribbean. In: Yanez-Arancibia, A., (Ed.), *Fish community ecology in estuaries and coastal lagoons,* pp. 535-556, UNAM Press, Mexico City.

Ley, J.A., and Montague, C.L., 1991. Influence of changes in freshwater flow on the use of mangrove prop root habitat by fish. Report to The South Florida Water Management District. University of Florida, Gainesville, 200pp.

Little, M.C., Reay, P.J., and Grove, S.J., 1988. The fish community of an east African mangrove creek. *Journal of Fish Biology* **32**:729-747.

Longhurst, A.R., and Pauly, D., 1987. *Ecology of tropical oceans,* Academic Press, Inc, San Diego, 407pp.

McFarlane, J.W., 1980. Surface and bottom sea currents in the Gulf of Papua and western Coral Sea. Department of Primary Industry (Papua New Guinea) Research Bulletin No. 27, 1-128.

MacNae, W., 1974. Mangrove forests and fisheries. Food and Agriculture Organization of the United Nations, Rome, IOFC/DEV/74/34, 1-35.

Martin, T.J., and Blaber, S.J.M., 1983. The feeding ecology of Ambassidae (Osteichthyes:Perciformes) in Natal estuaries. *South African Journal of Zoology* **18**:353-362.

Martosubroto, P.D., and Naamin, N., 1977. Relationship between tidal forests (mangroves) and commercial shrimp production in Indonesia. *Marine Research in Indonesia* **18**:81-86.

Marumo, R., Laoprasert, S., and Karnjanagesorn 1985. Plankton and near- bottom communities of the mangrove regions in Ao Khung Kraben and the Chantaburi River, Thailand. In: *Mangrove estuarine ecology in Thailand; Thai-Japanese Cooperative research project on mangrove productivity and development 1983-1984.* Jap. Ministry of Education, Science and Culture, pp. 55-74.

Minello, T.J., and Zimmerman, R.J., 1983. Fish predation on juvenile brown shrimp, *Penaeus aztecus* Ives: the effect of simulated *Spartina* structure on predation rates. *Journal of Experimental Marine Biology and Ecology* **72**:211-231.

Morton, R.M., 1990. Community structure, density and standing crop of fishes in a subtropical Australian mangrove area. *Marine Biology* **105**:385-394.

Nateewathana, A., and Tantichodok, P., 1984. Species composition, density and biomass of macrofauna of a mangrove forest at Ko Yao Yai, Southern Thailand. In: Soepadmo, E., Rao, A.N., and Macintosh, D.J. (Eds.), *Proceedings of the Asia symposium on mangrove environment - research and management,* pp. 258-285, UNESCO, Kuala Lumpur.

Odum, W.E., and Heald, E.J., 1972. Trophic analyses of an estuarine mangrove community. *Bulletin of Marine Science,* **22**:671-737.

Odum, W.E., and Heald, E.J., 1975. The detritus-based food web of an estuarine mangrove community. In: Cronin, L.E., (Ed.), *Estuarine Research,* pp. 265-286, Academic Press Inc., New York.

Odum, W.E., McIvor, C.C., and Smith, T.J. III, 1982. The ecology of the mangroves of South Florida: a community profile. U.S. Fish and Wildlife Service, Office of Biological Services, Washington D.C. FWS/OBS - 81/24, 154 pp.

Omori, M., 1969. Weight and chemical composition of some important oceanic zooplankton in the north Pacific Ocean. *Marine Biology* **3**:4-10.

Ong, J-E., Gong, W-K., Wong, C-H., and Dhanarajan, G., 1984. Contributions of aquatic productivity in a managed mangrove ecosystem in Malaysia. In: Soepadmo, E., Rao, A.N., and Macintosh, D.J., (Eds.), *Proceedings of the Asian symposium on mangrove environment: research and mangement,* pp. 209-215, University of Malaya and UNESCO, Kuala Lumpur.

Pages, J., Lemasson, L., and Dufour, P., 1981. Primary production measurement in a brackish tropical lagoon. Effects of light, as studied at some stations by the ^{14}C method. *Revue Hydrobiologie Tropicale* **14**:3-15.

Pant, A., Dhargalkar, V.K., Bhosale, N.B., and Untawale, A.G., 1980. Contribution of phytoplankton photosynthesis to a mangrove ecosystem. *Mahasagar* **13**:225-234.

Paranagua, M.N., and Eskinazi-Leca, E., 1985. Ecology of a northern tropical estuary in Brazil and technological perspectives in fish culture. In: Yanez-Arancibia, A., (Ed.), *Fish community ecology in estuaries and coastal lagoons,* pp. 595-614, UNAM Press, Mexico City.

Parrish, J.D., 1987. Characteristics of fish communities on coral reefs and in potentially interacting shallow habitats in tropical oceans of the world. *UNESCO Reports in Marine Science* **46**:171-218.

Pauly, D., 1975. On the ecology of a small West African lagoon. *Berichtder Deutschen Wissenschaftlichen Kommission fur Meeresforschung* **24**:46-62.

Pauly, D., and Ingles, J., 1986. The relationship between shrimp yields and intertidal vegetation (mangrove) areas: a reassessment. In: IOC/FAO Workshop on recruitment in tropical coastal demersal communities - submitted papers, pp. 227-284, Cuidad de Carman, Campeche, Mexico, 21-25 April, 1986, IOC, UNESCO, Paris.

Penn, J.W., and Stalker, R.W., 1975. A daylight sampling net for juvenile penaeid prawns. *Australian Journal of Marine Freshwater Research* **26**:287-291.

Percival, M., and Womersley, J.S., 1975. Floristics and ecology of the mangrove vegetation of Papua New Guinea. Department of Forestry, Division of Botany, Lae, Papua New Guinea, Botany Bulletin No. **8**, 96pp.

Peters, K.M., and McMichael, R.H., 1987. Early life history of the red drum *Sciaenops ocellatus* (Pisces:Sciaenidae), in Tampa Bay, Florida. *Estuaries* **10**:92-107.

Pinto, L., 1986. *Mangroves of Sri Lanka.* Natural Resources, Energy and Science Authority of Sri Lanka, 54 pp.

Pinto, L., 1988. Population dynamics and community structure of fish in the mangroves of Pagbilao, Philippines. *Journal of Fish Biology* **33**:(Suppl. A)35-43.

Pollard, D.A., 1980. Family Megalopidae. Oxeye herring. In: McDowall, R.M. (Ed.), *Freshwater Fishes of southeastern Australia*, pp. 53-54, A.H. and A.W. Reed, Sydney.

Qasim, S.Z., 1973. Productivity of backwaters and estuaries. In: Zietschel, B., (Ed.), *Ecological Studies,* pp. 143-154, Springer, Berlin, Vol 3.

Qasim, S.Z., Wellershaus, S., Bhattathiri, P.M.A., and Abidi, S.A.H., 1969. Organic production in a tropical estuary. *Proceedings of the Indian Academy of Science Section B* **59**:51-94.

Qasim, S.Z., Wellershaus, S., Bhattathiri, P.M.A., and Abidi, S.A.H., 1972. The influence of salinity on the rate of photosynthesis and abundance of some tropical phytoplankton. *Marine Biology* **12**:200-206

Quinn, N.J., 1980. Analysis of temporal changes in fish assemblages in Serpentine Creek, Queensland. *Environmental Biology of Fishes* **5**:117-133.

Quinn, N.J., and Kojis, B.J., 1985. Does the presence of coral reefs in proximity to a tropical estuary affect the estuarine fish assemblage. *Proceedings of the Vth International Coral Reef Congress* **5**:445-450.

Quinn, N.J., and Kojis, B.J., 1986. Annual variation in the nocturnal nekton assemblage of a tropical esturary. *Estuarine, Coastal and Shelf Science* **22**:63-90.

Ricard, M., 1984. Primary production in mangrove lagoon waters. In: Por, F.D., and Dor, I., (Eds.), *Hydrobiology of the mangal*, pp. 163-178, Dr W. Junk, The Hague.

Ridd, P.V., Wolanski, E., and Mazda, Y., 1990. Longitudinal diffusion in mangrove-fringed tidal creeks. *Estuarine, Coastal and Shelf Science* **31**:541-554.

Rimmer, M.A., and Merrick, J.A., 1982. A review of reproduction and development in the fork-tailed catfishes (Ariidae). *Proceedings of the Linnean Society of New South Wales* **107**:41-50.

Robertson, A.I., 1988. Abundance, diet and predators of juvenile banana prawns, *Penaeus merguiensis* in a tropical mangrove estuary. *Australian Journal of Marine and Freshwater Research* **39**:467-478.

Robertson, A.I., Alongi, D.M., Daniel, P..A., and Boto, K.G., 1989. How much mangrove detritus enters the Great Barrier Reef lagoon. *Proceedings of the VIth International Coral Reef Conference,* **2**:601-606.

Robertson, A.I., Alongi, D.M., Christoffersen, P., Daniel, P.A., Dixon, P., and Tirendi, F.,1990. The influence of freshwater and detrital export from the Fly River system on adjacent pelagic and benthic systems. Australian Institute of Marine Science Report No. 4, 199pp.

Robertson, A.I., Daniel, P.A., and Dixon, P., 1991. Mangrove forest structure and productivity in the Fly River estuary, Papua New Guinea. *Marine Biology* **111**:147-155.

Robertson, A.I., Daniel, P.A., Dixon, P., and Alongi, D.M.,1992. Pelagic biological processes along a salinity gradient in the Fly estuary and adjacent river plume (Papua New Guinea). *Continental Shelf Research* (in press).

Robertson, A.I., Dixon, P., and Daniel, P.A., 1988. Zooplankton dynamics in mangrove and other nearshore habitats in tropical Australia. *Marine Ecology Progress Series* **43**:139-150.

Robertson, A.I. and Duke, N.C., 1987. Mangroves as nursery sites: comparisons of the abundance and species composition of fish and crustaceans in mangroves and other nearshore habitats in tropical Australia. *Marine Biology* **96**:193-205.

Robertson, A.I. and Duke, N.C., 1990a. Mangrove fish communities in tropical Australia: spatial and temporal patterns in densities, biomass and community structure. *Marine Biology* **104**:369-379.

Robertson, A.I. and Duke, N.C., 1990b. Recruitment, growth and residence time of fishes in a tropical Australian mangrove system. *Estuarine, Coastal and Shelf Science* **31**:725-745.

Rothlisberg, P.C., Staples, D.J., and Crocos, P.J.,1985. A review of the life history of the banana prawn, *Penaeus merguiensis* in the Gulf of Carpentaria. In: Rothlisberg, P.C., Hill, B.J., and Staples, D.J., (Eds.), *Second Australian National Prawn Seminar,* pp. 125-136, NPS2, Cleveland, Australia.

Saenger, P., Hegerl, E.G., and Davie, J.D.S., 1983. Global status of mangrove ecosystems. *The Environmentalist* **3**:1-88.

Salini, J.P., Blaber, S.J.M., and Brewer, D.T., 1990. Diets of piscivorous fishes in a tropical Australian estuary with particular reference to predation on penaeid prawns. *Marine Biology* **105**:363-374.

Salini, J.P., Blaber, S.J.M., and Brewer, D.J., 1992. Diets of sharks from estuaries and nearshore waters of the northeastern Gulf of Carpentaria. *Australian Journal of Marine and Freshwater Research* **43**:87-96.

Salini, J.P., and Shaklee, J.B., 1988. Genetic structure of Barramundi *(Lates calcarifer)* stocks from northern Australia. *Australian Journal of Marine and Freshwater Research* **39**:317-329.

Sambasivam, S., 1985. Species composition and distribution of prawn juveniles in Pichavaran mangrove. In: Bhosale, L., (Ed.), *The Mangroves*, pp. 481-491, Shivaji University, Kolhapur.

Sarkar, S., Baidya, A., Bhunia, A., and Choudury, A., 1984. Zooplankton studies in the Hooghly estuary around Sagar Island, Sunderbans, India. In: Soepadmo, E., Rao, A.N., MacIntosh, D.J., (Eds.), *Proceedings of the Asia symposium on mangrove environment: research and management,* pp. 286-297, University of Malaysia and UNESCO, Kuala Lumpur.

Sasekumar, A., Ong, T-L., and Thong, K.L., 1984. Predation on mangrove fauna by marine fishes. In: Soepadmo, E., Rao, A.N., and MacIntosh, D.J., (Eds.), *Proceedings of the Asian symposium on mangrove environment: research and management,* pp. 378-384, University of Malaysia and UNESCO, Kuala Lumpur.

Sevrin-Reyssac, J., 1980. Chlorophyll *a* et production primaire dans les eaux de le baie du Levrier et le Parc National du Banc d'Arguin (Sept-Nov 1984). *Bulletin Centre Nationale Recherche en Oceanographic, et Peches, Nouadhibon (R.I. Mauritainie)* **9**:56-65.

Shanmugam, A., Kasinathan, R., and Maruthamuthu, S.,1986. Biomass and composition of zooplankton from Pitchavaram mangroves, southeast coast of India. *Indian Journal of Marine Science* **15**:111-113.

Sherr, B., and Sherr, E., 1989. Trophic impacts of phagotrophic protozoa in pelagic foodwebs. In: Hattori, T., Ishida, Y., Maruyama, Y., Morita, R.Y., and Uchida, A.,(Eds.), Recent Advances in Microbial Ecology, pp. 388-396, Japan Scientific Press, Tokyo.

Springer, V.G., 1982. Pacific Plate biogeography with special reference to shorefishes. *Smithsonian Contributions, Zoology* **367**:1-182.

Staples, D.J., Vance, D.J., and Heales, D.S., 1985. Habitat requirements of juvenile penaeid prawns and their relationship to offshore fisheries. In: Rothlisberg, P.C., Hill, B.J., Staples, D.J., (Eds.), *Second Australian National Prawn Seminar,* pp. 47-54, NPS2, Cleveland, Australia.

Stephenson, W., and Dredge, M.L.C., 1976. Numerical analysis of fish catches from Serpentine Creek. *Proceedings of the Royal Society of Queensland* **87**:33-43.

Stoner, A.W., 1986. Community structure of the demersal fish species of Laguna Joyuda, Puerto Rico. *Estuaries* **9**:142-152.

Stoner, A.W., 1988. A nursery ground for four tropical *Penaeus* species: Laguna Joyuda, Peurto Rico. *Marine Ecology Progress Series* **42**:133-141.
Sundararaj, V., and Krishnamarthy, K., 1973. Photosynthetic pigments and primary production. *Current Science* 42:1-6.

Teixeira, C., and Gaeta, S.A., 1991. Contribution of picoplankton to primary production in estuarine, coastal and equatorial waters of Brazil. *Hydrobiologia* **209**:117-122.

Teixeira, C., Tundisi, J., and Kutner, M.B., 1965. Plankton studies in a mangrove environment. II. The standing stock and some ecological factors. *Bolletin Instituto de Oceanografico, Sao Paulo* **14**:13-42.

Teixeira, C., Tundisi, J., and Santoro-Ycaza, J., 1969. Plankton studies in a mangrove environment. VI. Primary production, zooplankton standing stock and some environmental factors. *International Revue der Gesamten Hydrobiologie* **54**:289-301.

Thayer, G.W., Colby, D.R., and Hettler, W.F., 1987. Utilization of the red mangrove prop root habitat by fishes in South Florida. *Marine Ecology Progress Series* **35**:25-38.

Thayer, G.W., and Sheridan, P.F., 1992. Fish and aquatic invertebrate use of the mangrove prop-root habitat in Florida: a review. In: Yanez-Acancibia, A., (Ed.), *Marine ecosystems in tropical America: structure, function and management,* EPOMEX Serie Cientifica, Mexico City (in press).

Thollot, P., and Kulbicki, M., 1989. Overlap between the fish fauna inventories of coral reefs, soft bottoms and mangroves in Saint-Vincent Bay (New Caledonia). *Proceedings of the VIth International Coral Reef Symposium* **2**:613-618.

Thompson, R.B., and Hackman, B.D., 1969. Some geological notes on areas visited by the Royal Society Expedition to the British Solomon Islands, 1965. *Philosophical Transactions of the Royal Society B* **2**:189-202.

Thresher, R.E., 1984. *Reproduction in reef fishes.* T.F.H. Publications, New Jersey, U.S.A, 399pp.

Tundisi, J., Tundisi, T.M., and Kutner, M., 1973. Plankton studies in a mangrove environment. VIII. Further studies on primary production, standing stock of phyto- and zooplankton and some environmental factors. *Internationale Revue der Gesamten Hydrobiologie* **58**:925-940.

Turner, R.E., 1977. Intertidal vegetation and commercial yields of penaeid shrimp. *Transactions of the American Fisheries Society* **106**:441-416.

Vance, D.J., Haywood, M.D.E., and Staples, D.J., 1990. Use of a mangrove estuary as a nursery area by postlarval and juvenile banana prawns, *Penaeus merguiensis* de Man, in northern Australia. *Estuarine, Coastal and Shelf Science* **31**:689-702.

Warburton, K., 1978. Community structure, abundance and diversity of fish in a Mexican coastal lagoon system. *Estuarine and Coastal Marine Science* **7**:497-519.

Wium-Andersen, S., 1979. Plankton primary production in a tropical mangrove bay at the south-west coast of Thailand. *Ophelia* **18**:53-60.

Wolanski, E., 1986. An evaporation driven salinity maximum zone in Australian tropical estuaries. *Estuarine, Coastal and Shelf Science* **22**:415-424.

Wolanski, E., 1989. Measurements and modelling of the water circulation in mangrove swamps. COMARAF Regional Project for Research and Training on Coastal Marine Systems in Africa - RAF/87/038. Serie Documentaire No. **3**:1-43.

Wolanski, E., Jones, M., and Bunt, J.S., 1980. Hydrodynamics of a tidal creek - mangrove swamp system. *Australian Journal of Marine and Freshwater Research* **31**:431-50.

Wolanski, E., and Ridd, P., 1986. Tidal mixing and trapping in mangrove swamps. *Estuarine, Coastal and Shelf Science* **23**:759-771.

Wolanski, E., Mazda, Y., King, B., and Gay, S., 1990. Dynamics, flushing and trapping in Hinchinbrook Channel, a giant mangrove swamp, Australia. *Estuarine, Coastal and Shelf Science* **31**:555-580.

Yanez-Arancibia, A., (Ed.), 1985. *Fish community ecology in estuaries and coastal lagoons.* Direccion General de Publicaciones, Universidat National Autonoma de Mexico, UNAM Press, Mexico City, 653 pp,

Yanez-Arancibia, A., Lara-Dominguez, A.L., Rojas-Galaviz, J.L., Sanchez-Gil, P., Day, J.W., and Madden, C.J., 1988. Seasonal biomass and diversity of estuarine fishes coupled with tropical habitat heterogeneity (southern Gulf of Mexico). *Journal of Fish Biology* **33**(Suppl. A):191-200.

Youngbluth, M.J., 1980. Daily, seasonal and annual fluctuations among zooplankton populations in an unpolluted tropical embayment. *Estuarine, Coastal and Shelf Science* **10**:265-287.

Zimmerman, R.J., and Minello, T.J., 1984. Densities of *Penaeus aztecus, Penaeus setiferus* and other rotant macrofauna in a Texas salt marsh. *Estuaries* **7**:421-433.

Zoppi de Roa, E., 1974. Comparison de algunas caracteruticas del plankton entre las lagunas costeras de Tacarigua y Unare, Venezuela. *Boletin del Instituto Oceanografico de la Universidad de Oriente Cumana* **13**:129-146.

8

Primary Productivity and Growth of Mangrove Forests

B.F. Clough

8.1 Introduction

It is widely stated that mangrove ecosystems are highly productive. General statements to this effect, equally common in many texts and in the introductory remarks of research papers, can be misleading. In some cases the basis for such statements, whether it be biomass accumulation, primary productivity or secondary production, has not been clearly stated. While there is certainly good evidence that some mangrove systems are highly productive, both in terms of primary and/or of secondary production, there is equally strong evidence that others are not. The issue is further clouded by an often poor correlation between primary production, biomass accumulation and secondary production.

Mangroves occur along coastlines lying roughly between latitudes 35°N and 38°S. Within this broad latitudinal band they grow over a very wide variety of climates, ranging from extremely arid coast of the Persian Gulf or the cool-temperate coast of southern Australia on the one hand, to the wet equatorial coasts of Asia, Africa and Latin America on the other. Mangroves also grow on a very wide range of soil types, including heavy consolidated clays, unconsolidated silts, calcareous and mineral sands, coral rubble, and organic peats, with salinities ranging from close to 0‰ (parts per thousand) to about 90‰ (seawater = 35‰). The point to be made here is that the range of conditions over which mangroves occur naturally encompasses most environmental extremes apart from the low temperatures of cold-temperate, sub-arctic and arctic climates. It would be surprising indeed if mangrove systems across such a broad spectrum of environments were all of uniformly high productivity, irrespective of the basis for its assessment.

This chapter is concerned with these issues. It will focus on primary production and biomass accumulation by mangrove ecosystems, and on the way in which environmental constraints and ecophysiological responses interact to regulate primary productivity and growth.

Table 1. Coefficients for allometric relationships between diameter at breast height (DBH, cm) or girth at breast height (GBH, cm) and above-ground dry weight (kg) for various parts of mangrove trees.

Species	Equation	Variable	A	B	r^2	Source
B. gymnorrhiza	$DWT = A \cdot DBH^B$	Leaf	0.0679	1.4914	0.854	(1)
n=17; 2-21 cm DBH		Branch	0.0315	2.2789	0.926	
		Stem	0.2248	2.1407	0.977	
		Total	0.1858	2.3055	0.989	
B. parviflora	$DWT = A \cdot DBH^B$	Leaf	0.0268	1.407	0.621	(1)
n=16; 2-21 cm DBH		Branch	0.0115	2.4639	0.885	
		Stem	0.1361	2.4037	0.922	
		Total	0.1679	2.4167	0.993	
C. australis	$DWT = A \cdot DBH^B$	Leaf	0.0117	2.1294	0.927	(1)
n=26; 2-18 cm DBH		Branch	0.0197	2.5516	0.938	
		Stem	0.1468	2.3393	0.977	
		Total	0.1885	2.3379	0.989	
R. candeleria	$DWT = A \cdot DBH^B$	Leaf	0.8389	0.3999	0.846	(2)
n=270; 0.4-7 cm DBH		Branch	1.6267	0.9349	0.917	
		Stem	0.9482	1.1411	0.877	
		Root	0.8429	0.4647	0.919	
		Total	3.3512	0.7652	0.888	
R. apiculata	$DWT = A \cdot DBH^B$	Total	0.1709	2.516	0.98	(3)
n=NA; DBH range NA						
R. apiculata	$DWT = A \cdot GBH^B$	Total	0.0277	2.1668	0.899	(4)
n=30; 7-88 cm GBH						
R. apiculata	$DWT = A \cdot DBH^B$	Leaf	0.0139	2.1072	0.857	(1)
R. stylosa		Branch	0.0127	2.6844	0.912	
n=23; 3-23 cm DBH		Stem	0.0886	2.5621	0.991	
		Root	0.0068	3.1353	0.968	
		Total	0.1050	2.6848	0.995	
R. mangle	$DWT = A \cdot e^{B \ DBH}$	Stem	0.94	0.27	0.97	(5)
n=10; 3-11 cm DBH		Total	1.41	0.30	0.93	
X. granatum	$DWT = A \cdot DBH^B$	Leaf	0.0058	2.3966	0.951	(1)
n=15; 3-17 cm DBH		Branch	0.0047	3.0975	0.959	
		Stem	0.0817	2.4624	0.988	
		Total	0.0823	2.5883	0.994	

Sources: 1. Clough and Scott (1989); 2. Aksornkoae (1976); 3. Putz and Chan (1986); 4. Ong *et al.*,(1985); 5. Silva *et al.*, (1991).

8.2 Biomass accumulation and growth

8.2.1 Above ground biomass

Mangrove forest biomass above ground is usually estimated indirectly from measurements of stem diameter at a height of 1.3 m (usually abbreviated to DBH, the diameter at breast height). Allometric relationships between DBH as the independent variable and the dry weights of different parts of the tree as dependent variables are then used to calculate biomass from measurements of DBH (Whittaker and Marks, 1975; Ong *et al.*, 1984; Putz and Chan, 1986; Clough and Scott, 1989). These allometric relationships may take several forms, but are usually based on one of the following:

$$\text{Biomass (or volume)} = a \cdot DBH^{b} \quad (1)$$

or

$$\text{Biomass (or volume)} = a(DBH^{2}H)^{b} \quad (2)$$

or

$$\text{Biomass (or volume)} = a + b(DBH^{2}H)^{c} \quad (3)$$

where B = biomass, V = volume, H = height, and a, b, and c are constants. Experience with terrestrial trees suggests that Eqns. 2 and 3 may give a better fit to data than Eqn. 1 (Causton, 1985). However, both Eqn. 2 and Eqn. 3 require the measurement of tree height in addition to DBH, adding considerably to the time needed to obtain biomass estimates. In many cases, Eqn. 1 provides an acceptable estimate of above-ground biomass without the need for the measurement of tree height (Ong *et al.*, 1984; Putz and Chan, 1986; Clough and Scott, 1989).

Published coefficients and equations for allometric relationships between DBH or GBH (girth at breast height) and the dry weight of various parts of a number of mangrove tree species are summarised in Table 1. Correlation coefficients of better than 0.95 indicate that stem and total above-ground biomass can be estimated with confidence from measurements of DBH (Table 1). Allometric relationships between DBH and the biomass of leaves and branches are generally less robust than those for stem or total biomass because leaves and branches are more easily be broken off the tree by strong winds. Leaf biomass may also vary seasonally.

None of the allometric relationships summarised in Table 1 have included trees with a DBH much above 30 cm. Extrapolation of these allometric relationships to trees of larger diameter should be done with caution, and only after checking that they also hold for larger trees in the size class for which they will be used. Furthermore, in addition to obvious differences in the values of corresponding coefficients between species (Table 1), there may also be differences between the same species at different localities, depending on site-dependent factors such as tree density, whether there is a monoculture or a mixed forest, and management practices. Figure 1, for example, shows the total above-ground biomass for *Rhizophora apiculata* trees of different DBH, using the three sets of allometric coefficients given in Table 1 for this species. Curves (A) and (B) were derived from natural, mixed forest stands, whereas Curve (C) was obtained in a monoculture managed for charcoal production.

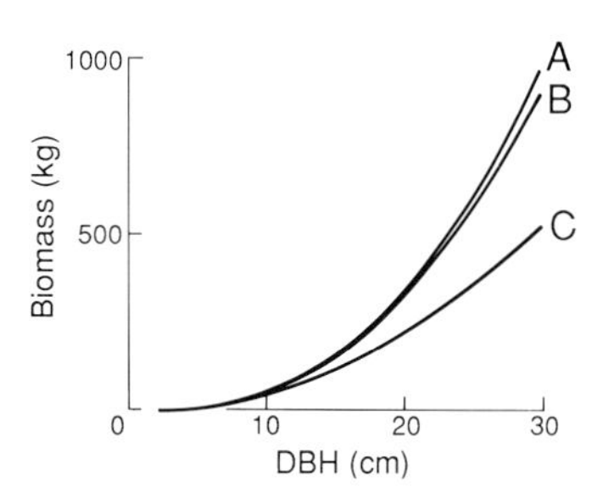

Figure 1. Total above-ground biomass of *Rhizophora apiculata* calculated from three different sets of allometric coefficients (Table 3). A, Clough and Scott (1989); B, Putz and Chan (1986); C, Ong *et al.*, (1984). See text for explanation.

The highest published estimates of above-ground biomass in old-aged mangrove forests in Asia and the Pacific are of the order of 500-550 $t.ha^{-1}$ (Paijmans and Rollet, 1977; Putz and Chan, 1986). However, some undisturbed *Rhizophora* forests in northern Australia are estimated to have an above ground biomass of up to 700 $t.ha^{-1}$ and basal areas of up to 70 $m^2.ha^{-1}$ (Table 2). These values may be close to the maximum for natural, undisturbed mangrove forest in the warm, humid tropics. With less favourable conditions of lower temperatures, arid climates, hypersaline soils, or nutrient limitation, above-ground biomass is usually somewhat less than these maximum values. In addition, major episodic events like severe coastal storms, hurricanes or cyclones, as well as the dynamics of erosion and sediment accretion can also have a significant impact on the accumulation of above-ground biomass by mangrove forests. Indeed, as will be discussed later, the complexity of the interactions between the hydrodynamics of mangrove swamps, climatic conditions and soil factors on the one hand, and forest structure and growth on the other, often make it extremely difficult to identify which are the main factors influencing the accumulation of biomass at any given site.

The rate of accumulation of above-ground biomass is usually estimated by applying allometric relationships (Table 1) to consecutive measurements of DBH made at intervals of at least 1 year apart. Such estimates vary widely. On the one hand, rates of net above-ground biomass accumulation may be close to zero in some mangrove communities growing on hypersaline saltpans in the strongly seasonal monsoonal climate of north-eastern Australia (Clough, unpublished) while, on the other, mangrove ecosystems growing in areas of high rainfall and low to moderate salinity in the wet humid tropics of the Indo-Pacific may have annual rates of accumulation of above-ground biomass of up to 45 $t.ha^{-1}$ (Table 2).

The most reliable estimates of biomass accumulation come from managed mangrove plantations in South East Asia for which the ages are also known. Ong *et al.*, (1984), for example, reported a mean annual biomass increment (above-ground) of 18 $t.ha^{-1}$ in a 10 year old stand of *Rhizophora apiculata* on the west coast of Peninsula Malaysia. Similarly, Aksornkoae (1976) found that the mean annual increase in biomass ranged from 14 $t.ha^{-1}$ to 33 $t.ha^{-1}$ in plantations of *Rhizophora candelaria* aged between 6 and 14 years in Thailand. In the latter study, there was no consistent trend of growth rate with age between 6 and 14 years.

Table 2. Stand characteristics and accumulation of above-ground dry weight in well-developed *Rhizophora apiculata / R. stylosa* forests on the Daintree River in north-eastern Australia.

Soil Salinity at 20 cm depth ppt	**Stems ha^{-1}**	**Mean DBH cm**	**Biomass t ha^{-1}**	**Basal Area m^{2}ha^{-1}**	**Mean DBH Increment cm yr^{-1}**	**Biomass Accumulation t ha^{-1}yr^{-1}**
27	2,600	14.2	461	48	0.31	26.5
23	2,150	12.6	264	31	0.13	6.3
25	1,825	13.8	297	32	0.31	16.1
32	2.725	16.6	711	69	0.48	45.4
35	1,600	20.6	686	59	0.51	41.2
32	2,400	15.4	497	51	0.32	20.1

Fewer published data exist for the growth rates of unmanaged mixed-species mangrove forests. Putz and Chan (1986) have provided a detailed account of the long-term (over 60 years) changes in forest structure and growth of an old-aged mixed-species mangrove forest in Peninsula Malaysia. They reported wide fluctuations in above-ground biomass during the 31 years up to 1981, owing to the death of older trees and their replacement by new trees, often of more shade tolerant species. Annual diameter increases in all size classes also varied from one 5-year growth period to another, with a long-term average of 0.26 cm.year^{-1} for trees ranging from 10 cm DBH to 60 cm DBH. Putz and Chan (1986) estimated the mean annual increment in above-ground biomass of this unmanaged, mixed-species forest to be 6 t. ha^{-1}.year^{-1}, a relatively low figure by comparison with that for managed plantations in the same area of Malaysia (Ong *et al.*, 1984) and the estimates shown in Table 2 for unmanaged, well-developed *Rhizophora* forests in north-eastern Australia.. This low figure is largely a consequence of the high rate of tree mortality (Putz and Chan, 1986).

Stem wood densities (kg dry weight per m^3 volume) for mangroves vary from 486 kg.m^{-3} to 980 kg.m^{-3}, depending on the the species (Putz and Chan, 1986; Clough and Scott, 1989; A. Robertson, unpublished).

8.2.2 Below-ground biomass

In terrestrial forests below-ground biomass usually represents less than 30% of the total forest biomass (Ulrich *et al.*, 1974). Unfortunately, few estimates are available for below-ground biomass in mangrove forests, in part owing to the difficulty of obtaining them. Of those available, most suggest that below ground biomass may represent 40-60% of the total biomass (Saenger, 1982; Lugo, 1990). For a number of reasons, however, some of these estimates must be viewed with caution. Firstly, few published reports describe the method by which the estimates were obtained; secondly, data are often reported in such a way that no clear distinction can be made between below-ground and above-ground roots; and thirdly, it is usually very difficult to distinguish between living and dead roots, particularly in the case of fine, fibrous roots, where live and dead roots often form a dense, intertwined mat (Komiyama *et al.*, 1987). The inclusion of dead roots and other organic detritus as below-ground root biomass would clearly lead to an overestimate of below-ground root biomass.

Recently, Gong and Ong (1990) reported that below-ground roots constituted an average of only 8.5% of the total biomass for *Rhizophora apiculata* trees of varying age in Malaysia. Root biomass as a proportion of the total biomass varied over the range 3.3-19.7% for individual trees, smaller trees in general being at the higher end of the range, and larger trees at the lower end of the range (J.-E. Ong and W.-K. Gong, pers. comm.). These data are interesting on three counts. Firstly, the measured root biomass refers only to live roots below ground (J.-E. Ong and W.-K. Gong, pers. comm.); secondly, the below ground root mass consisted only of fine, fibrous roots (J.-E. Ong and W.-K. Gong, pers. comm.); and thirdly, below-ground roots represented a much smaller proportion of total biomass than has been reported previously. While this proportion might at first seem too low by comparison with other published figures, it should be remembered that unlike most other mangrove species, *R. apiculata* does not have an extensive underground cable root system, instead being supported by an extensive prop root system above the ground. The absence of woody roots below ground (Gill and Tomlinson, 1977) may explain the relatively small contribution of below-ground roots to total tree biomass. It is likely that root biomass makes up a greater proportion of total biomass in other mangroves with a more extensive underground cable root system, such as *Avicennia, Sonneratia* and *Bruguiera* spp.. Even within any given species, the ratio of above to below ground biomass is likely to vary widely from site to site, particularly where soil and other environmental conditions differ between sites. For example, the published data on above- and below-ground biomass suggest that the balance shifts progressively to an increasing allocation to below-ground biomass with increasing latitude. This trend might be a response to low temperature.

8.2.3 Partitioning of dry matter between parts of the tree

The partitioning of photosynthate for growth amongst different tree parts varies with species and with time. Factors like age, environmental conditions (particularly those inducing stress), forest structure and competitive interactions all have an influence on the the proportion of photosynthate allocated for the growth of roots, stems, branches and leaves (Landsberg, 1986; Koslowski *et al.*, 1991). While little is known, either qualitatively or quantitatively, about such effects in mangroves, analysis of allometric data for trees of different sizes at the same (or similar) site(s) gives some insight into the effect of tree size on the allocation of carbon for growth.

Figure 2 shows the changes in the proportion of leaves, branchwood, stem, and above-ground (prop) roots with stem diameter for *Rhizophora* species at several sites in north-eastern Australia. Noteworthy in Figure 2 is that as the size of the tree increases, there is an increasing allocation to above-ground prop roots at the expense of that to the stem. This presumably reflects the greater support requirements of larger trees, since the prop roots are the major support structures for *Rhizophora* species. The architecture of the prop roots and the rest of the tree, as well as their relative dimensions, are amenable to structural analysis using standard engineering analytical techniques; such an analysis, however, is beyond the scope of this chapter.

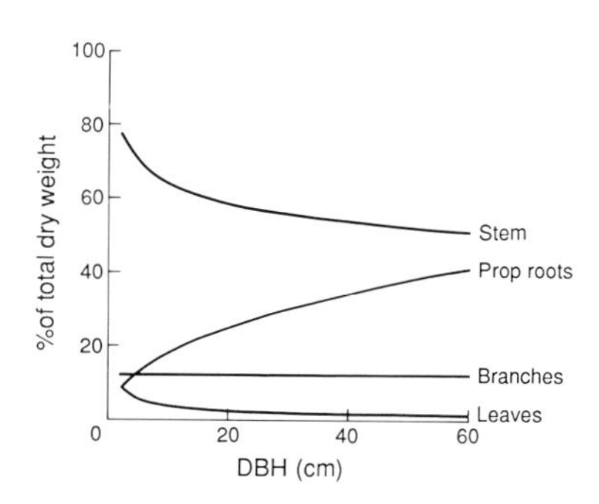

Figure 2. Partitioning of above-ground biomass in relation to tree size for *Rhizophora apiculata* and *R. stylosa* in north-eastern Australia.

Since both the prop roots and the stem are relatively long-lived, stable structures (unless the tree dies), the accumulation of biomass in each provides at least some clue to the way in which carbon is partitioned between them as the tree grows. This of course underestimates the amount of photosynthate allocated to each because it neglects respiration. Nevertheless, it is a useful first approximation in the absence of reliable measurements of respiratory losses from the prop roots and stem.

By contrast, leaves, and to a lesser extent twigs and branches, are shed as litter throughout the year. In consequence, carbon must be allocated to replace leaves, twigs and branches that are lost as litter, in addition to that allocated for expansion of the leaf canopy.

Carbon allocation for the production of below-ground roots is more difficult to estimate. As pointed out earlier in this chapter (see also Gill and Tomlinson, 1977), the large woody structural roots of *Rhizophora* are mainly above ground; the roots below ground comprise mainly soft, non-woody, first- and second-order laterals of less than 10 mm in diameter, and fine, fibrous roots of less than 1 mm in diameter. Gong and Ong (1990) estimated that below-ground roots represented only about 8.5% of total tree biomass in *Rhizophora*. This is clearly a significant underestimate of the total amount of carbon allocated to below-ground roots which, like those of other tree species (Kozlowski *et al*., 1991), probably have a high rate of turnover.

8.3 Primary production

In addition to the trees themselves, seedlings, algae and other periphyton also contribute to the primary productivity of mangrove ecosystems, in some cases significantly (Lugo, 1990). An important distinction therefore needs to be made between net primary production of the whole mangrove ecosystem and that of mangrove trees themselves. In the former case, net primary production includes the contribution to gross primary production of the trees, seedlings and periphyton, as well as losses due to respiration by these components of the ecosystem. Many of these respiratory losses are difficult to quantify, even in the short term. The main problem in measuring the gas exchange of components such as the branches, stem and above-ground roots is contamination by non-photosynthetic organisms on the surface of these structures, which can lead to overestimation of respiration by primary producers. Furthermore, accurate measurements of the respiration of below-ground roots *in situ* are almost impossible, owing to the high population density and metabolic activity of sedimentary bacteria (Alongi, 1989).

The contribution of seedlings and periphyton to mangrove ecosystem net primary production is likely to be highly variable from one site to the next. It will depend, for example, on the relative proportions of trees, seedlings and periphyton in the ecosystem. It will also depend on the structural characteristics of the ecosystem, in particular on the leaf area index of the upper canopy and the degree of canopy cover, which will determine the amount of light reaching the forest floor. The minimum light flux density required for gross photosynthesis to balance the daily respiratory losses from the whole organism (seedlings or periphyton) is likely to be about 200 μmol photons.m^{-2}.s^{-1}, a reasonable value considering that the light compensation for photosynthesis of mangrove leaves is in the range 30-50 mmol photons.m^{-2}.s^{-1}. Figure 3 shows the approximate photon flux density reaching the forest floor as a function of leaf area index, assuming a photon flux density of 2000 μmol photons.m^{-2}.s^{-1} above the canopy. For a fairly typical extinction coefficient, *k*, of 0.5 (Clough *et al.*, unpublished), the light flux density on the forest floor falls below 200 μmol photons.m^{-2}.s^{-1} at a leaf area index above 4.5. Even at leaf area indices between 3.5 and 4.5, photosynthesis by seedlings and periphyton may not contribute significantly to total ecosystem primary production. In mangrove ecosystems with lower leaf area index, or where the canopy is very open, however, photosynthesis by seedlings and periphyton may make a greater contribution to ecosystem net primary production, in part because of the greater light flux densities reaching seedlings and periphyton, and in part because canopy photosynthesis is reduced as a consequence of the low leaf area index.

Notwithstanding the difficulty of accurately measuring rates of photosynthesis and respiration by different compartments of mangrove forests, gas exchange techniques, coupled with estimates of canopy light absorption and a modelling approach (e.g. Miller, 1972), offer some promise of giving reliable estimates of gross and net primary production and providing

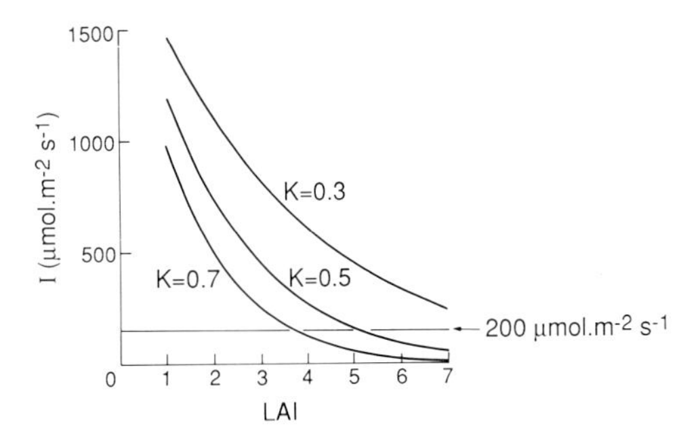

Figure 3. Photon flux density (I) reaching the forest floor as a function of leaf area index (LAI), for three values of the extinction coefficient, k. The photon flux density above the canopy is taken to be 2000 μmol photons.m^{-2}.s^{-1}. The dotted line parallel to the lower axis represents a photon flux density of 200 μmol photons.m^{-2}.s^{-1}, which is assumed to be necessary for photosynthesis of seedlings and periphyton to meet their daily respiratory requirement.

further insight into the factors regulating net primary production. As an alternative to these techniques, which require expensive instrumentation, many workers in Asia have estimated potential net production using the technique described by Bunt *et al.,* (1979) and elaborated further by Boto *et al.,* (1984). Unfortunately, this technique does not measure net production, and estimates derived from it are generally much too low to account for observed rates of accumulation of above-ground biomass (Gong *et al.*, 1991; Clough *et al.*, unpublished).

Estimates of net forest primary production can be obtained by summing net biomass accumulation, both above and below ground, plus all losses of dry matter. The latter include losses due to shedding of leaves, flowers, propagules and branches, to death of trees, to root turnover, to extrusion of organic compounds from the roots, and to direct losses from herbivory on leaves and other plant parts. To this author's knowledge, no studies have included concommitant estimates of all these components of net primary production, mainly because of the difficulty of obtaining estimates of all of them over a suitable time interval. Estimates of the below-ground components of net primary production, such as biomass accumulation by roots or root turnover, are entirely lacking. Furthermore, leaching of polyphenolics, amino acids and other organic compounds from roots has been observed in laboratory culture of mangrove seedlings (Clough, unpublished). This is also likely to occur in the field. Leaching of these organic materials from mangrove roots could, hypothetically, amount to a sizeable fraction of net primary production.

In contrast to the lack of data on the below-ground components of net primary production, there is an increasing body of data on the above-ground components, derived from studies in many parts of the world. The ease with which litter fall can be measured and its important contribution to estuarine detritus-based food chains (see Chapter 10, this volume) has encouraged many studies of this component of net primary production (e.g. Pool *et al.*, 1975; Duke *et al.,* 1981; Twilley *et al.*, 1986). Unfortunately, there has also been a tendency by some to erroneously equate litter fall with net primary production. Furthermore, there is no convincing evidence that litter fall and net primary production are necessarily correlated.

Generally there appears to be a degree of consistency in the annual rates of litter fall across broad geographic boundaries, with an average annual rate of 8-10 t dry weight.ha^{-1}.yr^{-1} (Golley *et al.*, 1962; Lugo and Snedaker, 1974; Pool *et al.*, 1975; Duke *et al.*, 1981; Singh *et al.*, 1987; Kyuma *et al.,* 1989; Mall *et al.*, 1991. Highest rates, up to 20 t.ha^{-1}.yr^{-1}, have been reported in north-eastern Australia (Duke *et al*, 1981). Litter fall usually takes place throughout the year, with a peak generally just prior to, or during, the monsoon season (Pool *et al*, 1975; Williams *et al.*, 1981).

8.4 Edaphic Factors influencing primary production and growth

8.4.1 Anaerobiosis (Anoxia)

Mangroves grow in waterlogged soils that are commonly anaerobic. Their anaerobic nature stems from the slow rate of diffusion of oxygen in water and the biological activity of microorganisms that use oxygen. Redox potentials of mangrove sediments vary considerably (see Chapter 9, this volume), depending on the frequency and duration of tidal inundation, drainage, sediment organic matter content, and the availability of electron acceptors such as nitrate, Fe^{3+} and Mn^{4+}. In general, redox potential would be expected to increase with decreasing frequency and duration of tidal inundation, and hence with higher elevation. However, this trend may be reversed by the presence of mangroves, owing to diffusion of oxygen from roots into the surrounding soil (e.g. Boto and Wellington, 1984). Similar findings in a salt marsh system were reported by Howes *et al.*, (1981).

Anoxic conditions can influence plant growth in three ways. Firstly, the absence of oxygen in the soil means that below-ground roots must rely on gas transport internally to satisfy their oxygen requirement. Secondly, low oxidation-reduction potentials favour chemical transformations of a number of essential elements, in some cases improving their availability, in others restricting their availability. Finally, extreme degrees of anaerobiosis can lead to the formation of H_2S and other compounds that may be toxic to plants.

Mangroves are shallow rooted, the roots being concentrated mainly in the top 50-100cm of the soil and seldom penetrating to depths beyond 2 m. Water and nutrients are absorbed by a dense mat of fibrous roots that usually originate from underground cable roots or smaller secondary roots in close proximity to the aerial "breathing" roots or, in the case of the genus *Rhizophora,* at the points of entry of the prop roots to the soil (Gill and Tomlinson, 1977). The shallowness of the root system, together with the close physical coupling of the smaller metabolically active roots and the aerial roots minimises the distance through which gases such as oxygen must diffuse in order to reach the extremities of the root system.

Transport of oxygen and other gases between the atmosphere and below-ground roots occurs via the intercellular space in a spongy, aerenchymatous cortex of both the aerial and below-ground roots (Scholander *et al.*, 1955; Gill and Tomlinson, 1977; Curran, 1985). Generally the rate at which oxygen is supplied to underground roots exceeds that required to satisfy their respiration, resulting in diffusion of surplus oxygen from the root into the surrounding soil and leading to the development of an oxidised rhizosphere around the roots and (Nickerson and Thibodeau, 1985; Thibodeau and Nickerson, 1986). The extent of this oxidised rhizophere, which may be up to 5 mm thick, presumably depends on the rate of oxygen diffusion to the underground roots, their respiratory requirements, and the oxygen demand of the surrounding soil as reflected in its oxidation-reduction potential. Hence, in strongly anaerobic soils, the oxidised rhizosphere tends to be smaller than that in moderately anaerobic soils.

In view of the fact that most of the fine absorbing roots are found near the soil surface, the physical and chemical properties of the top metre of the soil are probably more important than those at greater depth.

8.4.2 Nutrient availability and net primary production

As pointed out above, the physical and chemical properties of the soil, in particular redox potential, have a significant effect on the availability of both macro and micro nutrients. These have been summarised by Clough *et al.*, (1983) and will not be detailed here. It is sufficient to point out that much of the inorganic phosphorus in mangrove soils is adsorbed or otherwise incorporated within hydrated iron and aluminium colloidal sesquioxides. Calcium-bound phosphorus may be an important component of the inorganic phosphate pool in calcareous soils. Generally, low redox potentials in anaerobic mangrove soils can be expected to lead to the release of phosphate, particularly at pH<7 (Patrick and Mahapatra, 1968).

Ammonium is the main form of inorganic nitrogen in anaerobic mangrove soils because nitrification of organic nitrogen stops at ammonium due to the lack of oxygen to oxidise it further to nitrate (Ponnamperuma, 1972). Like other cations, ammonium is capable of occupying cation exchange sites, but the high level of sodium in most mangrove soils tends to swamp the cation exchange sites, thereby displacing ammonium. Thus ammonium occurs mainly in the interstitial phase, where it is highly mobile and susceptible to being leached by heavy rain or lost through drainage following tidal inundation. While ammonium is the main form of inorganic nitrogen in mangrove soils, it is not certain that all mangrove species can utilise ammonium as their primary nitrogen source. Boto *et al.*, (1985) compared the growth of *Avicennia marina* seedlings with either nitrate or ammonium ions as their sole source of nitrogen, and found that nitrate appeared to be the preferred nitrogen source. Prolonged exposure to ammonium ions in the absence of nitrate lead to a marked reduction in growth and eventual death of *A. marina* seedlings. This observation, as well as more general considerations of nutrient transformations in response to redox potential, points to the significance of the oxidised rhizosphere surrounding mangrove roots in influencing nutrient availability. Other essential nutrients whose availability is likely to be influenced by this oxidised rhizosphere include phosphorus, iron and manganese. For these nutrients, however, the oxidised rhizophere is likely to reduce availability.

From the foregoing discussion it is clear that interactions between redox potential and the availability of essential nutrients are very complex, and this complexity is further increased by the heterogeneous nature of many mangrove soils. For this reason the extent to which net primary production and growth of mangroves is limited by nutrients is not at all clear. For example, Boto and Wellington (1984) studied the changes in redox potential, salinity, soil nitrogen, soil phosphorus and stem volume along a pronounced topographic gradient in north-eastern Australia. They found significant correlations between soil elevation, soil salinity, soil phosphorus (both total and extractable), redox potential and stem volume along the transect. In another study using the same transect, Boto and Wellington (1983) enriched plots with N and P by applying fertilizers. They found an increase in the rate of production of new leaves in response to both N and P in some areas, but not others. Similarly, the new leaves produced in fertilized plots in most cases, but not always, had higher N or P contents than those in control plots. Other workers have also observed an increase in foliar N and P in response to enrichment by sewage or guano, in some cases associated with an apparent increase in growth rate (Onuf *et al.*, 1977), in others with no apparent increase in growth (Clough *et al.*, 1983). In a recent survey of thirty different sites located in five geographically separated localities in Peninsula Malaysia and East Malaysia, Gong *et al.*, (1991) found no correlation between potential primary production and soil characteristics. This contrasts with the findings of Boto *et al.*, (1984) who reported that low availability of soil phosphorus was a major factor contributing to low values of potential primary production in northern Australia and the Gulf of Papua. Nevertheless, it is difficult to interpret with confidence such observations in terms of a nutrient limitation on growth owing to the complexity of interactions between all factors likely to influence it. In general, soils in actively accreting deltaic areas receiving inputs of silt and other materials from terrigenous sources are likely to have higher nutrient levels than sandy, rocky or calcareous soils of predominantly marine origin.

8.4.3 Salinity

While some species, notably *Avicennia marina*, do not seem to survive well in non-saline soils and may be obligate halophytes (Downton, 1982; Clough, 1984), others such as *Bruguiera* spp., *Xylocarpus granatum*, *Sonneratia caseolaris, Heritiera littoralis* and *Nypa fruticans* grow well in non-saline soils (personal observation) and do not appear to have an obligatory requirement for NaCl beyond the trace element level shown by non-halophytes (cf. Brownell, 1979). Laboratory studies with seedlings of generally less than 18 months in age suggest that soil salinities of 10-20 ‰ are optimal for the growth of *Avicennia marina* (Connor, 1969; Downton, 1982; Burchett *et al.*, 1984; Clough, 1984), *Rhizophora stylosa* (Clough, 1984), *R. apiculata, R. mangle, Lumnitzera racemosa* and *Xylocarpus granatum* (Clough, unpublished). These laboratory experiments, all with seedlings of less than 2 years in age, are consistent with field observations in Malaysia, Thailand and Australia which show that growth is faster at low than at high salinities. However, it is not yet clear whether the optimum salinity range for seedling growth is the same as that for more mature shrubs and trees.

In the field, salt tolerance appears to vary widely between species (Table 3). While some species appear to be restricted to salinities well below that of seawater (35 ‰), others can survive in hypersaline conditions. *Rhizophora mangle*, for example, has been reported to grow at a soil salinity of 65‰ (Teas, 1979), while *Avicennia marina* and *Lumnitzera racemosa* have been found at salinities as high as 90 ‰ (Macnae, 1968). Such exteme values might be associated with the drier periods of the year, and corresponding salinities during the wetter months could be lower. It is likely that most species can withstand brief periods when soil salinity is much greater than the annual average. Soil water content, seasonal variation in rainfall/evapotranspiration, extremes of temperature, and other environmental conditions may also have a major effect on the capacity of any particular species to withstand hypersaline regimes. The upper salinity limit for survival of a particular species is therefore likely to vary from one location to another.

High salinity poses two major problems for mangrove growth. Firstly, the low osmotic potential of saline soils sets constraints on the water relations of mangroves, leading to physiological responses that are analagous to those of terrestrial plants experiencing drought.

Table 3. Relative tolerances of some mangrove species to salinity, aridity, low temperature and tidal level. Range of tolerances from, +++++ = very tolerant, to + = not tolerant; for tidal levels, H = high, M = mid and L = low.

Species	Salinity	Aridity	Low Temperature	Intertidal Level
Acanthus ilicifolius	++	+	++	H,M,L
Aegialitis annulata	++++	++++	++	M,H
Aegiceras corniculatum	+++	+++	++++	M,L
Avicennia germinans	+++++	+++++	+++++[A]	H,M,L
Avicennia marina	+++++	+++++	+++++[A]	H,M,L
Bruguiera cyclindrica	++	++	+	H,M
Bruguiera exaristata	+++	+++	++	H,M
Bruguiera gymnorrhiza	+++	+++	+++	H,M
Bruguiera parviflora	++	++	++	H,M
Bruguiera sexangula	+	+	+	H,M
Camptostemon schultzii	?	?	?	H,M
Ceriops decandra	++	++	+	H,M
Ceriops australis[B]	++++	++++	+++	H,M
Ceriops tagal[C]	+++	+++	++	H,M
Cynametra iripa	++	++	++	H
Diospyros ferrea	+	+	+	H
Excoecaria agallocha	+++	+++	+++	H,M
Heritiera formis	?	?	?	?
Heritiera littoralis	++	++	+	H,M
Kandelia candel	+++?	+++?	+++	H,M
Laguncularia racemosa	?	?	?	?
Lumnitzera littorea	++++	++++	++	H
Lumnitzera racemosa	++++	++++	+++	H
Nypa fruticans	+	+	+	M,L
Osbornia octodonta	+++	+++	+++	H
Rhizophora apiculata	+++	+++	++	M,L
Rhizophora lamarckii	+++	+++	++	M,L
Rhizophora mangle	++++	++++	?	M,L
Rhizophora mucronata	++	++	++	M,L
Rhizophora stylosa	++++	++++	+++	M,L
Scyphiphora hydrophyllacea	?	?	+	H,M
Sonneratia alba	+++	+++	++	M,L
Sonneratia caseolaris	+	+	+	M,L
Xylocarpus granatum	+++	+++	++	H,M
Xylocarpus mekongensis[D]	+++	+++	++	H,M

A Wide ecotypic variation
B Formerly *Ceriops tagal* var *australis* (Ballment *et al.*, 1988)
C Formerly *Ceriops tagal* var *tagal* (Ballment *et al.*, 1988)
D syn. *X. australasicus* (Tomlinson, 1986)

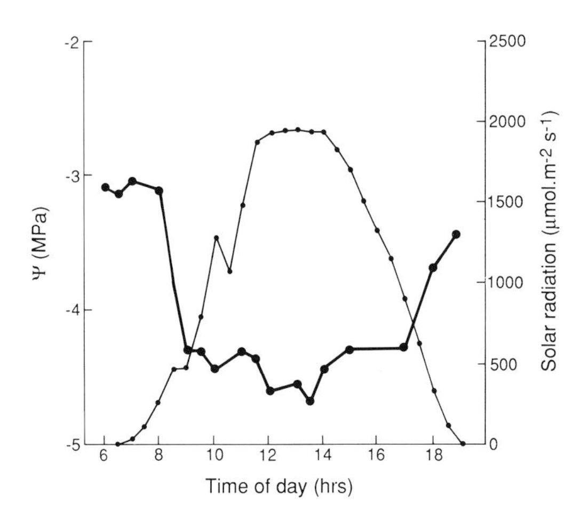

Figure 4. Relationship between photon flux density and daily variation of shoot water potential in *Rhizophora apiculata* during the dry season in southern Thailand (Taken from Unesco, 1991).

The water potential of saturated soils is largely determined by the osmotic potential of the soil solution. At any given soil salinity, the corresponding osmotic potential represents the highest water potential that can be attained by a plant overnight. For soils with salinities equivalent to seawater this is about -2.5 MPa. During the daytime, however, when water loss exceeds water uptake, plant water potentials fall well below the overnight maximum, giving rise to a diurnal variation in shoot water potential like that shown in Figure 4. The rate of water loss from mangrove leaves depends chiefly on stomatal conductance and the vapour pressure gradient between the leaf and surrounding air (VPD). The latter in turn depends on leaf temperature and ambient relative humidity, both of which are influenced by the radiation flux density. As shown in Figure 4, shoot water potential recovers quite rapidly with declining irradiance late in the afternoon. Similar responses were reported by Smith *et al.*, (1989).

Maintenance of a water balance that is acceptable physiologically requires that cellular osmotic potentials be lower (more negative) than the osmotic potential of the soil solution. Hence, the second problem posed by high salinity for mangrove net primary production and growth is related to the need to take up sufficient inorganic ions to maintain osmotic balance while at the same time avoiding the adverse metabolic effects of high cytoplasmic concentrations of these ions. Most evidence suggests that inorganic ions are stored in the vacuole where they can do little physiological damage, while osmotic balance in the cytoplasm is maintained through the synthesis of a range of compatible organic solutes (Wyn-Jones and Storey, 1981; Popp, 1984a, 1984b; Popp *et al.*, 1984). Little is known of the energetics of the synthesis of compatible organic solutes or of the maintenance of high ion concentrations in the vacuole against a steep concentration gradient. However, both processes presumably require the expenditure of metabolic energy that might otherwise be used to give higher rates of net primary production and growth.

8.5 Climatic factors influencing net primary production and growth

Net primary production and growth of mangrove forests, like that of other forest and woodland communities, is greatly influenced by climatic conditions. The solar radiation regime, daylength, temperature, rainfall, potential evapotranspiration, as well as the seasonal variability of these factors, all play an important role in influencing net primary production and growth rate. These factors operate at all levels of plant organisation, from the cellular to the whole tree.

Collectively, mangroves seem to grow best in equatorial climates where temperature is moderately high, rainfall is moderate to high, conditions are persistently cloudy, the ratio of precipitation to evapotranspiration is low, solar radiation fluxes are not extreme, and seasonal variability in climate is small. Growth and net primary production are lower in arid climates, such as the coasts of Pakistan or the Arabian Pensinsula, and those with a pronounced seasonal asymmetry in rainfall, such as the strongly seasonal monsoonal climate of north-western Australia. Differences in growth and net primary production between humid equatorial climates and arid or seasonally monsoonal climates can be explained at least partly by the effect of climatic conditions on the carbon, water and salt balance of mangroves.

8.5.1 Carbon balance

Mangroves are C_3 plants, notwithstanding some biochemical evidence to the contrary (Joshi *et al.*, 1974; Joshi, 1976). Their carbon isotope ratio lies between -24‰ and -32‰ (Andrews *et al.*, 1983), well within the range reported for other C_3 plants and outside that reported for C_4 plants (Smith and Epstein, 1971); they do not have the specialised leaf anatomy associated with the C_4 pathway; they have a measurable, temperature-dependent CO_2 compensation point of 40-90 $\mu l.l^{-1}$ (Moore *et al.*, 1972; Ball and Critchley, 1982; Andrews *et al.*, 1984; Ball and Farquhar, 1984a); and a temperature optimum for photosynthesis below 35°C (Andrews *et al.*, 1984). These are all C_3 characters.

Rates of net photosynthesis vary widely depending on the species, the position of the leaf in the canopy, and environmental conditions, particularly soil salinity, solar radiation flux density and the vapour pressure difference (VPD) between the leaf and its environment. Under favourable conditions of low VPD (<22 mbar) and low salinity (<15‰), the rate of net photosynthesis may exceed 25 $\mu mol\ CO_2.m^{-2}.s^{-1}$ (Clough and Sim, 1989). More commonly, however, rates of net photosynthesis in the field lie in the range 5-20 $\mu mol\ CO_2.m^{-2}.s^{-1}$ (Moore *et al.*, 1972, 1973; Attiwill and Clough, 1980; Ball and Critchley, 1982; Andrews *et al.*, 1983; Andrews and Muller, 1985; Clough and Sim, 1989; Cheeseman *et al.*, 1991). In a wide-ranging study of 19 mangrove species in diverse environments, Clough and Sim (1989) found that maximum rates of net photosynthesis under saturating light flux densities (>800 μmol photons.$m^{-2}.s^{-2}$) decreased with both increasing soil salinity and increasing VPD. Of the 19 species sampled, *Avicennia marina* consistently had higher rates of net photosynthesis than other species at those sites where it was present, while members of the genus *Rhizophora* consistently had higher rates than their close relatives in the genus *Bruguiera* at the same site (Clough and Sim, 1989). It is not known whether these interspecific differences in photosynthetic performance are also reflected in differences in growth rate, since it is possible that species with lower photosynthetic rates may have other compensatory responses that offset their low photosynthetic capacity.

At saturating light flux densities, photosynthetic performance may be limited by stomatal conductance, carboxylation processes, or both (Ball and Critchley, 1982; Andrews *et al.*, 1984; Cheeseman *et al.*, 1991). In consequence, leaves exposed to direct sunlight often receive excess visible radiation beyond that which can be utilised effectively in photosynthesis (Clough *et al.*, 1982). In some species excess visible light leads to damage to the electron transport chain of photosystem II, and may result in photoinhibition (Bjorkman and Demmig, 1987; Bjorkman *et al*, 1988). In other species the xanthophyll cycle appears to provide an effective mechanism for protecting against excess visible radiation (Lovelock, 1991).

8.5.2 Water balance

Mangrove leaves exposed to bright direct sunlight absorb enough radiation for their temperature to rise substantially above that of the atmosphere (Andrews and Muller, 1985; Ball *et al.*, 1988; Smith *et al.*, 1989). Moreover, leaves that are perpendicular to the sun's rays are hotter than those that are inclined more acutely. Such leaves may have temperatures >40°C (Andrews and Muller, 1985; Ball *et al.*, 1988). High leaf temperatures may reduce the biochemical capacity to fix carbon (Moore *et al.*, 1972, 1973; Andrews *et al.*, 1984; Ball *et al.*, 1988). In addition, however, high leaf temperatures increase the vapour pressure difference (VPD) between the leaf and its environment, leading to an increase in the rate of water loss which is seldom fully offset by the compensatory reduction in stomatal aperture (Clough *et al.*, 1982; Ball *et al.*, 1988). The leaves of many mangrove species also possess morphological adaptations that aid in reducing water loss. These include pubescence, thick cuticles, wax coatings and sunken or otherwise hindered stomata (Sidhu, 1975; Saenger, 1982).

Since the stomata also control the influx of CO_2 for photosynthesis, reduced stomatal conductance resulting from partial stomatal closure also incurs the penalty of reduced photosynthetic carbon fixation, provided that carboxylation and electron transport processes are not severely co-limiting photosynthetic capacity. Linear relationships between photosynthetic rate and stomatal conductance have been reported by a number of workers (eg. Ball and Critchley, 1982; Andrews *et al.*, 1984; Andrews and Muller, 1985; Ball *et al.*, 1988). The slope of this relationship is also an index of relative water use efficiency (Schulze and Hall, 1982; Ball *et al.*, 1988; Clough and Sim, 1989). In addition, the linear relationship between photosynthetic rate and stomatal conductance is often of a form that results in a low and more or less constant intercellular CO_2 concentration of 160-200 $\mu l.l^{-1}$ over a wide range of conditions (eg. Ball and Critchley, 1982; Andrews *et al.*, 1984; Andrews and Muller, 1985; Ball *et al.*, 1988). Recent studies of these characteristics in the field by Ball *et al.*, (1988) and by Clough and Sim (1989) have confirmed earlier laboratory studies (Ball and Farquhar, 1984a, 1984b) that water use efficiency increases with increasing salinity and VPD between the leaf and its environment. Thus, while the actual photosynthetic carbon gain at high salinity and high VPD was less than that under less adverse conditions, it was achieved for a relatively lower cost in water use. This implies some degree of acclimation of metabolism in response to saline and/or arid conditions.

These ecophysiological responses to climatic conditions and the high water use efficiency of mangroves help to explain why they grow better in persistently wet equatorial climates than in seasonally monsoonal or arid climates. In wet equatorial climates, persistent cloud cover prevents exposure of the forest canopy to high levels of direct sunlight, with the consequence that the leaves remain relatively cool. This, coupled with high ambient relative humidity, places less stress on the water and salt balance of the trees and minimises the need for an unusually conservative use of water. In contrast, mangroves growing in arid climates are exposed to high solar insolation and in consequence, experience high leaf temperatures and low ambient relative humidities, which in turn leads to high rates of water loss, partial stomatal closure and very conservative use of water. High soil salinities simply exacerbate this situation. Mangroves growing in seasonally monsoonal climates often experience similar conditions during the dry season, depending on its length and whether or not the dry season is broken by occasionally cloudy conditions and rain.

8.5.3 Temperature

Mangroves grow over a wide range of temperatures. At the lower end of their temperature range they occur in climates with average minimum air and soil temperatures of about 15°C and 20°C, respectively, during the coldest months, with episodic minima reaching as low as 4°C for air temperature. At the upper end of their temperature range, average maximum air and soil temperatures may be as high as 35°C and 30°C, respectively, during the warmest months, with episodic maxima for air temperature reaching 40°C. Given this wide temperature range, it is surprising that so little quantitative information of a comparative nature is available for the effect of temperature on net primary production and growth.

General observation indicates an overall trend of decreasing growth rate of mangroves with increasing latitude along the eastern coastline of Australia, in association with a decrease in average and minimum temperatures of both air and soil. This progressive reduction in growth rate with decreasing temperature is undoubtedly a consequence of complex responses of many growth-regulating processes. These could include an overall reduction in growth potential, reduced photosynthetic carbon fixation, increased respiration, or a reduced capacity to achieve osmoregulation. Further speculation is unwarranted in the absence of supporting evidence.

8.6 Effect of forest structure on net primary production

The structural characteristics of forests have a marked influence on growth and net primary production (Oliver and Larson, 1990). By comparison with tropical rainforest, tropical mangrove forest is relatively simple structurally, having a comparatively low diversity of tree species and generally lacking an understorey of shrub or herbaceous species. While a number of structural attributes may have some influence on net primary production, the most important structural attributes influencing net primary production and growth are likely to be stem density, basal area, canopy leaf area index (LAI), and the distribution and orientation of foliage. Comprehensive and comparative data sets for these structural parameters, which also include an estimate of net primary production or growth, are generally lacking. Recently, Gong *et al.*, (1991) found a positive correlation between potential net primary production, measured by the method of Bunt *et al.*, (1979), and both stand density and basal area in a wide range of mangrove forests in both Eastern and Peninsular Malaysia. However, as outlined earlier and recognized by Gong *et al.*, (1991), the method of Bunt *et al.*, (1979) for estimating potential net production is suspect. Other data sets, where net accumulation of above-ground biomass has been measured, do exist for the west coast of Peninsular Malaysia (J.-E. Ong and W.-K. Gong, pers. comm.) and for north-eastern Australia (Clough, unpublished), but have not yet been analysed in terms of effects due to stand density, basal area and size class distribution.

The photosynthetic production of mangrove canopies depends on the leaf area available to absorb light, how that leaf area is arranged spatially in order to utilise light efficiently, and on the photosynthetic rate per unit leaf area. Like the other structural attributes of mangrove forests the leaf area index and the way the foliage is displayed varies betweem species, and with environment. The leaf area index of mangrove forests is often low by comparison with other tropical forests. As pointed out above, this is partly attributable to the absence of understorey plants with very efficient light harvesting characteristics. Apart from this, however, the amount of foliage per unit ground area, or Leaf area index, of mangrove forest, as well as other vegetation types, generally decreases with increasing environmental stress of one kind or another. This includes both high and low temperature, high salinity, an arid climate, severe nutrient deficiency, excessive wind, and others. Stress in one form or another probably accounts for the relatively low leaf area indices reported from the new world mangroves, where LAI seems to fall between about 2 and 5 (Golley *et al.*, 1962; Miller, 1972; Lugo *et al.*, 1975). Similarly Clough and coworkers (unpublished) recorded leaf area indices of less than 5 in mangrove canopies in southern Thailand that were subjected to intense anthropogenic activity. However, in other less stressful environments, such as the west coast of Peninsula Malaysia or north-eastern Australia, leaf area index can reach 7 (e.g. Cheeseman *et al.*, 1991).

Both the amount of light absorbed by the canopy and its distribution through the canopy depend on the spatial arrangement of the canopy foliage. One obvious example is the change of leaf angle with position in the canopy shown by members of the genus *Rhizophora.* In this genus, leaf inclination from almost vertical at the top of the canopy to progressively more horizontal with depth in the canopy. This allows the upper leaves that are more exposed to the sun to avoid absorption of excessive radiation, while at the same time allowing light to penetrate further into the canopy. Figure 5 shows the light profile with depth in two canopies of *Rhizophora apiculata* in southern Thailand. Nothwithstanding the steep inclination of leaves near the top of the canopy, in each case about 60% of the incoming photon flux density was absorbed in the top metre of the canopy. The response function shown in Figure 5 is a least squares fit to the data using the relationship

$$I = I_o \exp^{-kL}$$

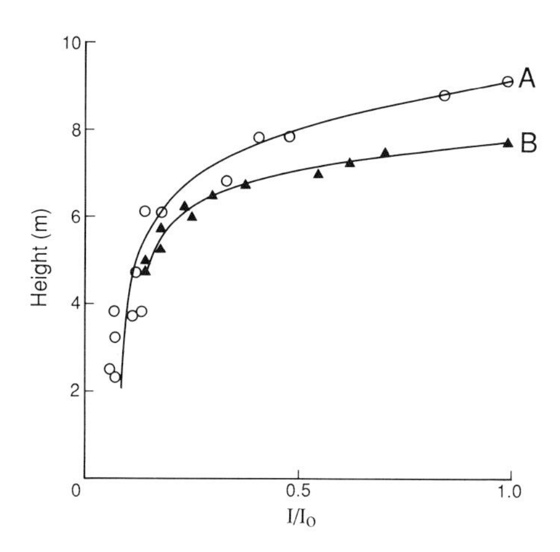

Figure 5. Attenuation of photosynthetically active radiation with increasing depth in two canopies (A and B) of *Rhizophora apiculata* (Taken from Unesco, 1991).

where I is the photon flux density at a given height beneath leaf area index, L; I_0 is the photon flux above the canopy; and k is the canopy light extinction coefficient. The latter, which varies with leaf angle and the spatial arrangement of the leaves, had values between 0.4 and 0.6 for the two canopies.

8.7 Conclusions

Given the diversity of environmental and geomorphic settings in which mangroves grow it is not surprising that their primary productivity and rate of growth is highly variable, both locally and regionally. At one end of the spectrum, growth rate and net primary productivity may be close to zero when climatic conditions are severe and/or soil salinity is high. At the other end of the scale, under favourable climatic conditions and moderate salinities, the net primary productivity of mangrove forests compares favourably with that of terrestrial forests.

Although the environmental constraints on primary productivity and growth are known in a general, qualitative sense, there is still a lack of good quantitative information on these effects and of comparative data on primary productivity and growth under differing environmental conditions. In particular, there is a paucity of data on soil-plant interactions and other processes below ground that could influence primary productivity and growth. For example, reliable data on below ground biomass and its rate of turnover is crucial to assessing mangrove primary productivity. Information is also needed on the relationship between soil characteristics, nutrient availability and nutrient uptake. Another obvious gap is information on the leaching or loss of soluble organic materials from root systems. This may well represent a significant part of the carbon economy of mangrove forests.

This lack of quantitative data has hampered attempts to develop predictive models of mangrove forest dynamics and growth. Such models serve not only as an aid in conceptualizing the functional relationships of mangrove systems and in identifying gaps in understanding, but could also be of great value as a management tool.

8.8 References

Aksornkoae, S., 1976. Mangrove plantation productivity at Amphoe Khlung, Changwat Chantaburi. Paper presented at the First National Seminar on Mangrove Ecology, Phuket, Thailand.

Alongi, D.M., 1989. The role of soft-bottom benthic communities in tropical mangrove and coral reef ecosystems. *Reviews in Aquatic Sciences* **1**:243-279.

Andrews, T.J., and Muller, G.J., 1985. Photosynthetic gas exchange of the mangrove, *Rhizophora stylosa* Griff., in its natural environment. *Oecologia* **65**:449-455.

Andrews, T.J., Clough, B.F., and Muller, G.J., 1984. Photosynthetic gas exchange properties and carbon isotope ratios of some mangroves in North Queensland. In: Teas, H.J. (Ed.), *Physiology and Management of Mangroves, Tasks for Vegetation Science 9*, pp. 15-23, Dr. W. Junk, The Hague, .

Attiwill, P.M., and Clough, B.F., 1980. Carbon dioxide and water vapour exchange in the white mangrove. *Photosynthetica* **14**:40-47.

Ball, M.C., 1988. Ecophysiology of mangroves. *Trees* **2**:129-142.

Ball, M,C., and Critchley, C., 1982. Photosynthetic responses to irradiance by the grey mangrove, *Avicennia marina*, grown under different light regimes. *Plant Physiology* **70**:1101-1106.

Ball, M.C., Cowan, I.R., and Farquhar, G.D., 1988. Maintenance of leaf temperature and optimisation of carbon gain in relation to water loss in a tropical mangrove forest. *Australian Journal of Plant Physiology* **15**:263-276.

Berry, J., and Bjorkman, O., 1980. Photosynthetic response and adaptation to temperature in higher plants. *Annual Review of Plant Physiology* **31**:491-543.

Bjorkman, O., and Demmig, B., 1987. Photon yield of O_2 evolution and chlorophyll fluorescence characteristics at 77°K among vascular plants of diverse origins. *Planta* **170**:489-504.

Bjorkman, O, Demmig, B., and Andrews, T.J., 1988. Mangrove photosynthesis: response to high-irradiance stress. *Australian Journal of Plant Physiology* **15**:43-61.

Blasco, F., 1984. Climatic factors and the biology of mangrove species. In: Snedaker, S.C. & Snedaker, J.G., (Eds.), *The Mangrove Ecosystem: Research Methods*, pp 18-34, Unesco, Paris.

Boto, K.G., Bunt, J.S., and Wellington, J.T., 1984. Variations in mangrove forest productivity in Northern Australia and Papua New Guinea. *Estuarine, Coastal and Shelf Science* **19**:321-329.

Boto, K., Saffigna, P., and Clough, B., 1985. Role of nitrate in nitrogen nutrition of the mangrove *Avicennia marina. Marine Ecology Progress Series* **21**:259-265.

Boto, K.G., and Wellington, J.T., 1983. Phosphorus and nitrogen nutritional status of a northern Australian mangrove forest. *Marine Ecology Progress Series* **11**:63-69.

Boto, K.G., and Wellington, J.T., 1984. Soil characteristics and nutrient status in a northern Australian mangrove forest. *Estuaries* **7**:61-69.

Brownell, P.F., 1979. Sodium as an essential micronutrient element for plants and its possible role in metabolism. *Advances in Botanical Research* **7**:117-224.

Bunt, J.S., Boto, K.G., and Boto, G., 1979. A survey method for estimating potential levels of mangrove forest primary production. *Marine Biology* **52**:123-128.

Burchett, M.D., Field, C.D., and Pulkownik, A., 1984. Salinity, growth and root respiration in the grey mangrove *Avicennia marina. Physiologia Plantarum* **60**:113-118.

Causton, D.R., 1985. Biometrical, structural and physiological relationships among tree parts. In: Cannell, M.G.R., and Jackson, J.E., (Eds.), *Attributes of Trees as Crop Plants*, pp. 137-159, Institute of Terrestrial Ecology, Huntingdon, UK.

Chapman, V.J., 1976. *Mangrove Vegetation.* Cramer, Vaduz. 425 pp.

Cheeseman, J.M., Clough, B.F., Carter, D.R., Lovelock, C.E., Ong Jin Eong, and Sim, R.G., 1991. The analysis of photosynthetic performance in leaves under field conditions: A case sutdy using Bruguiera mangroves. *Photosynthesis Research* **29**:11-22.

Clough, B.F., 1984. Growth and salt balance of the mangroves, *Avicennia marina* (Forsk.) Vierh. and *Rhizophora stylosa* Griff., in relation to salinity. *Australian Journal of Plant Physiology* **11**:419-430.

Clough, B.F., Andrews, T.J., and Cowan, I.R. 1982. Physiological processes in mangroves. In: Clough, B.F., (Ed.), *Mangrove Ecosystems in Australia: Structure, Function and Management*, pp. 193-210, Australian National University Press, Canberra.

Clough, B.F., Boto, K.G., and Attiwill, P.M., 1983. Mangroves and sewage: a re-evaluation. In: Teas, H.J. (Ed.), *Biology and Ecology of Mangroves, Tasks for Vegetation Science 8,* pp 151-161, Dr. W. Junk, The Hague.

Clough, B.F., and Scott, K., 1989. Allometric relationships for estimating above-ground biomass in six mangrove species. *Forest Ecology and Management* **27**:117-127.

Clough, B.F., and Sim, R.J., 1989. Changes in gas exchange characteristics and water use efficiency of mangroves in response to salinity and vapour pressure deficit. *Oecologia* **79:**38-44.

Connor, D.J., 1969. Growth of grey mangrove (*Avicennia marina*) in nutrient culture. *Biotropica* **1**:36-40.

Curran, M., 1985. Gas movements in the roots of *Avicennia marina* (Forsk.) Vierh. *Australian Journal Plant Physiology* **12**:97-108.

Downton, W.J.S., 1982. Growth and osmotic relations of the mangrove *Avicennia marina*, as influenced by salinity. *Australian Journal of Plant Physiology* **9**:519-528.

Duke, N.C., Bunt, J.S., and Williams W.T., 1981. Mangrove litter fall in north-eastern Australia. I. Annual totals by component in selected species. *Australian Journal of Botany* **29**:547-553.

Gill, A.M., and Tomlinson, P.B., 1977. Studies of the growth of red mangrove (*Rhizophora mangle* L.) 4. The adult root system. *Biotropica* **9**:145-155.

Golley, F., Odum, H.T., and Wilson R.F., 1962. The structure and metabolism of a Puerto Rican red mangrove forest in May. *Ecology* **43**:9-19.

Gong, W.-K., and Ong, J.-E., 1990. Plant biomass and nutrient flux in a managed mangrove forest in Malaysia. *Estuarine, Coastal and Shelf Science* **31**:519-530.

Gong, W.-K., Ong, J.-E., and Wong, C.-H., 1991. The light attenuation method for measuring potential primary productivity in mangrove ecosystems: an evaluation. In: Alcala, A.C., 1991(Ed.), *Proceedings of the Regional Symposium on Living Resources in Coastal Seas.* pp. 399-406,University of the Philippines, Quezon City, Philippines, 597pp.

Goulter, P.F.E., and Allaway, W.G., 1979. Litter fall and decomposition in a mangrove stand, *Avicennia marina* (Forsk.) Vierh., in Middle Harbour, Sydney. *Australian Journal of Marine and Freshwater Research* **30**:541-546.

Howes, B.L., Howarth, R.W., Teal, J.M., and Valiela, I., 1981. Oxidation-reduction potentials in a salt marsh. Spatial patterns and interactions with primary production. *Limnology and Oceanography* **26**:350-360.

Joshi, G.V., 1976. *Studies in Photosynthesis in Saline Conditions.* Shivaji University Press, Kolhapur, India. 195 pp.

Joshi, G.V., Karekar, M.D., Jowda, C.A., and Bhosale, L., 1974. Photosynthetic carbon metabolism and carboxylating enzymes in algae and amangrove under saline conditions. *Photosynthetica* **8**:51-52.

Komiyama, A., Ogino, K., Aksornkoae, S., and Sabhasri, S., 1987. Root biomass of a mangrove forest in southern Thailand. 1. Estimation by the trench method and the zonal structure of root biomass. *Journal of Tropical Ecology* **3**:97-108.

Kozlowski, T.T., Kramer, P.J., and Pallardy, S.G., 1991. *The Physiological Ecology of Woody Plants.* Academic Press, San Diego. 657 pp.

Kyuma, K., Oya, K., Ishimura, K., Hirai, H., and Chan, H.T., 1989. Report on the field studies conducted by the soil group of the mangrove workshop held on Iriomote Island, Okinawa, in 1987. *Galaxea* **8**:31-41.

Landsberg, J.J., 1986. *Physiological Ecology of Forest Production.* Academic Press, London. 198 pp.

Lovelock, C.E., 1991. Adaptation of Tropical Mangroves to High Solar Radiation. Ph.D. Thesis, James Cook University of North Queensland. 148 pp.

Lugo, A.E., 1990. Fringe wetlands. In: Lugo, A.E., Brinson, M., and Brown,S., (Eds.), *Ecosystems of the World, 15, Forested Wetlands,* pp 143-169, Elsevier, Amsterdam.

Macnae, W., 1968. A general account of the fauna and flora of mangrove swamps and forests in the Indo-West Pacific region. *Advances in Marine Biology* **6**:73-270.

Mall, L.P., Singh, V.P., and Garge, A., 1991. Study of biomass, litter fall, litter decomposition and soil respiration in monogeneric and mixed mangrove forests of Andaman Islands. *Tropical Ecology* **32**:144-152.

Miller, P.C., 1972. Bioclimate, leaf temperature, and primary production in red mangrove canopies in South Florida. *Ecology* **53**:22-45.

Moon, G.J., Clough, B.F., Peterson, C.A., and Allaway, W.G., 1986. Apoplastic and symplastic pathways in *Avicennia marina* (Forsk.) Vierh. roots revealed by fluorescent tracer dyes. *Australian Journal of Plant Physiology* **13**:637-648.

Moore, R.T., Miller, P.C., Ehleringer, J., and Lawrence, W., 1973. Seasonal trends in gas exchange characteristics of three mangrove species. *Photosynthetica* **7**:387-394.

Moore, R.T., Miller, P.C., Albright, D., and Tieszen, L.L., 1972. Comparative gas exchange characteristics of three mangrove species during the winter. *Photosynthetica* **6**:387-393.

Nickerson, N.H., and Thibodeau, F.R., 1985. Association between pore water sulfide concentrations and the distribution of mangroves. *Biogeochemistry* **1**:183-192.

Oliver, C.D., and Larson, B.C., 1990. *Forest Stand Dynamics.* McGraw-Hill, New York. 467 pp.

Ong, J.-E., Gong, W.-K., Wong, C.-H., and Dhanarajan, G., 1984. Contribution of aquatic productivity in managed mangrove ecosystem in Malaysia. In: Soepadmo, E., Rao, A.N., and Macintosh, D.J., (Eds), *Proceedings of the Asian Symposium on Mangrove Environment Research and Management,* pp. 209-215, University of Malaya, Kuala Lumpur.

Onuf, C.P., Teal, J.M., and Valiela, I., 1977. Interactions of nutrients, plant growth and herbivory in mangrove ecosystem. *Ecology* **58**:514-526.

Paijmans, K., and Rollet, B., 1977. The mangroves of Galley Reach, Papua New Guinea. *Forest Ecology and Management* **1**:141-147.

Patrick, W.H. Jr., and Mahapatra, I.C., 1968. Transformation and availability to rice of nitrogen and phosphorus in waterlogged soils. *Advances in Agronomy* **20**:323-359.

Ponnamperuma, F.N., 1972. The chemistry of submerged soils. *Advances in Agronomy* **24**:29-96.

Pool, D.J., Lugo, A.E., and Snedaker, S.C., 1975. Litter production in mangrove forests of southern Florida and Puerto Rico. In: Walsh, G.E., Snedaker, S.C., and Teas, H.J., (Eds.), *Proceedings of the International Symposium on Biology and Management of Mangroves*, pp 213-237, University of Florida, Gainesville, Florida.

Popp, M., 1984a. Chemical composition of Australian mangroves I. Inorganic ions and organic acids *Zeitsechrift fÅr Pflanzenphysiologie* **113**:395-409.

Popp, M. 1984b. Chemical composition of Australian mangroves II. Low molecular weight carbohydrates. *Zeitsechrift fÅr Pflanzenphysiologie* **113**:411-421.

Popp, M., Larher, F. and Weigel, P. 1984. Chemical composition of Australian mangroves III. Free amino acids, total methylated onium compunds and total nitrogen. *Zeitsechrift fÅr Pflanzenphysiologie* **114**:15-25.

Putz, F.E., and Chan, H.T., 1986. Tree growth, dynamics, and productivity in a mature mangrove forest in Malaysia. *Forest Ecology and Management* **17**:211-230.

Saenger, P., 1982. Morphological, anatomical and reproductive adaptations of Australian mangroves. In: Clough, B.F., (Ed.), *Mangrove Ecosystems in Australia: Structure, Function and Management,* pp 153-191, Australian National University Press, Canberra.

Scholander, P.F., 1968. How mangroves desalinate seawater. *Physiologia Plantarum* **21**:251-261.

Scholander, P.F., van Dam, L., and Scholander, S.I., 1955. Gas exchange in the roots of mangroves. *American Journal of Botany* **42**:92-98.

Schulze, E.-D., and Hall, A.E., 1982. Stomatal responses, water loss and CO2 assimilation rates of plants in contrasting environments. In: Lange, O.L., Nobel, P.S., Osmond, C.B., and Zeigler, H., (Eds.), *Encyclopedia of Plant Physiology (New Series), Physiological Plant Ecology II Vol 12B*, pp 182-230, Springer-Verlag, Heidelberg.

Singh, V.P., Garge, A., Pathak, S.M., and Mall, L.P., 1987. Pattern and process in mangrove forests of the Andaman Islands. *Vegetatio* **71**:185-188.

Smith, B.N., and Epstein, S., 1971. Two categories of $^{13}C/^{12}C$ ratios for higher plants. *Plant Physiology* **47**:380-384.

Smith, J.A.C., Popp, M., Luttge, U., Cram, W.J., Diaz, M., Griffiths, H., Lee, H.S.J., Medina, E., Schafer, C., Stimmel, K.H., and Thonke, B., 1989. Ecophysiology of xerophytic and halophytic vegetation of a coastal alluvial plain in northern Venezuela. VI. Water relations and gas exchange of mangroves. *New Phytologist* **111**:293-307.

Teas, H.J., 1979. Silviculture with saline water. In: Hollaender, A., Aller, J.C., Epstein, E., San Pietro, A., and Zaborsky, O.R., (Eds.), *The Biosaline Concept. An Approach to the Utilization of Underexploited Resources,* pp 117-161, Plenum Press, New York.

Thibodeau, F.R., and Nickerson, N.H., 1986. Differential oxidation of mangrove substrate by *Avicennia germinans* and *Rhizophora mangle*. *American Journal of Botany* **73**:512-516.

Tomlinson, P.B., 1986. *The Botany of Mangroves.* Cambridge University Press, Cambridge, U.K. 413 pp.

Tsilemanis, C., 1989. The Effect of Salinity on Photosynthesis in Mangroves. M.Envir.Sc. Thesis, Monash University. 49 pp.

Twilley, R.R., Lugo, A.E., and Patterson-Zucca, C., 1986. Litter production and turnover in basin mangrove forests in southern Florida. *Ecology* **67**:670-683.

Ulrich, B., Mayer, R., and Heller, H., (Eds.), 1974. *Data Analysis and Data Synthesis of Forest Ecosytems*. Gîttinger Bodenkundliche Berichte, Vol. 30, 459 pp.

Unesco, 1991. Soils and forestry studies. In: Final Report of the Integrated Multidisciplinary Survey and Research Programme of the Ranong Mangrove Ecosystem. Unesco in cooperation with the National Research Council of Thailand, Paris. pp. 35-81.

Walsh, G.E., 1977. Exploitation of the mangal. In: Chapman, V.J. (Ed.), *Ecosystems of the World 1: Wet Coastal Ecosystems*, pp 347-362, Elsevier, Amsterdam.

Whittaker, R.H., and Marks, P.L., 1975. Methods of assessing terrestrial productivity. In: Lieth, H., and Whittaker, R.H., (Eds.), *Primary Productivity of the Biosphere, Ecological Studies 14*, pp 55-118, Springer-Verlag, New York.

Williams, W.T., Bunt, J.S., and Duke, N.C., 1981. Mangrove litter fall in north-eastern Australia. II. Periodicity. *Australian Journal of Botany* **29**:555-563.

Wyn-Jones, R.B., and Storey, R., 1981. Betaines. In: Paleg, L.G., and Aspinall, D., (Eds.), *Physiology and Biochemistry of Drought Resistance in Plants*, pp 171-204,. Academic Press, Sydney .

9

Nitrogen and Phosphorus Cycles

D.M. Alongi, K.G. Boto and A.I. Robertson

9.1 Introduction

Nitrogen and phosphorus are elements essential to a variety of biological and chemical processes, both at the organismic level (e.g. somatic growth, reproduction) and on the scale of ecosystems. In this chapter, we summarise nutrient cycling in mangrove ecosystems by examining the distribution, transformation and fluxes of various nitrogenous and phosphorus species in mangrove sediments and in tidal creek waters. Nitrogen and phosphorus availability and their role in controlling mangrove plant growth has been dealt with in the previous chapter.

Several excellent reviews and books have presented a comprehensive picture of nitrogen and phosphorus cycling in temperate estuarine and marine ecosystems, including specific processes, such as nitrogen fixation and denitrification (Fenchel and Blackburn, 1979; Carpenter and Capone, 1983). These sources also summarise techniques currently in use to measure these processes.

In general, the dynamics and standing stocks of various nitrogenous and phosphorus compounds (organic and inorganic) in tropical waters and sediments have been poorly studied compared to their temperate counterparts (Furnas, 1991). Most available information for the tropics has come from coral reefs (D'Elia and Wiebe, 1990). Data on standing amounts and fluxes of nitrogen and phosphorus in tropical mangrove waters and sediments are available, but are mainly found in technical bulletins, as are short reviews of the topic (Wiebe, 1987; Agate *et al.*, 1988; Field and Vannucci, 1988). This chapter provides a comprehensive overview of nitrogen and phosphorus dynamics in tropical mangroves and concludes with a nitrogen budget for a tropical mangrove ecosystem (Hinchinbrook Island, northern Australia) where nutrient dynamics have been studied in most detail.

Table 1. Mean concentrations of dissolved nitrogen and phosphorus (μM) in mangrove creeks and estuaries.

Location	NH_4^+	NO_3^-	PO_4^{3-}	DON	DOP	Reference
Vatuwaga, Fiji						
unpolluted	0.60	0.65	1.02	-	-	Nedwell, 1975
polluted	50.94	36.58	13.39	-	-	
Purari River, Papua New Guinea	-	-	1.0-4.16	-	-	Viner, 1982
Kerala estuaries, India	5.50-13.54	4.31-6.23	1.15-3.15	-	-	Sarala Devi *et al.*, 1983
West Coast, Malaysia	1.5-24.7	0-13.8	1.9-17.9	20-50	0-0.8	Nixon *et al.*, 1984
Phuket, Thailand	-	4.0-16.0	0.5-20.0	-	-	Limpsaichol, 1984
Kadinamkulam, India	-	0.0-15.0	0.1-13.0	-	-	Balakrishnan Nair *et al.*, 1984
Kedah, Malaysia	0.0-50	0.1-6.0	0.1-6.0	-	-	Wong, 1984
Pitchaivaram, India	-	-	-	9.7-39.3	0.5-2.75	Balasubramanian and Venugopalan, 1984
Porto Novo, India	-	3.16-30.5	0.37-1.38	-	-	Kannan and Krishnamurthy, 1985
Hinchinbrook Is., Australia	0.1-0.65	0-0.22	0-0.45	2-8	0.1-0.6	Boto and Wellington, 1988
Magdalena Bay, Mexico	-	9-19	1.1-2.0	-	-	Guerrero *et al.*, 1988
Galley Reach, Papua New Guinea	0-5	0-1.0	0-3.9	-	-	Liebezeit and Rau, 1988
Fly River, Papua New Guinea	0.1-1.42	1.79-11.75	0.54-5.26	-	-	Robertson, *et al.*, 1992

9.2 Nitrogen

9.2.1 Concentrations

Waters

Concentrations of dissolved inorganic nitrogen species are low in tropical mangrove waters, within the μM range, and are dominated by ammonium with usually lower amounts of nitrate and nitrite (Table 1). Dissolved organic nitrogen has been measured less frequently, but, as in temperate waters, is often the major form of N (Table 1).

Variations in concentrations among estuaries and mangrove waterways can be ascribed to differences in the extent of freshwater and groundwater input, degree of solar insolation, oxygen availability, and standing stocks and productivity of phytoplankton and bacterioplankton. Dissolved nitrogen concentrations decrease with increasing salinity at the seaward end in mangrove estuaries of the wet tropics (Southeast Asia, India) due to dilution

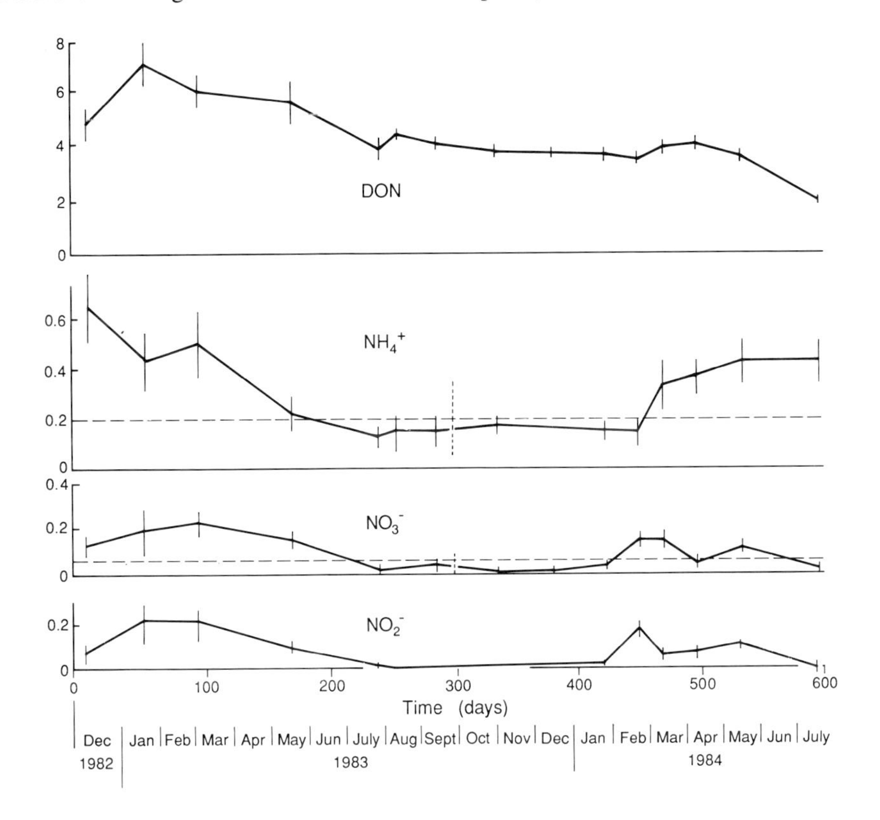

Figure 1. Variations in mean concentrations of dissolved nitrogen concentrations (μM) in Coral Creek, tropical Queensland, Australia over a 20 month sampling period. (Adapted from Boto and Wellington, 1988.)

from monsoonal rains (Nixon *et al.*, 1984; Wong, 1984). Lowest concentrations are generally recorded in the premonsoon season, coincident with high rates of primary productivity (Sarala Devi *et al.*, 1983). In the dry tropics, variations in estuarine nutrient concentrations are greatest over a tidal cycle with highest concentrations occurring at high tide and decreasing during ebb tide (Guerrero *et al.*, 1988; Ovalle *et al.*, 1990).

In northern Australia, some mangrove waterways are influenced almost entirely by tidal action with little input from groundwater or terrestrial runoff, and thus nitrate concentrations are generally lower than in other tropical waterways (Figure 1).

Few investigations have examined particulate nitrogen or dissolved organic nitrogen levels in mangrove waters (Balasubramanian and Venugopalan, 1984; Nixon *et al.*, 1984; Boto and Wellington, 1988). In Malaysian creek waters, Nixon *et al.*, (1984) found low and variable concentrations of PON, ranging from < 10 to 131 µM. DON was not clearly related to salinity but rather to total suspended load. In contrast, they found a sharp decline in DON concentrations with increasing salinity, with values ranging from 20-50 µM. Balasubramanian and Venugopalan (1984) recorded DON concentrations within the same range, with peak concentrations correlating with chlorophyll *a*, suggesting that decomposing plant matter may be responsible. In a mangrove creek in Missionary Bay, Australia, DON

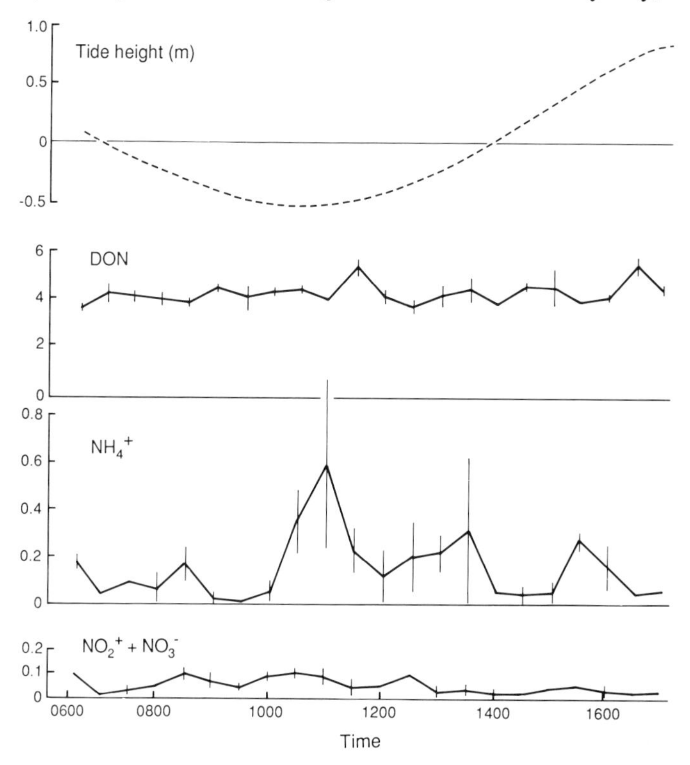

Figure 2. Typical tidal cycle variations in water level and of dissolved nitrogen concentrations in (µM).in Coral Creek, tropical Queensland, Australia (Adapted from Boto and Wellington, 1988.)

concentrations are lower (range 2-8 μM), due to less freshwater input, varying irregularly with time of year but not over tidal cycles (Figures 1 and 2). The lack of large temporal and spatial variations of DON were ascribed to the refractory nature of most of the dissolved organic material in mangrove waters (Boto and Wellington, 1988).

Sediments

As in other tropical marine deposits, dissolved inorganic nitrogen concentrations are low in mangrove soils, typically within the μM range and composed mostly of ammonia with lesser amounts of nitrate and nitrite (Table 2). In contrast, DON concentrations are high (> total DIN) and comparable to concentrations in temperate aquatic sediments.

Concentrations of NH_4^+ are higher in mud than in muddy sand and sand and are influenced by degree of tidal wetting, plant uptake and seasonal changes in microbial decomposition, temperature and rainfall (Boto, 1982, 1984; Boto *et al.*, 1985). The effects of mangrove growth on dissolved inorganic nitrogen concentrations is best illustrated in Figures 3 and 4 . Figure 3 shows typical depth profiles for ammonium-N during periods of high and low plant productivity. The rate of NH_4^+ uptake by mangroves exceeds regeneration rates (Boto and Wellington, 1984). Figure 4 shows differences in ammonium-N and nitrate + nitrite as affected by the presence and absence of mangrove roots (Alongi, unpublished data).

Table 2. Mean concentrations of dissolved nitrogen and phosphorus (μM) in mangrove porewaters.

Location	**Depth range (cm)**	**NH_4^+**	**NO_3^-**	**PO_4^{3-}**	**DON**	**DOP**	**Reference**
Florida Bay, USA	0-60	3-760	-	0.1-13	-	-	Rosenfeld, 1979
Mangrove Bay, Bermuda	0-14	<10-114	-	0.2-2.1	-		Hines and Lyons, 1982
Indian River, Florida	NA	1-23	<2	12-176	-	-	Carlson *et al.*, 1984
Hinchinbrook Is. Australia	0-100	0.5-120	<0.01-0.3	0.1-35.2	0-300	0.1-3.5	Boto and Wellington, 1984
							Iizumi, 1986 Stanley *et al.*, 1987 Alongi, unpublished data
Phuket, Thailand	0-10	0-75	0-21	-	-	-	Kristensen *et al.*, 1988

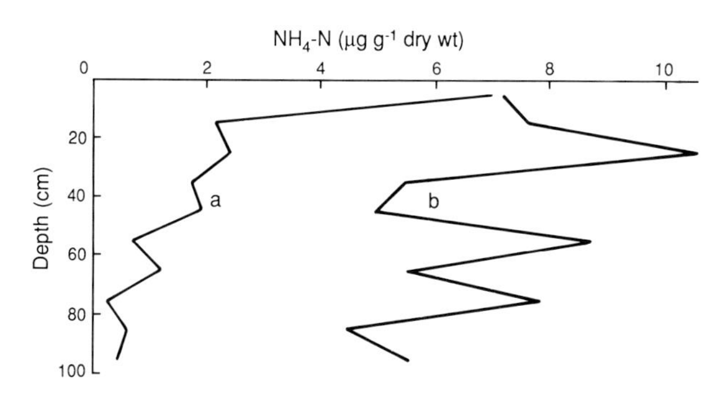

Figure 3.Variations in ammonium-N with sediment depth in Missionary Bay mangroves, northeastern Australia: a = a period of rapid plant growth; b = a period of little plant growth. (Adapted from Boto and Wellington, 1984.)

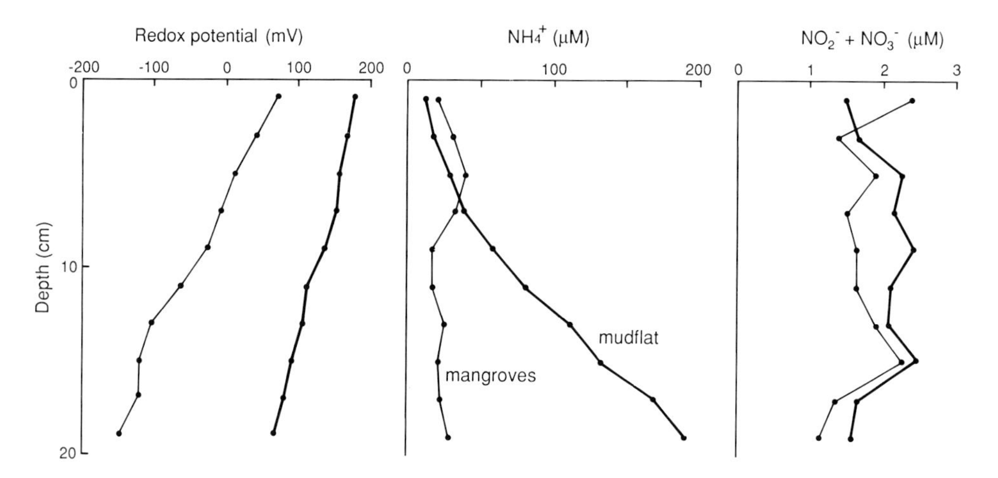

Figure 4. Depth variations in redox potential, ammonium-N and nitrite + nitrate in sediment porewaters in Missionary Bay mangroves and in an adjacent mudflat, April 1990(Alongi, unpubl. data.)

Concentrations of NH_4^+ and nitrate + nitrite are significantly lower in sediments in which mangrove roots are found (Figure 4). This reduction probably reflects uptake by the plants. Redox potential is more negative within the mangroves, despite the translocation of oxygen by below-ground roots and rhizomes (see Chapter 8, this volume). Vertical profiles of porewater nutrients in tropical mangroves indicate similarly irregular depth patterns, (Rosenfeld, 1979; Hines and Lyons, 1982; Kristensen *et al.*, 1988). These patterns are contrary to most intertidal temperate deposits where dissolved nitrogen levels normally increase with sediment depth due to increasing anoxia.

Vertical profiles of porewater ammonium in aquatic sediments are usually affected by adsorption onto sediment solids, particularly clays (Mackin and Allen, 1984). On Hinchinbrook Island in tropical Australia, adsorption capacity of mangrove sediments is low (Alongi, unpublished data). In most marine and estuarine muds, the adsorption coefficient is usually ~ 1.2 (Mackin and Allen, 1984), whereas in Hinchinbrook mangrove muds, the coefficient is much lower averaging ~ 0.5. This means that of the total sedimentary ammonium, one half more is adsorbed onto the sediment than is dissolved in the interstitial water. The low values are in good agreement with the low k values obtained by Mackin and Allen (1984) for tropical mangrove and *Thalassia* bed sediments in Florida. They ascribed the anomalous low k values to sediments with high $CaCO_3$ content, which have low cation exchange capacity and low clay content. Hinchinbrook muds are nearly devoid of $CaCO_3$, and the mud fraction is dominated by less adsorbative silts than clay (Alongi, 1988), thus accounting for the low concentration of adsorbed NH_4^+. Whether or not low adsorbtive capacity of NH_4^+ is typical of other mangrove soils awaits further analysis from other sites.

Concentrations of total extractable nitrogen vary with sediment type (higher in finer deposits) and reflect the amount and distribution of fine roots (Boto and Wellington, 1984; Table 3). Total N concentrations generally range from 0.02 to 0.40% of sediment dry weight (Table 3). Concentrations of N in surface sediments in the wet tropics is greatly influenced by the intensity and duration of monsoonal rains (Japtap, 1987; Alongi, 1988). In Goa, Japtap (1987) found maximal concentrations during the monsoon season due to deposition of huge volumes of suspended plant debris. In contrast, Alongi (1988) found that monsoonal rains scoured surface sediments, lowering N values significantly in northern Australia.

Vertical profiles exhibit subsurface maxima in total extractable nitrogen usually between 20-40 cm depth, reflecting the contribution of fine roots (Boto and Wellington, 1984). The contribution of particulate organics to the total extractable N fraction has been examined only

Table 3. Mean concentrations of total extractable nitrogen and phosphorus in some tropical mangrove sediments. Values are percentage of dry weight.

Location	Depth range (cm)	TN(%)	TP(%)	Reference
Sierra Leone, West Africa	0-10	-	0.08-0.16	Hesse, 1961, 1963
Florida Bay, USA	0-50	0.2-0.4	-	Rosenfeld, 1979
Hinchinbrook Island, Australia	0-100	0.1-0.40	0.022-0.054	Boto and Wellington, 1984
Sunderban, India	0-25	0.06-0.09	0.01-0.048	Sahoo *et al.*, 1985
Queensland estuaries, Australia	0-2	0.02-0.40	0.011-0.067	Alongi, 1987; unpublished data
Goa, India	0-2	0.02-0.16	-	Japtap, 1987
Iriomote Island, Japan	0-5	0.05-0.09	0.022-0.038	Kuraishi *et al.*, 1985
Phuket, Thailand	0-10	0.08-0.10	-	Kristensen *et al.*, 1988

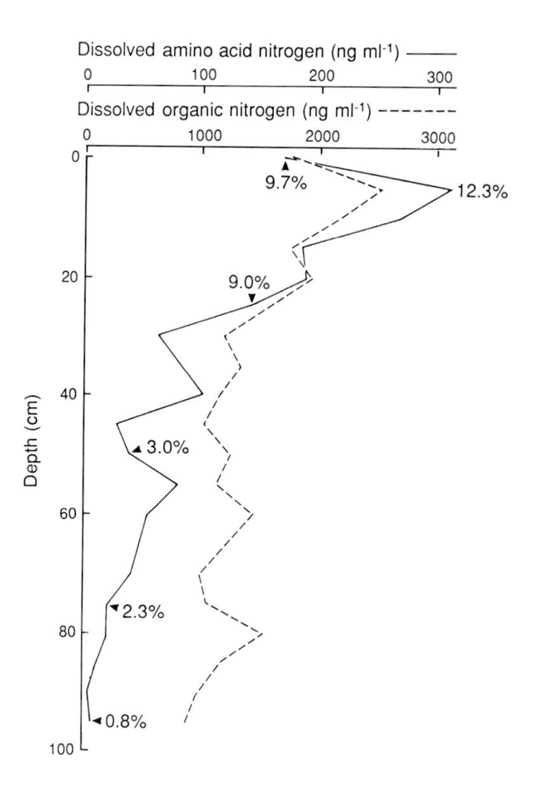

Figure 5. Depth variations in porewater concentrations of total DON and dissolved amino acid nitrogen and percentage contribution of dissolved free amino acids to DON in Missionary Bay mangroves. (From Stanley *et al.*, 1987.)

by Rosenfeld (1979) in Florida and Hatcher *et al.*, (1982) in Bermuda. They calculated that amino acids and humic materials constitute approximately 30% and nearly 25-75% respectively of the organic nitrogen content. In the less organic-rich mangrove forests of the Australasian region, it is less likely that amino acids constitute such a large fraction of the total organic nitrogen pool. In the mangrove sediments of Hinchinbrook Island, amino acids account for, on average, only 6% of the total DON pool (Figure 5). As in other marine deposits, humic nitrogen is likely to be a large proportion of total extractable N in most mangroves. Even though roots and other plant-derived detritus appear to be the major sources of nitrogen, the contribution of specific compounds such as the essential amino acids and protein remains unknown.

9.2.2 Transformation Processes

The nitrogen cycle is mediated primarily by biological, mostly microbial, processes rather than by chemical means. By virtue of a wide range of oxidation states (-3 for ammonia to +5 for nitrate), nitrogen can serve as a reductant or an oxidant in the diagenesis of organic matter in addition to its role as a nutrient in microbial and plant assemblages. Figure 6 summarises the major transformations and pools of nitrogenous compounds important in coastal pelagic and benthic energetics. General aspects of nitrogen cycling in the estuarine and marine environment have been summarized by Fenchel and Blackburn (1979) and Carpenter and Capone (1983).

Ammonification (organic N mineralization)

Ammonification is the process in which the end product of organic nitrogen mineralization (hydolyzation and catabolization of proteins and polynucleotides) is the liberation of ammonium (step 1 of Figure 6). Nearly all organic nitrogen breakdown occurs via this pathway in organisms ranging in size from bacteria to vertebrates in which ammonium is the end product of excreta.

Indication of ammonification in the water column comes mainly from estimates of ammonium excretion by protozoan and metazoan grazers, and from patches of locally elevated concentrations. Direct measurements of ammonium liberation by protozoa, zooplankton and fish in other habitats have been made (Taylor, 1982; Bidigare, 1983; Meyer and Schultz, 1985; Ip *et al.*, 1990), but not in mangrove creek waters. The studies of Ikeda *et al.*, (1982a, b) on ammonium excretion by coastal zooplankton of the Great Barrier Reef offer the only clues as to the contribution of excreta by planktonic organisms to nutrient cycles in mangrove waters. Ammonium excretion rates of various organisms ranged from 1 to 415 $ngN.animal^{-1}.h^{-1}$ over body sizes ranging from ~1 to 10,000 $\mu g.animal^{-1}$, respectively (Ikeda *et al.*, 1982a). These values are lower on an equivalent body weight basis than excretion rates for temperate zooplankton, but Ikeda *et al.*, (1982b) calculated that the excretion rates of net zooplankton and microzooplankton combined could supply from 9 to 29% of the nitrogen required for phytoplankton production. Excretion rate data for tropical

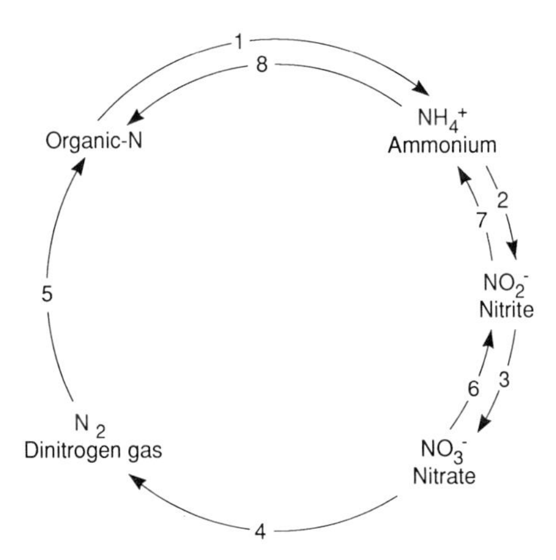

Figure 6. The nitrogen cycle with major pools and transformations. Major transformations processes as numbered are: 1 = ammonification; 2 = nitrification (ammonium oxidation); 3 = nitrification (nitrite oxidation); 4 = denitrification and dissimilatory nitrate reduction; 5 = nitrogen fixation; 6 = nitrate reduction; 7 = nitrite reduction; 8 = immobilization and assimilation. Steps 6 and 7 are assimilatory processes. (Adapted from D'Elia and Wiebe, 1990.)

coastal nekton such as fish and prawns are rare, but data are available for prawns similar to those inhabiting mangroves. Dall and Smith (1986) examined respiration and ammonium excretion in fed and immature *Penaeus esculentus*, a seagrass-associated prawn in subtropical and tropical Australia. Average daily excretion rates for ammonia-N ranged from 5-30μg.g wet wt^{-1} of prawn.

Zooplankton and nekton are likely to contribute substantially to the ammonium pool in mangrove creek-waters, but estimates of their standing stocks and excretion rates are too few to allow reasonable extrapolations at present. Bacterioplankton activities, as in other pelagic systems, must be the major pathway of nitrogen transformation via ammonification, but the process at the bacterial level has yet to be quantified in tropical mangrove waters.

In mangrove sediments, ammonium excretion by microbes, meiofauna and macrofauna must occur, but rates are almost wholly unknown. Bacterial activity must regulate most of the available ammonia pool, particularly in deeper subsurface sediments (>50 cm) devoid of other biota, as they do in other aquatic sediments. The presence of nematodes did not affect rates of NH_4^+ regeneration from mangrove (*A. marina, R. stylosa*) litter and Tietjen and Alongi (1990) attributed the lack of an effect to the inability of nematode populations to attain sufficient numbers because of poor nutritional quality of the litter. Poor food (low N) quality may also help to explain the low and erratic ammonium excretion rates of the mangrove crab, *Sesarma messa* fed a diet of fresh and aged *R. stylosa* leaf litter (Gissler, unpublished data).

Rates of ammonification have been measured directly in mangrove sediments (Hesse, 1961a, b; Rosenfeld, 1979; Iizumi, 1986; Shaiful *et al.*, 1986 and are within the range of values measured in salt marsh sediments). Rosenfeld (1979) reported net ammonium release in a Florida mangrove sediment of 1064 $ngN.g^{-1}$ sediment $DW.d^{-1}$. In Bermudan mangrove sediments, Hines and Lyons (1982) estimated rates of ammonium production to be on the order of 530 to 800 $ngN.g^{-1}.d^{-1}$, based on measurement of sulphate reduction. On Hinchinbrook Island, Iizumi (1986) measured ammonia production in anaerobic sediment slurries and found rates ranging from 420 to 1820 $ngN.g^{-1}.d^{-1}$ (Figure 7). In Malaysia, Shaiful *et al.*, (1986) measured highest rates in surface sediments, ranging from 0 to 0.375 $\mu M\text{-}N.g^{-1}.d^{-1}$, and being undetectable below the top 3-4 cm. Similar rates were measured earlier with a similar incubation method by Hesse (1961a, b) in Sierra Leone muds.

These ammonification rates (and standing amounts in the interstitial water) are net estimates, that is, gross production minus incorporation by plants and bacteria (see next paragraph). True production rates are likely to be much higher than those observed, but are more difficult to measure, involving the use of $^{15}NH_4^+$ in an isotope dilution technique (see Blackburn, 1986).

Ammonium immobilization and assimilation (step 8 in Figure 6), mainly by microbes (bacteria, fungi) and plants, always accompanies and counteracts the mineralization process. As pointed out by Boto (1982), the extent to which these processes balance each other is highly dependent on the C:N ratio of the decomposing organic matter. Substances rich in

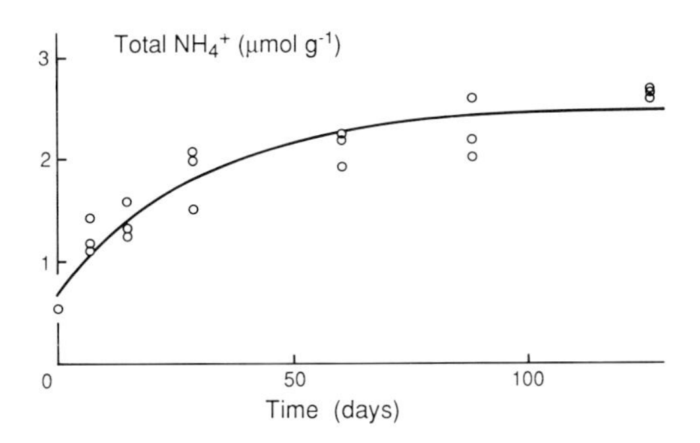

Figure 7. Changes in ammonium content (μmol $NH_4^+.g^{-1}$) in sediments (0-5 cm, mid-intertidal) incubated for over 120 days, Missionary Bay. (From Iizumi, 1986.)

nitrogen (low C:N) favor net mineralization, whereas those poor in nitrogen (high C:N) favor net immobilization. The generally high C:N ratio of mangrove tree components (e.g. leaves, wood), high rates of bacterial cell production (Alongi, 1988) and low NH_4^+ flux across the sediment-water interface (Kristensen *et al.,* 1988) indicate that rates of immobilization may be high in mangrove sediments (see section 9.6.3 and Melillo *et al.*, 1982 for terrestrial forest comparisons).

Nitrification

Nitrification is the process by which ammonia is oxidized to nitrite (ammonia oxidation, step 2 of Figure 6) which, in turn, is oxidized to nitrate (nitrate oxidation, step 3 of Figure 6). In sediments these processes occur close to the sediment-water interface in the oxidized lining of animal burrows and within the oxidized region of the rhizosphere (Boto, 1982). Comparatively few estimates have been made of nitrification in marine sediments, mainly because it is very difficult to measure (Kaplan, 1983). Nitrate, the final product of the process, disappears rapidly as it is involved in many biological reactions, such as mangrove plant growth (Boto *et al.,* 1985).

The only empirical measurements of nitrification in tropical mangrove sediments have been conducted by Iizumi (1986) in mangroves on Hinchinbrook Island and by Shaiful *et al.,* (1986) in Selangor, Malaysia. Iizumi (1986) incubated sediments under aerated conditions using N-serve to inhibit nitrification and measure net increase in ammonia over time. Nitrification was low (~ 14ng $N.g^{-1}.d^{-1}$), perhaps because nitrification was inhibited by soluble tannins, (Kimball and Teas, 1975) which are abundant in mangrove porewater (Boto *et al.*, 1989). Tannins are known to inhibit the growth and activity of nitrifying bacteria. In Malaysian mangrove muds, Shaiful *et al.,* (1986) measured nitrification rates (using N-serve) ranging from 0 to 0.22 μmol $N.g^{-1}.d^{-1}$.

Assimilatory uptake of N (steps 6 and 7; Figure 6) counterbalances oxidation processes, whereby uptake of nitrate occurs for the growth of mangroves and bacterial cells (see Chapter 8, this volume). The only empirical data of nitrate reduction in mangroves comes from a study conducted in low mangroves in Malaysia (Shaiful, 1987). Shaiful found highest

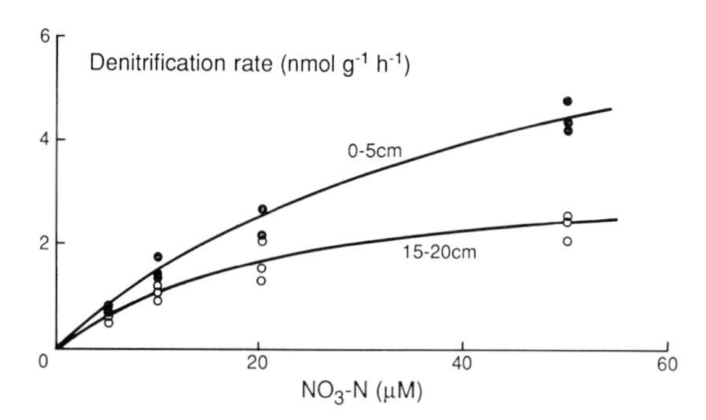

Figure 8. Denitrification activity (μmol.g^{-1}.h^{-1}) of sediments in the mid-intertidal mangroves of Missionary Bay in which Michaelis-Menten kinetics (line on graph) was applied. (From Iizumi, 1986.)

rates ranging from 0.46 to 1.34 mmol NO_3 reduced .m^{-2}.hr^{-1} in the top 2 cm soil layer. The soils were capable of reducing 88.7-93.5% of nitrate derived from nitrification.

Denitrification and dissimilatory nitrate reduction

Dissimilatory nitrate reduction refers to the process in which nitrate serves as the terminal election acceptor of respiration in facultative - and some obligate-anerobic bacteria, the energy yield of which is used for other cellular reactions. Many such bacteria reduce nitrate to nitrite as a terminal end-product, although a second pathway terminating in the production of ammonia has been recognized (Sorensen, 1978). As no data exists for this process in tropical mangroves, it will not be discussed further.

Denitrification (step 4 of Figure 6) the reduction of nitrate, in which N_2 and N_2O gases are end products, occurs in mangroves. Nitrate discharged from a sewage treatment plant in Fiji mangroves, and not taken up by phytoplankton, decreased by 30% by the time is passed through the estuary, indicating potentially large losses of N via denitrification (Nedwell, 1975). Nedwell reported 180-600 μM of nitrate for the K_m constant of denitrification, equating to a reaction rate of from 26.2-87.6 mgN.m^{-2}.d^{-1}. In an unpolluted mangrove stand on Hinchinbrook Island, Iizumi (1986) found low rates of denitrification (K_m=10-69μM NO_3^-) with a mean value of 0.18mgN.m^{-2}.d^{-1}. The activity showed Michaelis-Menten kinetics (Figure 8).

Nitrogen fixation

Nitrogen fixation (step 5 of Figure 6) is the best-studied transformation process of nitrogen in mangroves (Zuberer and Silver, 1975, 1978; Kimball and Teas, 1975; Gotto and Taylor, 1976; Zuberer and Smith, 1979; Potts, 1979; Uchino *et al.*, 1984; van der Valk and Attiwill, 1984; Hicks and Silvester, 1985; Iizumi, 1986; Myint, 1986; Mann and Steinke, 1989; Boto and Robertson, 1990). Nitrogen fixation is carried out by prokaryotes, mostly bacteria and cyanobacteria which can fix gaseous nitrogen to form ammonia via nitrogenase activity. These microbes are called diazotrophs and include bacteria of the genera *Clostridium, Azotobacter, Bacillus* and *Desulphovibrio* and cyanobacteria of the genera *Microcoleus, Oscillatoria* and *Voucheria* (Potts, 1984).

Table 4. Mean rates of nitrogen fixation (μgN.m^{-2}.d^{-1}) in a variety of microhabitats in some mangrove forests measured using the acetylene reduction technique. Values in parantheses are ranges.

Location	N_2 fixation rate	Reference
Southern Florida[1],		
low intertidal fringe	254.6	Kimball and Teas, 1975
high intertidal scrub	(27.1-310.3)	
Avicennia forest	118.3	
Laguncularia forest	189.4	
Southern Florida[2],		
anoxic sediment	262	Zuberer and Silver, 1970, 1975
prop root sediment	(383-725.8)	
pneumatophor sediment	120.9	
litter	35,784	
Southern Florida[3],		
fresh *R. mangle* leaves	0	Gotto and Taylor, 1976
aged *R. mangle* leaves	264.1	
Sinai[4]		
on pneumatophors	(120.6-126.2)	Potts, 1979
sediment surface	8.6	
0.5 cm below sediment surface	2.4	
2 cm below surface	1.2	
4 cm below surface	2.4	
Iriomote Is., Okinawa[2]		
bark of *B. gymnorrhiza*	(16.1-24.2)	Uchino *et al.*, 1984
Southern Australia[3]		
A. marina litter	(64-758)	van der Valk and Attiwill, 1984
dead *A. marina* roots	(11.0-22.6)	
live *A. marina* roots	7.0	
sediment surface (0-1 cm)	15.8±12.2	
New Zealand		
sediments, *A. marina* stand[5]	24.8	Hicks and Silvester, 1985
creek bank w/o mangroves[5]	9.9	
decomposing *A. marina* leaves[3]		
10°C	22.2(±12.4)	
20°C	39.2(±14.4)	
30°C	82.8(±16.6)	
Hinchinbrook Is., Australia[2]		
sediments	(1935.4-3427.2)	Iizumi, 1986
C. tagal roots	(2016-10,080)	

Table 4 continued/

Location	**N_2 fixation rate**	**Reference**
Batu Maung, Malaysia[2]		
mudflat	0.038(0.06-0.17)	Myint, 1986
mangrove sediments	0.127(0.10-0.18)	
mud w *Oscillatoria* bloom	0.164(0.14-0.24)	
mud w. *Voucheria Microcoleus* bloom	2.607(0.58-9.9)	
Mgeni estuary, S. Africa[5]		
cyanobacteria mats	(0-2520)	Mann and Steinke, 1989
pneumatophore, *A. marina*	(0-252)	
Hinchinbrook Is., Australia		
low intertidal sediments[5]	1.5(0-3.22)	Boto and Robertson, 1991
mid intertidal sediments[5]	0	
high intertidal sediments[5]	1.45(0-3.34)	
prop roots, low intertidal[6]	7.33(0-22.0)	
prop roots, mid intertidal[6]	58.5(34.9-72.7)	
prop roots, high intertidal[6]	2.77(0-8.3)	
aged logs, many wood-borers[7]	0.37±0.15	
aged logs, some wood-borers[7]	0.21±0.10	
aged logs, no live wood-borers[7]	0.25±0.10	

[1]sediments measured to depth of 104 mm
[2]$\mu gN.g^{-1}$(wet wt).d^{-1}
[3]$\mu gN.g^{-1}$(dry wt).d^{-1}
[4]$\mu gN.cm^{-3}.d^{-1}$
[5]$mgN.m^{-2}.d^{-1}$
[6]$\mu g.N.$(per whole prop root).d^{-1}
[7]$mg.kg^{-1}$(wet wt log).d^{-1}

Cyanobacterial mats on exposed sediments and cyanobacteria associated with prop roots and pneumatophores are the main loci for nitrogen-fixing activity (Potts, 1984; and see Table 4). Studies in Florida established that roots and litter have higher fixation rates than sediments (Gotto and Taylor, 1976; Zuberer and Silver, 1979). This is particularly true for aged leaf litter of *Rhizophora mangle* (Gotto and Taylor, 1976), which invariably supports an abundant microbial biocoenosis. Bacteria attached mainly to root surfaces may have access to root exudates and sloughed cell material as sources of organic matter for growth and nitrogen fixation.

Nitrogen-fixing ability of above-ground components, such as bark, logs, leaf litter, pneumatophores and prop roots, is related to the abundance of cyanobacteria (Potts, 1979). In the Sinai, Potts (1979) found high rates of N_2 fixation on pneumatophores covered with blue-greens whereas rates were significantly lower on and within the sediments without cyanobacteria. In Malaysian forests, highest nitrogen fixation rates were observed on cyanobacterial mats on mudflats adjacent to mangroves (Myint, 1986). Fixation rates on mangrove and mudflat sediment surfaces were equally low, although the mean rate in mangroves (0.127 $\mu gN.g^{-1}.d^{-1}$) was considerably greater than on the mudflat (0.038 $\mu gN^{-1}.d^{-1}$).

In contrast, Boto and Robertson (1990) observed rates of N_2 fixation on algal mats on a saltpan adjacent to mangroves equivalent to those on bare saltpan sediments and on the sediment surface within the forests on Hinchinbrook Island (Australia). The anomalously low fixation rates on the algal mats was attributed to a lack of nitrogen-fixing, heterocystous algae. Prop roots and their associated algae were the only components to exhibit significant diurnal activity, with night-time rates being much greater than those measured in the day. Most of the prop root-associated activity occurred in the low and mid-intertidal forests on the island, with rates ranging from 22.0-72.7 μgN.whole prop root.d^{-1}. Logs lying on the forest floor were not a source of significant N_2-fixation, suggesting that wood-boring teredinids do not contribute greatly to the fixation process. Rates of N_2-fixation in mangrove forests are low compared to other estuarine and marine habitats, probably due to low light levels under the forest conopy and low standing amounts of cyanobacteria (see review of Capone, 1983).

9.3 Phosphorus

9.3.1 Concentrations

Waters

Concentrations of dissolved inorganic and organic phosphorus in mangrove waters are low, within the μM range (Table 1). Dissolved inorganic phosphorus (soluble reactive phosphate) exists mainly as a nutrient salt (HPO_4^{2-}) at the pH of seawater. Values in unpolluted mangrove waterways range from < 0.1 to ~ 20 μM. Measurements of dissolved organic phosphorus are rare (Table 1), but values are very low, ranging from below detection limits to ~ 3 μM.

As in the case of nitrogen, differences in concentrations among mangrove estuaries can be ascribed to local characteristics, such as the extent of freshwater and groundwater input, and the productivity of the biota. Variations in dissolved P concentrations may mirror changes observed for dissolved N. For instance, in mangrove estuaries of the wet tropics, dissolved and total phosphorus concentrations decrease with increasing salinity (Nixon *et al.*, 1984; Wong, 1984; Liebezeit and Rau, 1988; Robertson *et al.*, 1992). Patterns of P in estuaries subjected to heavy monsoonal rainfall are also nearly identical to those for nitrogen. Lowest P concentrations are usually coincident with dry periods when primary production is highest (Sarala Devi *et al.*, 1983; Balakrishnan Nair *et al.*, 1984).

In some wet tropical river systems, the extremes of rainfall and runoff shift the importance for biological availability from dissolved orthophosphate to P tied to the suspended load. For instance, in the estuary of the Purari river of Papua New Guinea (where rainfall is extremely high) most of the available P is bound, either as particulates or as phosphates adsorbed to clays, in the suspended load, rather than dissolved free in the water column (Viner, 1982).

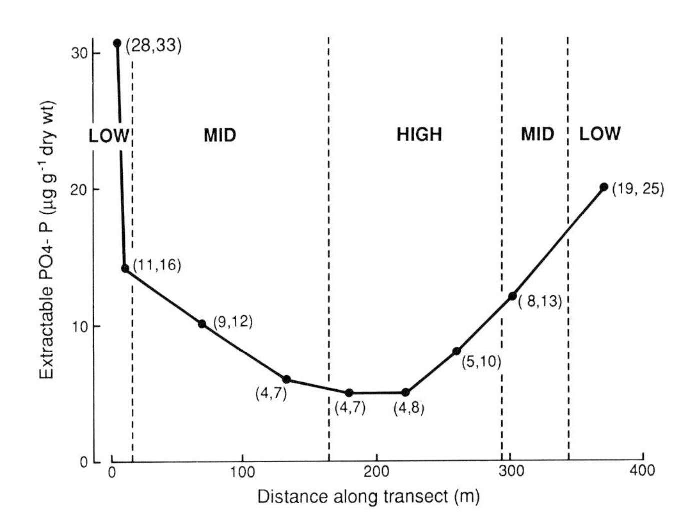

Figure 9. Variation of extractable PO_4-P (averaged over 1 m depth and 14 month period) with tidal height (low, mid, high intertidal zones) along a transect in the Missionary Bay mangroves, tropical Queensland. Numbers in parentheses refer to the maximum and minimum values of PO_4-P recorded at each site. (Modified from Boto and Wellington, 1984.)

In the dry tropics, concentrations vary mostly over a tidal cycle with highest concentrations occurring at high tide and decreasing over an ebb tide (Guerrero *et al.*, 1988). On Hinchinbrook Island, where there is little freshwater input, dissolved organic and inorganic P concentrations are lower than those in wet tropical estuaries (Boto and Wellington, 1988). Monthly variations were similar to those for nitrogen, although phosphate concentrations were significantly higher at night when photosynthesis shuts down. DOP levels were very low (Table 1), but Boto and Wellington (1988) estimated that DOP and DIP accounted for 17.9% and 6.3% of net forest primary production requirements, respectively.

Only two other investigations have been made which examine dissolved organic phosphorus in mangrove waters (Balasubramanian and Venugopalan, 1984; Nixon *et al.*, 1984). In a south Indian estuary, Balasubramanian and Venugopalan found that the seasonal pattern of DOP was similar to that of DOC and DON, being lowest in the monsoon and post monsoon periods. This pattern was attributed to mineralization of DOP and dilution of accumulated organics by flood waters. Concentrations in bottom waters were significantly higher than those at the surface, correlating positively with salinity and chlorophyll *a*. In a Malaysian mangrove estuary, particulate organic P levels ranged from 1 to 17.8 μM, but were related to total suspended load rather than to salinity. DOP concentrations ranged from 0 to ~ 0.8 μM with maximum levels within the salinity range of 5 to 22‰ (Nixon *et al.*, 1984).

Sediments

Dissolved and particulate phosphorus concentrates in mangrove sediment have been measured rarely (Tables 2 and 3), but the available data indicates low dissolved

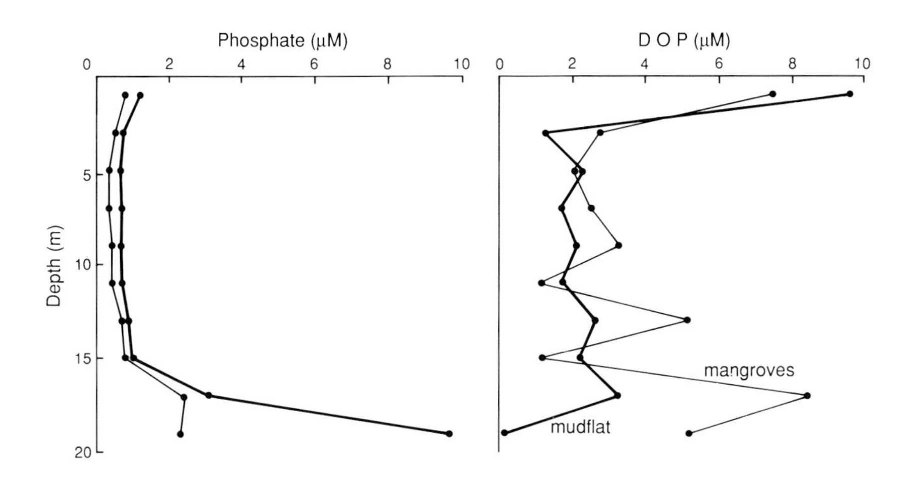

Figure 10. Variations in porewater PO_4^{3-} and DOP (μM) over 0-20 cm depth horizon in low intertidal mangroves and in adjacent mudflat, Missionary Bay, April 1990. (Alongi, unpubl. data.)

concentrations, generally <40 μM for DIP and <4 μM for DOP.

Concentrations vary over time and intertidal position reflecting seasonal effects of plant uptake and microbial growth, temperature, rainfall, degree of anoxia and sediment type (Boto, 1982, 1984). Mean concentrations of extractable phosphate across a mangrove forest gradient decreased with tidal height in tropical Australia (Figure 9), probably corresponding to increasing anoxia whereby available P is in the reduced form and bound to insoluble iron sulphides (Boto and Wellington, 1984). P becomes limiting only in elevated areas and fertilization with P results in a positive mangrove response (Boto and Wellington, 1983). No such response to P enrichment occurred in low intertidal forests, suggesting that P is not limiting in these areas. This conclusion is supported by porewater data from Hinchinbrook Island (Figure 10) in which differences in PO_4^{3-} and DOP concentrations between low intertidal mangrove and adjacent unvegetated mudflat sediments are not evident.

Total P content of mangrove sediments appears to fall within the range of 100-1600 $\mu g.g^{-1}$ (Table 3). Mangrove soils are expected to contain a high proportion of organic P compounds due to their generally high organic matter content (Boto, 1988). Hesse (1962, 1963), for instance, found that 75-80% of the total extractable P was organic. Boto (1988) has pointed out that much of this organic P is in the phytate form and bound to humic compounds, as has been found in lake sediments, and is probably not readily available for microbial and mangrove plant nutrition.

Although organic P is the major fraction, the inorganic phosphates probably represent the largest potential pool of plant-available, soluble reactive phosphorus (Boto, 1988). Most of the inorganic P in mangrove sediments is either bound in the form of Ca, Fe and Al phosphates or as soluble reactive phosphorus adsorbed onto, or incorporated into, hydrated Fe and Al sesquioxides (Table 5). Total organic P concentrations, proportionally greater in surface (0-25 cm) sediments, reflect the influence of roots, whereas the inorganic fractions,

Table 5. Forms of inorganic phosphorus (μgP.g^{-1} sediment dry wt) in mangrove sediments.

Fraction	**Victoria, Australia (Attiwill and Clough, 1978)**	**Sierra Leone (Hesse, 1962)**
Water-soluble P	0	2
Al phosphate	2-5	6
Fe phosphate	40-100	29
Ca phosphate	20-32	25
Fe phosphate (reductant soluble)	400-1100	0
Occluded Fe and Al phosphate	2-15	6

mainly Fe-P, proportionally and in real terms increase gradually with depth reflecting the influence of increasing anoxia particularly below the root layer (Figure 11).

9.3.2 Transformation Processes

Phosphorus in nature occurs mainly as phosphates with P having an oxidation number of +5, giving rise to a relatively simple cycle (Figure 12). General aspects of phosphorus and nitrage cycling in estuarine and marine environments can be found in Nixon (1980) and Valiela (1984).

Transformation of P in marine and estuarine systems, as summarized by D'Elia and Wiebe (1990) for coral reefs, can be categorized as two sets of processes; biotic (assimilation, and excretion and hydrolysis) and abiotic (precipitation, adsorption and chemisorption, and dissolution and desorption) activity.

Chemical processes (precipitation, sorption, and dissolution)

Available forms of P for organisms, mainly as soluble reactive phosphorus, are derived from several sources: (1) natural weathering and erosion of phosphate minerals (e.g. apatite), (2) solubilization of metal phosphate precipitates or phosphate adsorbed onto clays, (3) excretion from bacteria, protozoa and larger organisms, (4) mineralization of organic phosphates and (5) anthropogenic sources such as sewage effluent and agricultural runoff.

Mineral phosphates (e.g. the apatites, iron and aluminium phosphates) are very insoluble and provide only a limited supply of P unless solubilization is accelerated via chemical or biological mediation. Through chemical and physical interactions, even the "readily available" phosphate ions in solution ($H_2PO_4^-$, HPO_4^{2-}, PO_4^{3-}) are prone to be strongly adsorbed to clay particles, particular to clay containing iron or manganese oxyhydroxides. Clays such as kaolinite, abundant in tropical soils and sediments (Kennett, 1982), are particularly efficient in phosphate adsorption. Phosphate immobilization by precipitation as Ca, Fe and Al salts and/or very strong adsorption on clays, results in a significant net removal of P into the large mineral pool, limiting the available pool to living systems.

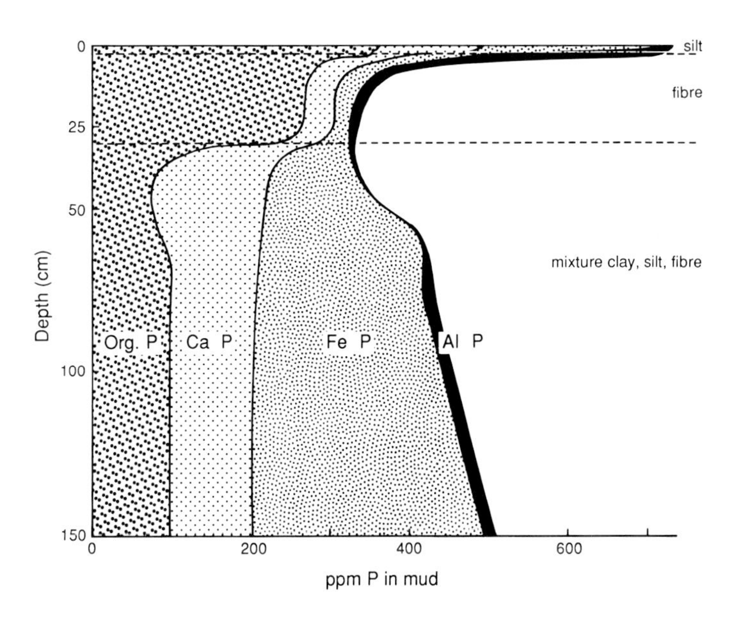

Figure 11. Vertical distribution of the various forms of particulate P (ppm) with depth and changes in roots and sediment type, Sierra Leone (From Hesse, 1963).

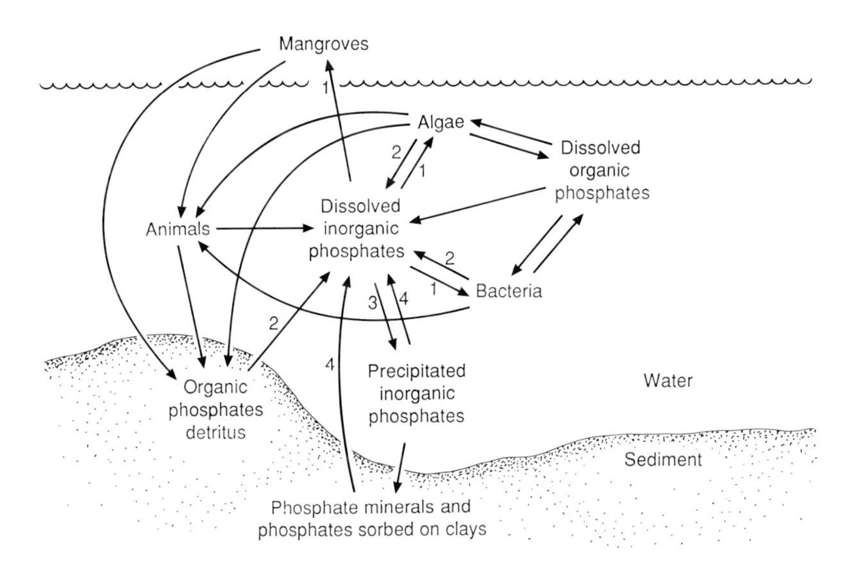

Figure 12. The phosphorus cycle with major pools and transformations. Major transformation processes as numbered are: 1 = assimilation; 2 = excretion and hydrolysis; 3 = precipitation, adsorption and chemisorption; 4 = dissolution and desorption. (Modified from Fenchel and Blackburn, 1979 and Boto, 1988.)

Soluble reactive phosphate is readily assimilated by bacteria, algae and higher plants, including mangroves. Most dissolved P in aquatic systems consists of various organic phosphates (primarily phosphate esters originating from living cells), which are often resistant to hydrolysis and therefore of limited availability.

Phosphate is immobilized in mangrove sediments (Hesse, 1962, 1963; Boto and Wellington, 1983). In experiments on mangrove muds from Sierra Leone, Hesse (1962, 1963) found that immobilization (as adsorption) of added inorganic phosphorus (30 ppm) by fresh (wet) and air-dried sediment was rapid, with equilibrium being reached within 1 hour. The dried mud adsorbed significantly more phosphorus, but ~ 90% of the adsorption took place within 5 min. In another experiment in which the distribution of immobilized phosphorus was monitored after addition of inorganic P, the iron-bound and aluminium-bound phosphorus concentrations doubled immediately, whereas calcium-bound P decreased by nearly one-third. After 6 months, the iron phosphate increased by about one-third, but the other two bound forms decreased sharply after about 1 month.

In similar experiments, Boto and Wellington (1983) found that superphosphate added to forest plots to a total quantity of 400 $kg.ha^{-1}$ over 12 months was rapidly immobilized to the extent that sediment P concentrations only increased to ~ 65 $\mu g.g^{-1}$ from ambient levels of 20-30 $\mu g.g^{-1}$. This increase was not large considering that 4 times the P concentrations normally employed in agricultural systems was used. In general, the capacity of mangrove soil to immobilize phosphate depends on the amount of organic matter, its C/P ratio, and the type and amount of clay minerals present. Dissolution of mineral phosphate also depends on physiochemical characteristics such as pH, available sulfides, alkalinity and redox state (Boto, 1988). These factors can, of course, be affected by the activity of microbes and larger organisms.

Uptake and Excretion by Organisms

Every organism participates to some extent in the phosphorus cycle by virtue of its need to assimilate organic and/or inorganic P for growth and maintenance, and by excreting P-containing byproducts. Bacteria, algae and higher plants, including mangroves, take up dissolved orthophosphate, and organic phosphates are either taken up directly or first hydrolyzed by extracellular alkaline phosphatases. As noted earlier, organic P may be very resistant to hydrolysis and not readily assimilable to organisms. Orthophosphate is coupled to ADP to form ATP in cells, and is essential for energy transfer and phosphorylations, and for synthesis of nucleic acids, phospholipids and phosphoproteins (Ingraham *et al.*, 1983).

In comparison with the release rates of phosphorus from mineral phosphates and refractory organic materials, the turnover time for P uptake, utilization and excretion by living organisms is very short, on the order of minutes to tens of hours, depending on the rate of biological activity and the amount of available P. Once P is taken up and used in cells as phosphate, it is eventually liberated via excretion or through mineralization of detritus as phosphate. This means that all organisms have evolved efficient uptake mechanisms for a very small and virtually constant proportion of the Earth's P in a very competitive cycle, and that growth of biomass may be limited by P on a localized level. Local P cycles can be very efficient in tropical mangroves, where it has been estimated that up to 88% of the forest P pool is retained within the system (Boto and Bunt, 1982).

The cycling of phosphorus through mangrove food webs is presumably similar to that in other aquatic systems. At the base of pelagic and benthic food webs, a 'microbial loop' exists in which interactions among bacteria, microalgae and nanoprotozoans and larger protists facilitates net release of P into the water column and porewater. Unfortunately, there are no data on P uptake or release by mangrove microbes.

At the intermediate steps in the food web (eg. zooplankton), the rate of P regenerated is not constant. For example, zooplankton convert a large amount of P to eggs and stored as phospholipids when algal food is abundant. When food is scarce, the stored phospholipids are used for maintenance and orthophosphate is excreted. P excretion by zooplankton in a mangrove-fringed bay in tropical Australia is expressed by the regression equation $\log y_p = -0.174 + 0.429 \log x$ in summer and by $\log y_p = 1.002 + 0.740 \log x$ in winter, where y_p = phosphate excretion ($ngP.animal^{-1}.h^{-1}$) and x = body dry weight (mg, Ikeda *et al.*, 1982a, b). The values obtained were within the range observed for boreal and temperate zooplankton. Hourly excretion of phosphate by microzooplankton and net zooplankton for the entire mangrove water-column were 0.55 and from 1.4 to 2.8 $\mu g.m^{-3}$, respectively. Combined P regeneration, on average, accounted for 6.6-25.6% of the phosphorus required to support coastal phytoplankton production. Large nekton (fish, prawns) excrete phosphate (Brabrand *et al.*, 1990), but no data are available for tropical coastal assemblages.

Similarly, no data exist for P assimilation and excretion for macrofauna in mangrove sediments. As shown in the culture experiments of Tietjen and Alongi (1990), regeneration of orthophosphate by mangrove nematodes was minimal, insufficient to raise concentrations above those resulting from leaching and bacterial uptake alone. This result was ascribed to the slow growth of the nematode populations because of the poor nutritional quality of mangrove leaf litter. Significant P regeneration by meiofauna, as for pelagic microbes and zooplankton, depends on nutritional quality of the foods tested (Andersen *et al.*, 1988; Berman *et al.*, 1987; Goldman *et al.*, 1987). Food of high nutritional value usually results in higher rates of P excretion.

9.4 Sediment-water exchange of nitrogen and phosphorus

Regeneration of nutrients is the major pathway through which the benthos can influence pelagic production (Dixon 1980). Direct measurements of fluxes of dissolved nitrogen and phosphorus between mangrove sediment and overlying water are few compared with the number of studies conducted in temperate intertidal habitats (Alongi, 1989, 1990). Diffusion from sediments may be calculated from porewater concentration gradients and diffusion coefficients using diagenetic models, but bioturbated sediments require more complex modelling, so only direct measurements are considered here.

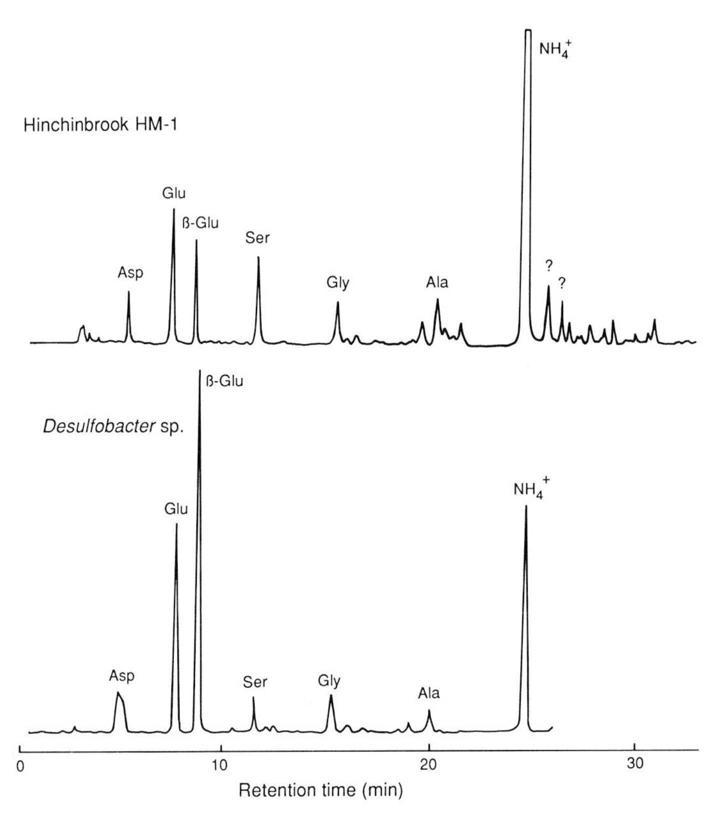

Figure 13. Chromatograms of the amino acids in (top) mangrove porewaters of Missionary Bay and (bottom) the spectrum of DFAA in the sulfate-reducing bacterium, *Desulfobacter* sp. (Modified from Stanley *et al.*, 1987.)

The data available indicate very low rates, in the $\mu mol.m^{-2}.d^{-1}$ range of flux of inorganic nutrients, from mangrove sediments (Kristensen *et al.*, 1988) and from subtidal sediments near mangroves (Hines and Lyons, 1982; Ullman and Sandstrom, 1987; Alongi, 1990). On Phuket Island in Thailand, Kristensen *et al.*, (1988) observed no clear differences in flux rates between light and dark chambers under the mangrove canopy but flux rates of NH_4^+ were significantly lower on a creek bank exposed to sunlight, suggesting the effect of algal uptake. Fluxes were into the sediment (for NH_4^+ = -104 to -29; for NO_3^- = -51 to -30 $\mu mol.m^{-2}.h^{-1}$) in all cases. On Hinchinbrook Island, flux rates are similarly low and erratic, and not related to intertidal position, rates of oxygen consumption, temperature or to solar radiation (Alongi, in prep). Fluxes are generally lower (total DIN:0 to +250; PO_4^{3-} = 0 to +100 $\mu mol.m^{-2}.d^{-1}$) in winter than in summer (total DIN:-55 to +2800; PO_4^{3-}-= - 50 to +522 $\mu mol.m^{-2}.d^{-1}$). Fluxes of DON and DOP are also low and erratic, exhibiting no seasonality or variation with intertidal position (DON = -20 to +45; DOP = -10 to +130 $\mu mol.m^{-2}.d^{-1}$). In most cases, there was no flux of dissolved inorganic and organic N and P. In subtidal sediments from embayments bordering mangroves, there are similarly low rates of DIN and DIP release (Hines and Lyons, 1982; Ullman and Sandstrom, 1987).

Low rates of nutrient flux, despite a concentration gradient between porewaters and overlying waters, are characteristic of other tropical shallow-water, sedimentary environments. Several reasons can be offered to account for this phenomenon, including adsorption onto clays, low rates of algal production, low nutritional quality of organic matter and sequestering via high rates of bacterial growth (Alongi, 1990a). Several studies offer evidence for one or all of these reasons, but the last factor appears to be the main cause (Stanley *et al.*, 1987; Ullman and Sandstrom, 1987; Alongi, 1989, 1990, 1991). Alongi (1990b) observed that mangrove litter added to boxcorer sediment samples did not affect rates of dissolved inorganic nutrient release. This result was ascribed to the very refractory nature (C:N:P ratio = 415:8:1) of this detritus.

High rates of sedimentary bacterial production were given as the reason for the retention of amino acids within sediments on Hinchinbrook Island (Stanley *et al.*, 1987). Flux chamber experiments showed negligible amino acid movement out of sediments in untreated chambers, but when sediments were poisoned to kill the biota, rates of dissolved free amino acid (DFAA) flux were measurable, ranging from 27 to 69 $mgN.m^{-2}.d^{-1}$, sufficient to account for 9 to 38% of the N required to support the measured rates of bacterial production. Moreover, experiments showed that ß-glutamic acid, an important component of the interstitial DFAA pool (Table 6), was not found in the flux waters, suggesting that this amino acid originates from sediment bacteria, and not as an exudate from plants or plant detritus. In fact, the composition of the DFAA pool is similar to that found within the intracellular pool of the common sulphate-reducing bacterium, *Desulfobacter* sp. (Figure 13).

9.5 Detritus decomposition

Nearly all of the organic detritus in mangrove forests is derived from above- and below-ground tree components, including leaves, fallen timber, roots, stipules, reproductive parts, twigs and bark. There are many studies detailing seasonal and geographical variations in litter production and/or elemental composition of these plants components (eg Bunt, 1982;

Table 6. Mean dissolved free amino acids ($ng.ml^{-1}$) in surface (0-5 cm) sediments and in creek water within the Hinchinbrook Island forests. Values are means ±1 S.D (from Stanley *et al.*, 1987).

Source				**Amino acids**			
	Asp	**Glu**	**B-Glu**	**AspN**	**Ser**	**Gly**	**Ala**
Low intertidal	21±11	81±62	90±60	tr	46±17	31±20	36±14
mid intertidal	125±68	328±157	167±139	40±22	156±175	88±122	64±53
high intertidal	25±16	224±10	380±92	6±8	40±57	19±13	28±17
creek water	5±5	3±3	nd	12±10	8±9	8±9	6±7

nd = not detected
tr = < 10 $ng\ ml^{-1}$

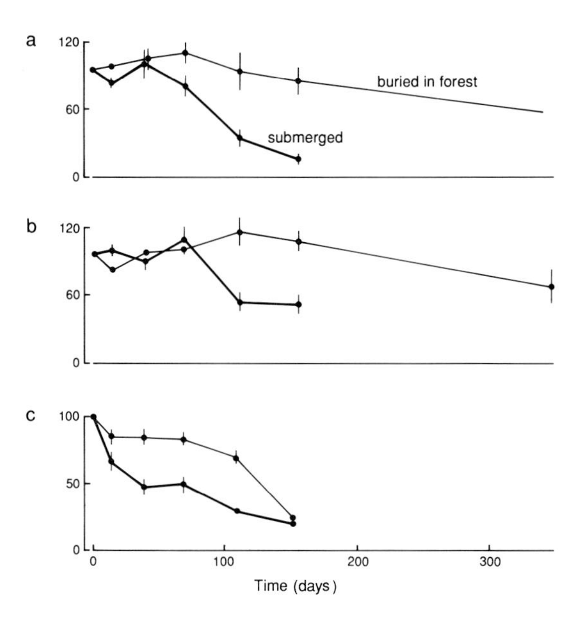

Figure 14. Changes in percentage of original nitrogen which occurs during the decomposition of leaves of (a) *Rhizophora stylosa*, (b) *Avicennia marina* and (c) *Ceriops tagal* in tidal creeks and forests of tropical Australia. (Modified from Robertson, 1988.)

Woodroffe, 1982; Boto and Wellington, 1983; Aksornkoae and Khemnark, 1984; Flores-Verdugo *et al.*, 1987; Woodroffe *et al.*, 1988).

An equally large literature exists for the rates of decomposition of mangrove detritus, particularly leaf litter (Fell *et al.*, 1975, 1980; Albright, 1976; Boonruang, 1978, 1980; Cundell *et al.*, 1979; Fell and Master, 1980; Sumitra-Vijayaraghavan *et al.*, 1980; Rice and Tenore, 1981; Poovachiranon and Chansang, 1982; Steinke *et al.*, 1983; Reice *et al.*, 1984; Zieman *et al.*, 1984; Benner and Hodson, 1985; Twilley *et al.*, 1986; Robertson, 1988; Robertson and Daniel, 1989), although most of these studies have not measured decompositional changes in nitrogen and/or phosphorus. Much of the information on mangrove litter decomposition is reviewed in the following chapter, and here we deal only with information relevant to nitrogen and phosphorus cycling.

Mangrove leaf decomposition studies indicate several common aspects: (1) mangrove detritus, in common with other vascular plant material, decomposes more slowly than detritus derived from non-vascular plants (e.g. seaweeds, macroalgae) because of decay-resistant components such as lignocelluloses; (2) among different mangrove species, *Avicennia* litter decomposes most quickly, because it is nutritionally more available to consumers and because it has lower concentrations of polyphenolic compounds, that inhibit microbial colonization and activity, (3) amounts of nitrogen either remain near 100% of original values, or increases during at least the first few months of decomposition and (4) as with other vascular plant detritus (Valiela, 1984), decomposition of mangrove detritus proceeds in phase: loss of labile water-soluble compounds, microbial colonization and utilization, and mechanical fragmentation.

For example, Robertson (1988) examined changes in mass nitrogen during the decomposition of mangrove leaves in litter bags. For *Rhizophora* and *Ceriops* leaves, the amount of nitrogen either remained constant or increased during the first 40-71 days of the experiment, with %N decreasing thereafter (Figure 14). After 348 days, ~ 80% of the original mass of nitrogen was still present for both species. Leaves of *Avicennia* lost nitrogen more rapidly throughout the experiment.

Rapid increases in bacterial numbers coincided with rapid loss of soluble tannins from leaves of all three species (Robertson, 1988) as observed in earlier studies of *Rhizophora* leaf decomposition (Cundell *et al.*, 1979; Benner *et al.*, 1986). Bacterial nitrogen contributed only a tiny fraction of the nitrogen content of decomposing leaves, in agreement with other studies of vascular plant decomposition (e.g. Rice and Hanson, 1984). Other microbes, particularly fungi, are undoubtedly important in the decomposition process (Fell *et al.*, 1975, 1980; Fell and Masters, 1980), but they are also unlikely to contribute any more to total detrital N than do bacteria (Rice and Hanson, 1984). Recent work suggests that the production of mucopolysaccharide exudates by bacteria and their incorporation into humic compounds, may account for much of the observed increase in total detritus nitrogen (Rice, 1982; Rice and Hanson, 1982).

The idea that the chemical composition of decomposing mangrove detritus is different than for marsh and seagrass detritus is supported by the study of Zieman *et al.*, (1984) in which temporal and spatial changes in stable isotope composition and in amino acid content of seagrass and mangrove detritus were compared. They found that while the amino acid composition of

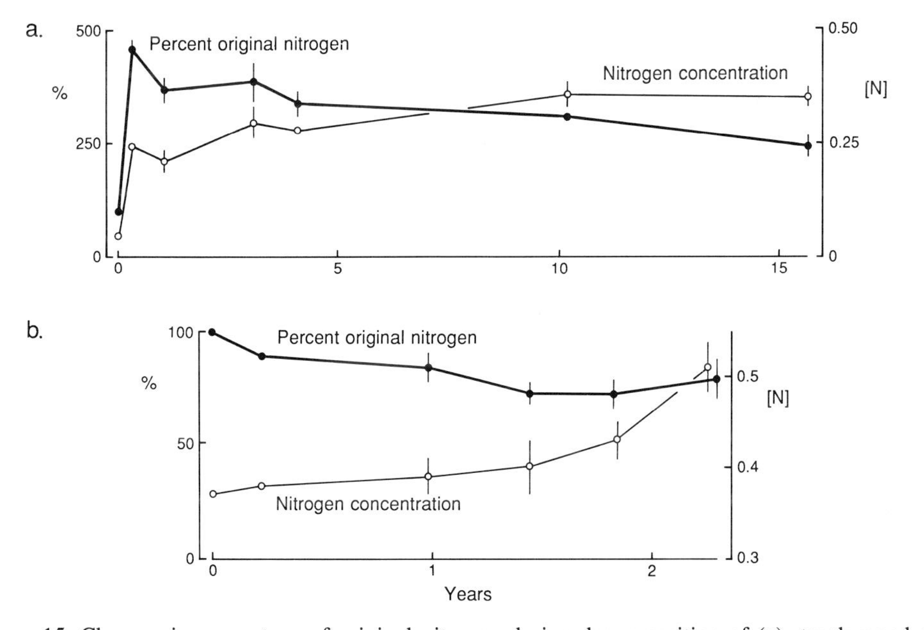

Figure 15. Changes in percentage of original nitrogen during decomposition of (a) trunk wood of *Rhizophora* and (b) sticks of *Rhizophora* over several years. (Modified from Robertson and Daniel, 1989.)

seagrass and mangrove (*R. mangle*) detritus was similar, the percentage of total N accounted for by amino acids decreased in the seagrasses, but increased for mangrove detritus, over a period of 6 weeks. Mangrove detritus showed little change in $\delta^{13}C$ but a large decrease in $\delta^{15}N$ and an increase in the D/L ratio of the amino acids. In contrast, seagrass detritus showed little change in either stable isotope and an increase in the D/L ratio of only some of the amino acids. Zieman *et al.,* hypothesized that detritivores obtain more nutrition from utilizing seagrass than mangrove detritus, which would require more microbial oxidation to become assimilable. Mangrove leaf litter has in fact been found to harbor large populations of bacteria (Robertson, 1988; Benner *et al.*, 1988) that appear to be highly productive (Benner *et al.*, 1988).

Only one study to date has considered nutrient dynamics during the decomposition of fallen timber (tree trunks, twigs and branches) in tropical mangrove forests (Robertson and Daniel, 1989). In mixed *Rhizophora* forests in tropical Australia, freshly fallen trunks had a N content of only 0.05% by dry weight, but N was immobilized rapidly during the first 2.4 months, increasing by five fold (%N) and then exhibiting only a gradual increase over the remaining 15.5 years (Figure 15a). Mass nitrogen also increased sharply during the first 8 weeks, but dropped steadily thereafter. Nitrogen concentration of small branches (sticks) (initially 0.37%) increased gradually during the initial 1.5 years, increasing to 0.54% after 2.2 years. Mass N decreased slowly throughout this period and was ~ 80% of the original after 2.3 years (Figure 15b). Most of this processed wood remains within the forest, serving as a mechanism to conserve N.

As suggested by Boto (1982, 1984) and Twilley *et al.*, (1986), mangroves are efficient at retaining and recycling nitrogen via several mechanisms that reduce export. These mechanisms include re-absorption or re-translocation of N prior to leaf fall, burial of fallen detritus by crabs, rapid and efficient uptake of dissolved materials by bacteria, and comparatively high N content invested in chemical defenses by the trees. Environmental factors also affect N retention, including among-forest variations in tidal inundation, sediment type, rainfall, climatic disturbances and topography (see Chapters 2, 3 and 8, this volume).

Phosphorus is probably also conserved, considering its important role in regulating plant growth. Only a few workers, however, have examined changes in P concentration in decomposing mangrove litter (Albright, 1976; Sumitra-Vijayaraghavan *et al.*, 1980; Steinke *et al.*, 1983). The study of Albright (1976) showed that addition of phosphorus did not stimulate root degradation, but did increase decomposition of pneumatophores. At the sediment surface, relative decomposition rates were: leaves > roots > pneumatophores. Sumitra-Vijayaraghavan *et al.*, and Steinke *et al.*, have shown that phosphorus content declines in *Rhizophora, Avicennia* and *Bruguiera* leaves during the first 4-6 weeks of decomposition, but either remains the same or gradually increases thereafter (~ 10-24 weeks), suggesting P immobilization.

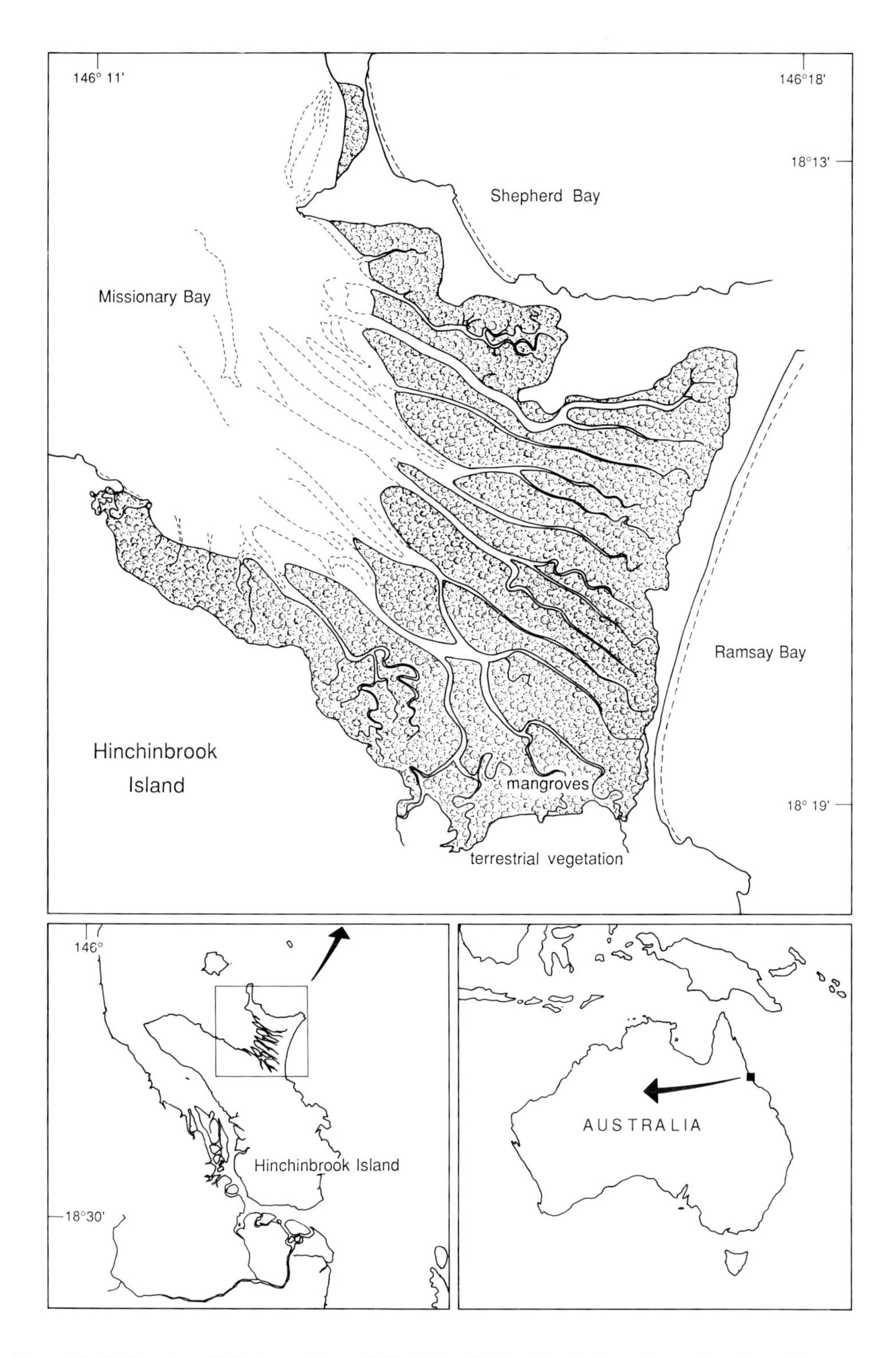

Figure 16. The location of Missionary Bay and Hinchinbrook Island in relation to the northern Australian coast.

9.6 The nitrogen budget of a tropical mangrove forest ecosystem (Hinchinbrook Island, Australia)

9.6.1 The mangrove ecosystem of Missionary Bay

For more than twelve years, the Australian Institute of Marine Science has conducted a long term program aimed at developing a comprehensive picture of mangrove ecology in the Australasian region. A main study site is Missionary Bay on Hinchinbrook Island in north Queensland, Australia. The island is a heavily forested and mountainous national park at 18°20'S and 146°10'E, adjacent to the Australian mainland and nearly 100 km NW of the city of Townsville (Figure 16). The western boundary of the island consists of mangrove forest and tidal creeks, separated from Missionary Bay at the northern end of the island by granite mountains (Grindrod and Rhodes, 1984). The island lies on a sharp climatic gradient receiving, on average, 2500 $mm.yr^{-1}$ of rainfall (Bunt, 1982). The western end of the system receives some additional freshwater input by runoff from the mountain watersheds, but the remainder of the system is tidally-dominated with no substantial groundwater input. Tides are semi-diurnal and range around 2 m, with spring tides exceeding 3 m. Even during heavy monsoonal rains, salinities in tidal waters are rarely less than 33‰. Hydrographic conditions within the Missionary Bay system have been described in detail by Wolanski *et al.*, (1980), Wolanski and Gardiner (1981) and Wolanski and Ridd (1990). Floristic surveys have been conducted by Bunt (1982) and Bunt *et al.*, (1982).

The total area of the Missionary Bay system is 64 km^2, and includes saltpans (7.5 km^2), eleven creeks (14 km^2) and mangrove forests (42.5 km^2). Most (75%) of the forest is within the mid-intertidal zone, with low- and high-intertidal mangroves constituting 17 and 8% of total forest area, respectively. The forest is mainly mixed *Rhizophora* spp, but 26 species of mangroves have been recorded (Bunt, 1982; Bunt *et al.*, 1982; Robertson and Daniel, 1990). Total volume of water exchanged over an average tidal run is estimated at 1.5 x 107 m^3 for all eleven creeks (Wolanski *et al.*, 1980).

9.6.2 Inputs

Nitrogen enters the mangroves of Missionary Bay through the fixation of atmospheric nitrogen by cyanobacteria growing on sediments, on prop roots and on timber lying on the forest floor, by tidal flushing and, to a much lesser extent, by monsoonal rainfall. The amount of 'groundwater outflow' from the land to the mangroves listed in Table 7 is the amount of average yearly rainfall within the mountainous barrier bordering the system and assumes that all of the precipitation runs down into Missionary Bay. The amounts of ammonium, nitrite and nitrate, and DON in the rainwater were determined (Alongi, unpublished data). No particulate material was found in gauges used to collect rainwater, so particulate N input is assumed to be zero. Half of the nitrogen derived from rain is DON, with nearly equal contributions from $NO_2^- + NO_3^-$ and ammonium (Table 7).

Table 7. Nitrogen budget for the Missionary Bay mangrove forest ($kgN.yr^{-1}$). Losses from the forest are shown as negative numbers in the net exchange column. The following physical characteristics of Missionary Bay were used in the calculations; saltpan area = 7.5 km^2 (~ 15% of total area); mangrove area = 42.5 km^2 (~ 66% of total area) with a) 17% low intertidal forest, 75% mid intertidal forest, 8% high intertidal forest, and where b) 50% is mature forest, 50% is young forest; creek area = 14 km^2 (~ 19% of total area); total water volume exchange over average tidal run (+0.7 m to -0.7 m) = 1.5 x 10^7 m^3 = 1.5 x 10^{10} liters; and, mean rainfall = 2500 $mm.yr^{-1}$

Processes	Input	Output	Net exchange
Precipitation[1]			
$NO_2^- + NO_3^-$ -N	8		
NH_4 - N NH_4^+ -N	6		
DON	15		
Particulate N	<<<1		29
Groundwater Flow[1]	29		29
N_2 fixation[2, 3, 4]			
saltpan	5594		
mangrove sediments			
a. low intertidal	3956		
b. mid intertidal	0		
c. high intertidal	1800		
prop roots			
a. low intertidal	327		
b. mid intertidal	13,961		
c. high intertidal	24		
logs			
a. young forests	853		
b. mature forests	10,316		36,831
Tidal water exchange[5, 6]			
$NO_2^- + NO_3^-$ -N	5250	6300	-1050
NH_4^+ - N	11,136	5362	5774
DON	152,212	104,446	47,766
particulate N (total)[7, 8]		76,321	-76,321
Denitrification[2]		2824	-2824
Totals	205,487	195,253	10,234

References:
1. Queensland Bureau of Meteorology Records, 1980-1990; Alongi, unpubl. data;
2. Iizumi, 1986;
3. Robertson and Daniel, 1989;
4. Boto and Robertson, 1990;
5. Wolanski *et al.*, 1980;
6. Boto and Wellington, 1989;
7. Boto and Bunt, 1981;
8. Robertson *et al.*, 1988.

Tidal flushing brings in nitrogen mainly as DON, with lesser amounts of DIN. Boto and Bunt (1981) indicate that no particulate matter is imported into the system on flood tides. Net tidal exchange of N is a loss to the system (see 9.6.3).

9.6.3 Outputs and net exchange

The major losses of N from the system occur via tidal flushing and, to a lesser extent, by denitrification (Table 7). Denitrification appears to be very low compared with nitrogen fixation and, as only mid-intertidal sediments were examined (Iizumi, 1986), the value (and extrapolation to the rest of the system) is clearly an underestimate.

It is, however, clear that significant outwelling occurs from Missionary Bay, mainly in the form of dissolved organic nitrogen and particulate N (as mangrove litter). These estimates are reasonable as tidal exchange of these materials has been well documented (Boto and Bunt, 1981;, 1982; Boto and Wellington, 1988; Robertson *et al.*, 1988; Alongi *et al.*, 1989; Alongi, 1990).

Other sources and losses from the system are less clear. Unaccounted losses such as volatilization of ammonium undoubtedly occur, but would probably be small at the pH range (6.5-8.2) measured in these sediments (Fenchel and Blackburn, 1979). Additional losses may occur with the emigration of juvenile fish and prawns from mangrove habitats to offshore (see Chapter 7), but as with the harvesting of marine products from Missionary Bay, these

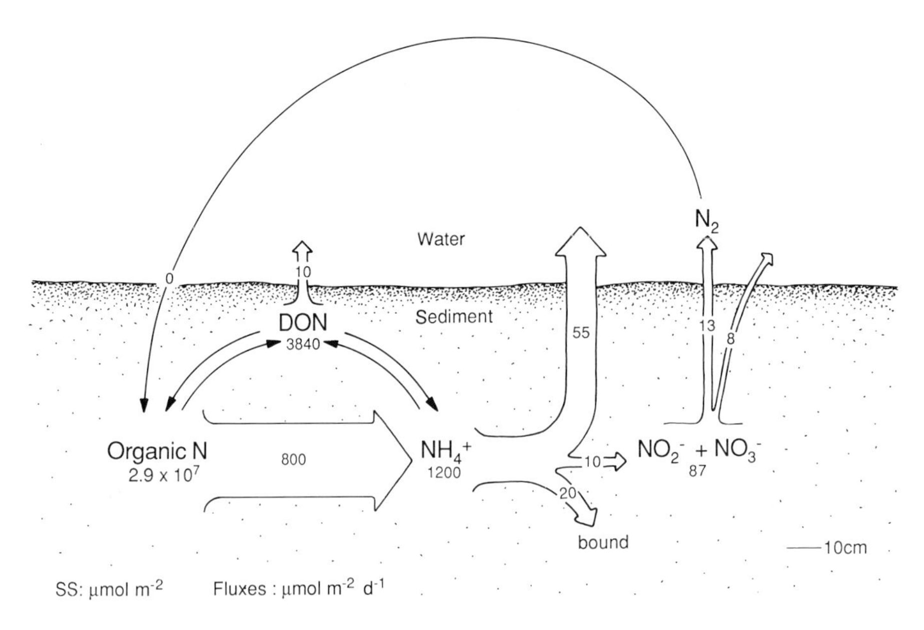

Figure 17. Nitrogen cycling in the mid-intertidal mangroves of Missionary Bay. Values are for a depth of 10 cm on a m2 basis. Standing amounts = µmolN.m^{-2}; fluxes = µmol N.m^{-2}.d^{-1}. Data from Boto and Wellington, 1984; Iizumi, 1986; Alongi, 1988; Boto and Wellington, 1990 and Alongi, unpublished data.

are likely to be relatively small (eg. see Zijlistra, 1988). Sedimentation may be the largest unaccountable loss (or source), but no data are available for this system and the adjacent seabed. Defecation by birds is another potential source of N but the numbers of birds (and bats) in the mangroves is poorly known, and their rates of defecation are not known at all.

Considering the extrapolations made from measurements over a small area and undocumented losses or inputs (sedimentation, volatization, etc.), the input and output sums are very close, and the estimated net loss is well within the range (~ 5% difference) of probable error. The budget is balanced. This balance conforms to the concept that in estuarine and marine ecosystems, nitrogen is conserved (Nixon and Pilson, 1983). Considering that these mangroves require about 1,147,500 kg $N.yr^{-1}$ for growth (Boto and Bunt, 1982), it is clear that very active recycling of nitrogen must take place within Missionary Bay.

The budget sheet does not reflect the high rate of N recycling, as calculated for transformations in mid-intertidal sediments (Figure 17). The low rate of N efflux from the sediments indicates very active uptake by sedimentary bacteria, as suggested by Stanley *et al.*, (1987) and Alongi (1988). Indeed, on the same scale (m^{-2} of forest floor to a depth of 10 cm) bacterial biomass (as N) is enormous (~ 2.2×10^6 $g.m^{-2}$), nearly 8% of the total remaining particulate organic N pool (Figure 17). This large biomass turns over rapidly (~ 10 days) with production rates exceeding 1.4×10^5 $\mu mol.m^{-2}.d^{-1}$ (Alongi, 1988). Even if the dissolved N pools contributed their entire stock, nearly all of the nitrogen required to sustain the measured rates of bacterial production must come from the PON pool, irrespective of assimilation efficiency. As suggested by Alongi (1989), bacteria in mangrove sediments are a sink for nutrients because of internal recycling (mortality, decomposition, uptake and growth) and thus serve as a mechanism for nutrient conservation.

Other biological mechanisms help to retain and conserve N in the forest. There is a large above- and below-ground mass of live trees and dead wood lying on the forest floor. Above-ground biomass is estimated to represent ~6000 tonnes N in Missionary Bay (Robertson and Daniel, 1989). The trees and dead wood also provide increased surface area for colonization of nitrogen-fixing biota. In addition, burial and processing of leaf litter by crabs in high intertidal forests has been estimated to supply between 11 and 64% of N required for forest primary production (See Chapter 10, this volume).

Further evidence of N retention in the system comes from the high C:N and C:P ratios for the dissolved compounds in the creek waters (C:N range = 18 to 35:1; C:P range = 250 to 700:1) suggesting that this material is refractory and in an advanced stage of diagenesis. Although not measured, microbial utilization of N and P in the water column is likely to be rapid. Finally, the forest appears to have evolved to the state in which N_2 fixation exceeds denitrification, causing a net import of N_2 from the atmosphere. Even if we assume that our denitrification rates are an underestimate, it is unlikely that they are low by a factor of 13 to equate with the nitrogen fixation estimate.

9.6.4 Outwelling and comparisons with other systems

It is clear that the forests of Missionary Bay, on average, import nitrogen in dissolved form and export it as particulate matter, mainly as leaves, twigs, flowering parts, etc. This outwelling supports adjacent coastal food chains and influences sediment nutrient chemistry in the central Great Barrier Reef Lagoon (Alongi *et al.*, 1989; Alongi, 1990, and see Chapter 10, this volume).

The extent of outwelling from a particular system is related to the geomorphology of the tidal basin, tidal amplitude and water motion, and the ratio of the areal extent of the vegetation to the receiving open ocean area (Nixon, 1980). These factors, in particular, the wide tidal range, are favorable for particulate export (and dissolved import) within this coastal area of Australia. Most other mangrove ecosystems appear to export nitrogen as litter, with the absolute amounts dependent upon the above-cited factors (Thailand, Aksornkoae and Khemmark, 1984; Florida; Fleming *et al.*, 1990; New Zealand, Woodroffe, 1982; Mexico, Flores-Verdugo *et al.*, 1987). Unfortunately, no other nitrogen budgets have been constructed for tropical mangrove forests.

Our budget can, however, be compared to those constructed for two salt marsh ecosystems, the Great Sippewissett Marsh in Massachusetts (Valiela and Teal, 1979) and the Sapelo Island marshes of Georgia (Pomeroy and Wiegert, 1981). The major similarities between these tropical and temperate systems are: (1) the dominance of physical processes controlling nitrogen exchange, (2) an overall balance of N, with outputs occurring via tides, (3) nitrogen fixation and tidal exchanges are major inputs and (4) tidal exchange is the major output. For instance, in both Missionary Bay and in the Great Sippewissett Marsh, biotic mechanisms contribute only 17% and 9% of the inputs and only 1% and 18% of the outputs, respectively.

However, there are clear differences between these tropical and temperate systems: (1) rates of N_2 fixation >>> denitrification rates in the mangroves, whereas in the marshes they are either nearly equal in magnitude (Great Sipperwissett) or are in reverse order (Georgia), (2) groundwater flow is negligible in the Missionary Bay mangroves (although it may be important in other tropical mangroves, e.g. Ovalle *et al.*, 1990). (3) N via rainfall is also less in the mangroves despite higher rates of rainfall and (4) there is little, if any, particulate N input into Missionary Bay, either from the atmosphere or via tidal exchange.

Other tropical mangrove systems may, of course, be more similar in terms of input and export pathways to temperate marshes than to the Missionary Bay system, but considering the larger amounts of nutrients tied up in above- and below-ground tree biomass, a characteristic of all mangrove forests, the inherent structural differences between mangroves and salt marshes undoubtedly outweigh their similarities.

9.7 References

Agate, A.D., Subramanian, C.V., and Vannucci, M., 1988. Mangrove Microbiology-Role of Microorganisms in Nutrient Cycling of Mangrove Soils and Waters. UNESCO Technical Report RAS/86/120, 118 pp.

Aksornkoae, S., and Khemmark, C., 1984. Nutrient cycling in mangrove forests in Thailand. In: Soepadmo, E., Rao, A.N., Macintosh, D.J. (Eds). *Proceedings of the Asian Symposium on Mangrove Environments: Research and Management,* pp. 545-557, University of Malaya and UNESCO, Kuala Lumpur.

Albright, L.J., 1976. In situ degradation of mangrove tissues. *New Zealand Journal of Marine and Freshwater Research* **10**:385-389.

Alongi, D.M., 1988. Bacterial productivity and microbial biomass in tropical mangrove sediments. *Microbial Ecology* **15**:59-79.

Alongi, D.M., 1989. The fate of bacterial biomass and production in marine benthic food chains. In: Hattori, T., Ishida, Y., Maruyama, Y., Monta, R.Y., and Achida, A. (Eds). *Recent Advances in Microbial Ecology*, pp. 355-359, Japan Scientific Societies Press, Tokyo.

Alongi, D.M., 1989. The role of soft-bottom benthic communities in tropical mangrove and coral reef ecosystems. *Reviews in Aquatic Sciences* **1**:243-280.

Alongi, D.M., 1990. The ecology of tropical soft-bottom benthic ecosystems. *Oceanography and Marine Biology Annual Reviews* **28**:381-496.

Alongi, D.M., 1990. Effect of mangrove detrital outwelling on nutrient regeneration and oxygen fluxes in coastal sediments of the Central Great Barrier Reef Lagoon. *Estuarine, Coastal and Shelf Science* **31**:581-598.

Alongi, D.M., 1991. The role of intertidal mudbanks in the diagenesis and export of dissolved and particulate materials from the Fly Delta, Papua New Guinea. *Journal of Experimental Marine Biology and Ecology* (in press).

Alongi, D.M., Boto, K.G., and Tirendi, F., 1989. Effect of exported mangrove litter on bacterial productivity and dissolved organic carbon fluxes in adjacent tropical nearshore sediments. *Marine Ecology Progress Series* **56**:133-144.

Andersen, O.K., Goldman, J.C., Caron, D.A., and Dennett, M.R., 1986. Nutrient cycling in a microflagellate food chain III. Phosphorus dynamics. *Marine Ecology Progress Series* **31**:47-55.

Attiwill, P.M., and Clough, B.F., 1978. Productivity and nutrient cycling in the mangrove and seagrass communities of Westernport Bay. Report to Ministry for Conservation, Victoria, Australia, 86 pp.

Balakrishnan Nair, N., Abdul Azis, P.K., Krishna Kumar, K., Dharmaraj, K., and Arunachalam, M., 1984. Ecology of Indian estuaries: Part VI - Physico-Chemical conditions in Kadinamkulam backwater, SW coast of India. *Indian Journal of Marine Sciences* **13**:69-74.

Balasubramanian, T., and Venugopalan, U.K., 1984. Dissolved organic matter in Pitchawaran mangrove environment, Tamil Nadu, South India. In: Soepadmo, E., Rao, A.N., and Macintosh, D.J. (Eds). *Proceedings of the Asian Symposium on Mangrove Environment: Research and Management,* pp. 496-513, University of Malaya and UNESCO, Kuala Lumpur.

Benner, R., and Hodson, R.E. 1985. Microbial degradation of the leachable and lignocellulosic components of leaves and wood from *Rhizophora mangle* in a tropical mangrove swamp. *Marine Ecology Progress Series* **23**:221-230.

Benner, R., Hodson, R.F., and Kinchman, D., 1988. Bacterial abundance and production on mangrove leaves during initial stages of leaching and biodegradation. *Ergebnisse der Limnologie* **31**:19-26.

Benner, R., Peele, E.R., and Hodson, R.E., 1986. Microbial utilization of dissolved organic matter from leaves of the red mangrove, *Rhizophora mangle*, in the Fresh Creek Estuary, Bahamas. *Estuarine, Coastal and Shelf Science* **23**:607-619.

Berman, T., Nawrocki, M., Taylor, G.T., and Karl, D.M., 1987. Nutrient flux between bacteria, bacterivorous nannoplanktonic protists and algae. *Marine Microbial Food Webs* **2**:69-82.

Bidigare, R.R., 1983. Nitrogen excretion by marine zooplankton. In: Carpenter, E.J. and Capone, D.G. (Eds). *Nitrogen in the Marine Environment*, pp. 385-409, Academic Press, New York.

Blackburn, T.H., 1986. Nitrogen cycle in marine sediments. *Ophelia* **26**:65-76.

Boonruang, P., 1978. The degradation rates of mangrove leaves of *Rhizophora apiculata* (B.L.) and *Avicennia marina* (Forsk) at Phuket Island, Thailand. *Phuket Marine Biological Center Research Bulletin, No.* **26**, 7 pp.

Boonruang, P. 1984. The rate of degradation of mangrove leaves, *Rhizophora apiculata* B.L. and *Avicennia marina* (Forsk) Vierh. at Phuket Island, western Peninsula of Thailand. In: Soepadmo, E., Rao, A.N. and Macintosh, D.J. (Ed.) *Proceedings of the Asian Symposium on Mangrove Environment: Research and Management*, pp. 200-208, University of Malaya and UNESCO, Kuala Lumpur.

Boto, K.G., 1982. Nutrient and organic fluxes in mangroves. In: Clough, B.F. (Ed.). *Mangrove Ecosystems in Australia,* Australian National University Press, Canberra, pp. 239-257.

Boto, K.G., 1984. Waterlogged saline soils. In: Snedaker, S.C. and Snedaker, J.G. (Eds). *The Mangrove Ecosystem: Research Methods.* pp. 114-130, UNESCO, Paris.

Boto, K.G., 1988. The phosphorus cycle. In: Agate, A.D., Subramanian, C.V. and Vannucci, M. (Eds). *Mangrove Microbiology*, pp. 85-100, UNDP/UNESCO Regional Project (RAS/86/1988), New Delhi.

Boto, K.G., and Bunt, J.S. 1981. Tidal export of particulate organic matter from a northern Australian mangrove system. *Estuarine, Coastal and Shelf Science* **13**:247-251.

Boto, K.G., and Bunt, J.S. 1982. Carbon export from mangroves. In: Galbally, I.E. and Freney, J.R. (Eds). *The cyclin of carbon, nitrogen, sulfur and phosphorus in terrestrial and aquatic ecosystems,* pp. 105-110, Australian Academy of Science, Canberra.

Boto, K.G., Alongi, D.M., and Nott, A.L.J., 1989. Dissolved organic carbon-bacteria interactions at sedimentwater interface in a tropical mangrove system. *Marine Ecology Progress Series* **51**:243-251.

Boto, K.G., and Robertson, A.I., 1990. The relationship between nitrogen fixation and tidal exports of nitrogen in a tropical mangrove system. *Estuarine, Coastal and Shelf Science* **31**:531-540.

Boto, K.G., Bunt, J.S., and Wellington, J.T., 1984. Variations in mangrove forest productivity in Northern Australia and Papua New Guinea. *Estuarine, Coastal and Shelf Science* **19**:321-329.

Boto, K.G., Saffigna, P., and Clough, B.F., 1985. Role of nitrate in the nitrogen nutrition of the mangrove *Avicennia marina*. *Marine Ecology Progress Series* **21**:259-265.

Boto, K.G., and Wellington, J.T., 1983. Phosphorus and nitrogen nutritional status of a northern Australian mangrove forest. *Marine Ecology Progress Series* **11**:63-69.

Boto, K.G., and Wellington, J.T., 1984. Soil characteristics and nutrient status in northern Australian mangrove forests. *Estuaries* **7**:61-69.

Boto, K.G., and Wellington, J.T., 1988. Seasonal variations in concentrations and fluxes of dissolved organic and inorganic materials in a tropical, tidally-dominated, mangrove waterway. *Marine Ecology Progress Series* **50**:151-160.

Brabrand, A., Faafeng, B.A., and Nilssen, J.P.M., 1990. Relative importance of phosphorus supply to phytoplankton production: Fish excretion versus external loading. *Canadian Journal of Fisheries and Aquatic Sciences* **47**:364-372.

Bunt, J.S., 1982. Studies of mangrove litter fall in tropical Australia. In: Clough, B.F. (Ed.). *Mangrove Ecosystems in Australia*, pp. 223-238, Australian National University, Canberra.

Bunt, J.S., Williams, W.T., and Duke, N.C., 1982. Mangrove distribution in north-east Australia. *Journal of Biogeography* **9**:111-120.

Capone, D.G., 1983. Benthic nitrogen fixation. In: Carpenter, E.J. and Capone, D.G. (Eds.), *Nitrogen in the Marine Environment* pp 105-137, Academic Press, New York.

Carlson, P.R., Yarbro, L.A., Zimmerman, C.F., and Montgomery, J.R., 1981. Pore water chemistry of an overwash mangrove island. *Estuaries* **4**:282 (Abstract).

Carpenter, E.J., and Capone, D.G. (Eds). 1983. *Nitrogen in the Marine Environment*, Academic Press, New York., 900 pp.

Chandramohan, D., 1988. The nitrogen cycle. In: Agate, A.D., Subramanian, C.V. and Vannucci, M. (Eds.), *Mangrove Microbiology*, pp. 61-84, UNDP/UNESCO Regional Mangroves Project RAS/86/1988, New Delhi.

Clough, B.F., Boto, K.G., and Attiwill, P.M., 1983. Mangroves and sewage: a re-evaluation. In: Teas, H.J. (Ed.). *Tasks for Vegetation Science, Vol. 8*. pp. 151-161, W. Junk, The Hague.

Cundell, A.M., Brown, M.S., Stanford, R., and Mitchell, R., 1979. Microbial degradation of Rhizophora mangle leaves immersed in the sea. *Estuarine and Coastal Marine Science* **9**:281-286.

Dall, W., and Smith, D.M., 1986. Oxygen consumption and ammonia-N excretion in fed and starved tiger prawns, *Penaeus esculentus* Haswell. *Aquaculture* **55**:2333.

D'Elia, C.F., and Wiebe, W.J., 1990. Biogeochemical nutrient cycles in coral-reef ecosystems. In: Dubinsky, Z. (Ed), *Coral Reefs,* pp. 49-74, Elsevier, Amsterdam.

Fell, J.W., and Newell, S.Y., 1980. Role of fungi in carbon flow and nitrogen immobilization in coastal marine plant litter systems. In: Wicklow, D.T. and Carroll, G.C. (Eds). *The Fungal Community: Its Organization and Role in the Ecosystem*, pp. 665-678. Marcel Dekker, New York.

Fell, J.W., Cefalu, R.C., Master, I.M., and Tallman, A.S., 1975. Microbial activities in the mangrove (*Rhizophora mangle*) leaf detrital system. In: Walsh, G.E., Snedaker, S.C., and Teas, H.J. (Eds.). *Proceedings of the International Symposium on the Biology and Management of Mangroves*, pp. 661-679, University of Florida, Gainesville.

Fell, J.W., Master, I.M., and Newell, S.Y., 1980. Laboratory model of the potential role of fungi (*Phytophthora* sp.) in the decomposition of red mangrove (*Rhizophora mangle*) leaf litter. In: Tenore, K.R., and Coull, B.C., (Eds.), *Marine Benthic Dynamics,* pp. 359-372, University of South Carolina Press. Columbia.

Fenchel, T., and Blackburn, T.H., 1979. *Bacteria and Mineral Cycling,* Academic Press, London, 225 pp.

Field, C.D., and Vannucci, M., (Eds), 1988. Symposium on New Perspectives in Research and Management of Mangrove Ecosystems. UNESCO Technical Report RAS/79/002, 267 pp.

Fleming, M., Lim, G., and da Silveria Lobo Sternberg, L., 1990. Influence of mangrove detritus in an estuarine ecosystem. *Bulletin of Marine Science* **47**:663-669.

Flores-Verdugo, F.J., Day, J.W., Ji., and Briseno-Duenas, R., 1987. Structure, litter fall, decomposition and detritus dynamics of mangroves in a Mexican coastal lagoon with an ephemeral inlet. *Marine Ecology Progress Series* **35**:8390.

Furnas, M.J., 1992. The behavior of nutrients in tropical aquatic ecosystems. In: Connell, D.W., and Hawker, D.W. (Eds.), *Pollution in tropical aquatic systems*, pp. 29-65, CRC Press Inc. Boca Raton.

Goldman, J.C., Caron, D.A., Andersen, O.K., and Dennett, M.R., 1985. Nutrient cycling in a microflagellate food chain: I. Nitrogen dynamics. *Marine Ecology Progress Series* **24**:231-242.

Goldman, J.C., Caron, D.A., and Dennett, M.R., 1987. Nutrient cycling in a microflagellate food chain IV. Phytoplankton-microflagellate interactions. *Marine Ecology Progress Series* **38**:75-87.

Goto, J.W., and Taylor, B.F., 1976. N_2 fixation associated with decaying leaves of the red mangrove (*Rhizophora mangle*). *Applied and Environmental Microbiology* **31**:781-783.

Grindrod, J., and Rhodes, E.G., 1984. Holocene sea-level history of a tropical estuary: Missionary Bay, North Queensland. In: Thom, B.G. (Ed.). *Coastal Geomorphology in Australia*, pp. 151-178, Academic Press, Sydney.

Guerrero, G.R., Cervantes, D.R., and Jimenez, I.A., 1988. Nutrient variation during a tidal cycle at the mouth of a coastal lagoon in the northwest of Mexico. *Indian Journal of Marine Science* **17**:235-237.

Hatcher, P.G., Smioneit, B.R.T., Mackenzie, F.T., Meuman, A.C., Thorstenson, D.C., and Gerchakov, S.M., 1982. Organic geochemistry and pore water chemistry of sediments from Mangrove Lake, Bermuda. *Organic Geochemistry* **4**:93-112.

Hesse, P.R., 1961. Some differences between the soils of *Rhizophora* and *Avicennia* mangrove swamps in Sierra Leone. *Plant and Soil* **14**:335-346.

Hesse, P.R., 1962. Phosphorus fixation in mangrove swamp muds. *Nature* **193**:295-296.

Hesse, P.R., 1963. Phosphorus relationships in a mangrove swamp mud with particular reference to aluminium toxicity. *Plant and Soil* **19**:205-218.

Hicks, B.J., and Silvester, W.B., 1985. Nitrogen fixation associated with the New Zealand mangrove (*Avicennia marina* (Forsk.)) Vierh. var. *resinifera* (Forsk.f.Bakl). *Applied and Environmental Microbiology* **49**:955-959.

Hines, M.E., and Lyons, W.B., 1982. Biogeochemistry of nearshore Bermuda sediments. I. Sulfate reduction rates and nutrient regeneration. *Marine Ecology Progress Series* **8**:87-94.

Iizumi, H., 1986. Soil nutrient dynamics. In: Cragg, S. and Polunin, N. (Eds). *Workshop on Mangrove Ecosystem Dynamics*, UNDP/UNESCO Regional Project (RAS/79/002) New Delhi, pp. 171-180.

Ikeda, T., Hing Fay, E., Hutchinson, S.A., and Boto, G.M., 1982a. Ammonia and inorganic phosphate excretion by zooplankton from inshore waters of the Great Barrier Reef, Queensland. I. Relationships between excretion rates and body size. *Australian Journal of Marine and Freshwater Research* **33**:55-70.

Ikeda, T., Carleton, J.H., Mitchell, A.W., and Dixon, P., 1982b. Ammonia and phosphate excretion by zooplankton from the inshore waters of the Great Barrier Reef. II. Their in situ contributions to nutrient regeneration. *Australian Journal of Marine and Freshwater Research* **33**:683-698.

Ingraham, J.L., Maaloe, O., and Neidhardt, F.C., 1983. *Growth of the Bacterial Cell,* Sinauer, Sunderland, 435 pp.

Ip, Y.K., Chew, S.F., and Lim, R.W.L., 1990. Ammoniagenesis in the mudskipper, *Periophthalmus chryospilos*. *Zoological Sciences* **7**:187-194.

Japtap, T.G., 1987. Seasonal distribution of organic matter in mangrove environment of Goa. *Indian Journal of Marine Science* **16**:103-106.

Jayasinghe, J.M.P.K., and De Silva, M.S.K.W., 1988. Development of mangrove swamps of the west coast of Sri Lanka for aquaculture. In: Field, C.W. and Vannucci, M. (Eds). *Symposium on New Perspectives in Research and Management of Mangrove Ecosystems,* pp. 99-103, UNDP/UNESCO Regional Project RAS/86/120, New Delhi.

Kannan, L., and Krishnamurthy, K., 1985. Nutrients and their impact on phytoplankton. In: Krishnamurthy, V., and Untawale, A.G. (Eds). *Marine Plants*, pp. 73-78, National Institute of Oceanography, Dona Paula, India.

Kaplan, W.A., 1983. Nitrification. In: Carpenter, E.J. and Capone, D.G. (Eds). *Nitrogen in the Marine Environment,* pp. 139-190, Academic Press, New York.

Kennett, J., 1982. *Marine Geology,* PrenticeHall, Englewood, 813 pp.

Kimball, M.C., and Teas, H.J., 1975. Nitrogen fixation in mangrove areas of southern Florida. In: Walsh, G.E., Snedaker, S.C., and Teas, H.J., (Eds). *Proceedings of the International Symposium on Biology and Management of Mangroves,* pp. 651-660, University of Florida, Gainsville.

Kristensen, E., Andersen, F.O., and Kofeod, L.H., 1988. Preliminary assessment of benthic community metabolism in a southeast Asian mangrove swamp. *Marine Ecology Progress Series* **48**:137-145.

Lakshmanan, P.T., Shynamma, C.S., Balchand, A.N., and Nambisan, P.N.K., 1987. Distribution and variability of nutrients in Cochin Backwaters, Southwest coast of India. *Indian Journal of Marine Science* **16**:99-102.

Liebeziet, G., and Rau, M.T., 1988. Nutrient chemistry of two Papua New Guinean mangrove systems. In: Field, C.D. and Vannucci, M. (Eds.), *Symposium on New Perspectives in Research and Management of Mangrove Ecosystems,* pp. 37-48, UNDP/UNESCO Regional Project RAS/86/120, New Delhi.

Limpsaichol, P., 1978. Reduction and oxidation properties of the mangrove sediment, Phuket Island, Southern Thailand. *Research Bulletin of Phuket Marine Biological Center, No.* **23**:1-13.

Limpsaichol, P., 1984. An investigation of some ecological parameters at Ao Nam Bor mangroves, Phuket Island, Thailand. In: Soepandmo, E., Rao, A.N. and Macintosh, D.J. (Eds), *Proceedings of the Asian Symposium on Mangrove Environments: Research and Management*, pp. 471-487, University of Malaya and UNESCO, Kuala Lumpur.

Mackin, J.E., and Aller, R.C., 1984. Ammonium adsorption in marine sediments. *Limnology and Oceanography* **29**:250-257.

Mann, F.D., and Steinke, T.D., 1989. Biological nitrogen fixation (acetylene reduction) associated with blue-green algal (cyanobacterial) communities in the Beachwood Mangrove Nature Reserve. I. The effect of environmental factors on acetylene reduction activity. *South African Journal of Botany* **55**:438-446.

Melillo, J.M., Aber, J.D., and Muratore, J.F., 1982. Nitrogen and lignin control of hardwood leaf litter decomposition dynamics. *Ecology* **63**:621-626.

Meyer, J.L., and Schultz, E.T., 1985. Migrating haemulid fishes as a source of nutrients and organic matter on coral reefs. *Limnology and Oceanography* **30**:146-156.

Myint, A., 1986. Preliminary study of nitrogen fixation in Malayan mangrove soils. In: Cragg, S. and Polunin, N. (Eds.), *Workshop on Mangrove Ecosystems Dynamics*, pp. 181-195, UNDP/UNESCO Regional Project RAS/79/002, New Delhi.

Nedwell, D.B., 1975. Inorganic nitrogen metabolism in a eutrophicated tropical mangrove estuary. *Water Research* **9**:221-231.

Nixon, S.W., 1980. Between coastal marshes and coastal waters - A review of twenty years of speculation and research on the role of salt marshes in estuarine productivity and water diversity. In: Hamilton, R. and MacDonald, K.B. (Eds), *Estuarine and Wetland Processes*, pp. 437-525, Plenum, New York.

Nixon, S.W., Furnas, B.N., Lee, V., Marshall, N., Ong, J.-E., Wong, C.-H., Gong, W.-K., and Sasekumar, A., 1984. The role of mangroves in the carbon and nutrient dynamics of Malaysia estuaries. In: Soepandmo, E., Rao, A.N., Macintosh, D.J. (Eds). *Proceedings of the Asian Symposium on Mangrove Environments: Research and Management*, pp. 496-513, University of Malaya and UNESCO, Kuala Lumpur.

Nixon, S.W., and Pilson, M.E.W., 1983. Nitrogen in estuarine and coastal marine ecosystems. In: Carpenter, E.J. and Capone, D.G. (Eds). *Nitrogen in the Marine Environment*, pp. 565-648, Academic Press, New York.

Onuf, C.P., Teal, J.M., and Valiela, I., 1977. Interaction of nutrients, plant growth and herbivory in a mangrove ecosystem. *Ecology* **58**:514-526.

Ovalle, A.R.C., Rezende, C.E., Lacerda, L.D., and Silva, C.A.R., 1990. Factors affecting the hydrochemistry of a mangrove tidal creek, Sepetiba Bay, Brazil. *Estuarine, Coastal and Shelf Science* **31**: 639-650.

Pomeroy, L.R., and Wiegert, R.G., (Eds) 1981. *The Ecology of a Salt Marsh*, Springer-Verlag, New York, 271 pp.

Poovachiranon, S., and Chansang, H., 1982. Structure of Ao Yon mangrove forest and its contribution to coastal ecosystem. In: Kasterman, A.V. and Sastroutoma, S.S. (Eds). *Proceedings of the Symposium on Mangrove Forest Productivity in South-east Asia, Biotrop Special Publication No.* **17**, pp. 101-111.

Potts, M., 1979. Nitrogen fixation (acetylene reduction) associated with communities of heterocystous and non-heterocystous bluegreen algae in mangrove forests of Sinai. *Oecologia* **39**:359-373.

Potts, M., 1984. Nitrogen fixation in mangrove forests. In: Por, F.D. and Dor, I. (Eds), *Hydrobiology of the Mangal*, pp. 155-162, Dr. W. Junk, The Hague,

Reice, S.R., Spira, Y., and Por, F.D., 1984. Decomposition in the mangal of Sinai - the effect of spatial heterogeneity. In: Por, F.D. and Dor, I. (Eds). *Hydrobiology of the Mangal*, pp. 193-199, Dr W. Junk Publ., The Hague.

Rice, D.L., 1982. The detritus nitrogen problem: new observations and perspectives from organic geochemistry. *Marine Ecology Progress Series* **9**:153-162.

Rice, D.L., and Hanson, R.B., 1984. A kinetic model for detritus nitrogen: role of the associated bacteria in nitrogen accumulation. *Bulletin of Marine Science* **35**:326-340.

Rice, D.L., and Tenore, K.R., 1981. Dynamics of carbon and nitrogen during the decomposition of detritus derived from estuarine macrophytes. *Estuarine, Coastal and Shelf Science* **13**:68-690.

Robertson, A.I., 1986. Leaf-burying crabs: their influence on energy flow and export from mixed mangrove forests (*Rhizophora* spp.) in northeastern Australia. *Journal of Experimental Marine Biology and Ecology* **102**:237-248.

Robertson, A.I., 1988. Decomposition of mangrove leaf litter in tropical Australia. *Journal of Experimental Marine Biology and Ecology* **116**:235-247.

Robertson, A.I., 1991. Plant-animal interactions and the structure and function of mangrove forest ecosystems. *Australian Journal of Ecology* **16**:433-443.

Robertson, A.I., Alongi, D.M., Daniel, P.A., and Boto, K.G., 1988. How much mangrove detritus enters the Great Barrier Reef Lagoon? *Proceedings of the VIth International Coral Reef Symposium* **2**:601-606.

Robertson, A.I., and Daniel, P.A., 1989. Decomposition and the annual flux of detritus from fallen timber in tropical mangrove forests. *Limnology and Oceanography* **34**:640-646.

Robertson, A.I., Daniel, P.A., Dixon, P., and Alongi, D.M., 1992. Pelagic biological processes along a salinity gradient in the Fly Delta and adjacent river plume (Papua New Guinea). *Continental Shelf Research* (in press).

Rodina, A.G., 1964. Nitrogen-fixing bacteria in the soils of the mangrove thickets of the Gulf of Tongking. *Doklady Akademii Nauk SSSR Biological Science Section* **155**:1437-1439.

Rosenfeld, J.K., 1979. Interstitial water and sediment chemistry of two cores from Florida Bay. *Journal of Sedimentary Petrology* **49**:989-994.

Ryther, J.H., and Dunstan, W.M., 1971. Nitrogen, phosphorus and eutrophication of the Gulf of Tonkin. *Doklady Akademii Nauk SSSR Biological Science Section* **155**:240-242.

Sahoo, A.K., Sah, K.D., and Gupta, S.K., 1986. Studies on nutrient status of some mangrove muds of the Sunderbans. In: Bhosale, L. (Ed.), *The Mangroves*: pp. 375-377, Shivaji University Kolhapur, India.

Sarala Devi, K., Venugopal, P., Remani, K., Zacharias, D., and Unnithan, R.V., 1983. Nutrients in some estuaries of Kerala. *Mahasagar* **16**:161-173.

Shaiful, A.A.A., 1987. Nitrate reduction in mangrove swamps. *Malaysian Applied Biology* **16**:361-367

Shaiful, A.A.A., Abdul Manan, D.M., Ramli, M.R., and Veerasamy, R., 1986. Ammonification and nitrification in wet mangrove soils. *Malaysian Journal of Science* **8**:47-56.

Smith, S.V., 1984. Phosphorus versus nitrogen limitation in the marine environment. *Limnology and Oceanography* **29**:1149-1160.

Sorensen, J., 1978. Capacity for denitrification and reduction of nitrate to ammonia in a coastal marine sediment. *Applied and Environmental Microbiology* **35**:301-305.

Stanley, S.O., Boto, K.G., Alongi, D.M., and Gillan, F.T., 1987. Composition and bacteria utilization of free amino acids in tropical mangrove sediments. *Marine Chemistry* **22**:13-30.

Steinke, T.D., Naidoo, G., and Charles, L.M., 1983. Degradation of mangrove leaf and stem tissues in situ in Mgeni estuary, South Africa. In: Teas, H.J., (Ed.) *Biology and Ecology of Mangroves, Tasks for Vegetation Science*, Vol. 8, pp. 141-149, Dr. W. Junk, The Hague.

Sumitra-Vijayaraghavan, Rhamadhas, B., Krishna, L., and Royan, J.P., 1980. Biochemical changes and energy content of the mangrove, *Rhizophora mucronata*, leaves during decomposition. *Indian Journal of Marine Science* **9**:120-123.

Taylor, G.T., 1982. The role of pelagic heterotrophic protozoa in nutrient cycling: a review. *Annales Institute Oceanographie, Paris* **58**:227-241.

Tietjen, J.H., 1980. Microbial-meiofaunal interrelationships: a review. In: Colwell, R.R. and Foster, J. (Eds). *Aquatic Microbial Ecology,* pp 130-140, University of Marylands, College Park.

Tietjen, J.H., and Alongi, D.M., 1990. Population growth and effects of nematodes on nutrient regeneration and bacteria associated with mangrove detritus from northeastern Queensland (Australia). *Marine Ecology Progress Series* **68**:169-180.

Twilley, R.R., 1985. The exchange of organic carbon in basin mangrove forests in a Southwest Florida estuary. *Estuarine, Coastal and Shelf Science* **20**:543-557.

Twilley, R.R., Lugo, A.E., and Patterson-Zucca, C., 1986. Litter production and turnover in basin mangrove forests in southwest Florida. *Ecology* **67**:670-683.

Uchino, F., Hambali, G.G., and Yatazawa, M., 1984. Nitrogen-fixing bacteria from warty lenticellate bark of a mangrove tree, *Bruguiera gymnorrhiza* (L.) Lamk. *Applied and Environmental Microbiology* **47**:44-48.

Ullman, W.J., and Sandstrom, M.W., 1987. Dissolved nutrient fluxes from the nearshore sediment of Bowling Green Bay, central Great Barrier Reef Lagoon. *Estuarine, Coastal and Shelf Science* **24**:289-303.

Valiela, I., 1984. *Marine Ecological Processes,* SpringerVerlag, New York, 546 pp.

Valiela, I., and Teal, J.M., 1979. The nitrogen budget of a salt marsh ecosystem. *Nature* **280**:652-656.

van der Valk, A.G., and Attiwill, P.M., 1984. Acetylene reduction in an *Avicennia marina* community in southern Florida. *Australian Journal of Botany* **32**:157-164.

Viner, A.B., 1982. A quantitative assessment of the nutrient phosphate transported by particles in a tropical river. *Revue Hydrobologie Tropicale* **15**:3-8.

Wiebe, W.J., 1987. Nutrient pools and dynamics in tropical marine coastal environments, with special reference to the Caribbean and Indo-West Pacific regions. *UNESCO Reports in Marine Science* **46**:19-42.

Wolanski, E., Jones, M., and Bunt, J.S., 1980. Hydrodynamics of a tidal creek-mangrove swamp system. *Australian Journal of Marine and Freshwater Research* **31**:431-450.

Wolanski, E., and Gardiner, R., 1981. Flushing of salt from mangrove swamps. *Australian Journal of Marine and Freshwater Research* **32**:681-683.

Wolanski, E., and Ridd, P., 1990. Mixing and trapping in Australian tropical coastal waters. In: Cheng, R.T. (Ed.). *Residual Currents and Long-Term Transport.* pp. 165-183. Springer-Verlag, New York.

Wong, C.-H., 1984. Mangrove aquatic nutrients. In: Ong, J.-E., and Gong, W.-K., (Eds). P*roductivity of the Mangrove Ecosystem: Management Implications,* pp. 60-68, Universiti Sains Malaysia, Penang.

Woodroffe, C.D., 1982. Litter production and decomposition in the New Zealand mangrove, *Avicennia marina* var *resinifera. New Zealand Journal of Marine and Freshwater Research* **16**:179-188.

Zieman, J.C., Macko, S.A., and Mills, A.L., 1984. Role of seagrasses and mangroves in estuarine food webs: temporal and spatial changes in stable isotope composition and amino acid content during decomposition. *Bulletin of Marine Science* **35**:380-392.

Zijlistra, J.J., 1988. Fish migrations between coastal and offshore areas. In: Jansson, B.O. (Ed.), *Coastal-offshore ecosystem interactions*, pp. 257-272, Springer-Verlag, Berlin.

Zuberer, D.A., and Silver, W.S., 1975. Mangrove-associated nitrogen fixation. In: Walsh, G.E., Snedaker, S.C., and Teas, H.J., (Eds). *Proceedings of the International Symposium on Biology and Management of Mangroves*, pp. 643-653, University of Florida, Gainesville.

Zuberer, D.A., and Silver, W.S., 1978. Biological dinitrogen fixation (acetylene reduction) associated with Florida mangroves. *Applied and Environmental Microbiology* **35**:567-575.

Zuberer, D.A., and Silver, N.S., 1979. Nitrogen fixation (acetylene reduction) and the microbial colonization of mangrove roots. *New Phytologist* **82**:467-472.

10

Food chains and carbon fluxes

A.I. Robertson, D.M. Alongi and K.G. Boto

10.1 Introduction

Following the pioneering work of W.E. Odum and E.J. Heald in southern Florida in the late 1960's a paradigm was established that stressed the importance of mangrove forests in supporting nearshore secondary production via detrital-based food chains. Although this paradigm has been subsequently modified to include alternative energy and carbon sources for consumers in mangrove ecosystems (e.g. Odum *et al.*, 1982), much of the argument for the preservation of mangrove forests rests on the belief that carbon and energy fixed by mangrove vegetation is the most important nutritive source for animal communities in and near mangrove wetlands (e.g. Saenger *et al.*, 1983).

In the same way that isotope ratios in producers and consumers were used to question the importance of saltmarsh carbon in the food chains of temperate nearshore regions (e.g. Haines, 1977), recent analyses of carbon isotopes in the biota of sub-tropical and tropical mangrove systems has also indicated that the number of consumer organisms in tropical coastal regions adjacent to mangrove forests that are dependent on mangrove carbon is smaller than suggested by the classic paradigm (Rodelli *et al.*, 1984; Zieman *et al.*, 1984). These studies have highlighted the need for a more detailed analysis of mangrove food chains.

In this chapter, we review recent research on the food chains of tropical mangrove systems with particular emphasis on work in tropical Australia. The review is in three sections. We first consider trophodynamics within mangrove habitats by following the fate of components of forest primary production (litter, wood, roots). We then consider the export of mangrove carbon and its influence on consumers and sediments in adjacent habitats. Finally, we identify the major gaps that still exist in our understanding of tropical mangrove ecosystem trophodynamics.

10.2 Food Chains Within Mangrove Habitats

10.2.1 Sources of Energy and Carbon

Earlier chapters in this book indicate that the relative contribution of mangrove carbon to total estuarine or wetland primary production varies widely with forest type and the amount of clear, open water in the system. Within most riverine mangrove systems, where the ratio of forested to open water habitats is high, mangrove production is the dominant source of carbon (see Chapter 7, this volume). Only in large lagoons, with fringing mangrove forests, does phytoplankton, benthic algal or other macrophyte production become a major source of carbon (e.g. Day *et al.*, 1982). Where shading is not severe, prop root epiphytes may be highly productive in some forests. Values for periphyton production on prop roots of 0.14 and 1.1 g $C.m^{-2}.d^{-1}$ have been reported in Florida mangroves (Lugo *et al.*, 1975; Hoffman and Dawes, 1980. Kristensen *et al.*, (1988) recorded benthic production of 110-180 mg $C.m^{-2}.d^{-1}$ in a Thai mangrove forest where mangrove production was estimated to range from 1900 to 2750 mg $C.m^{-2}.d^{-1}$. In most closed canopy mangrove forests light penetration to the mud surface is likely to be too low to support significant benthic algal production (see 10.4 below and Chapter 8, this volume). For instance, it is estimated that benthic primary production in the mangrove forests of Missionary Bay, North Queensland is only ~20mgC.$m^{-2}.d^{-1}$ (Alongi, unpub. data).

In the Indo-West Pacific region, the majority of mangrove forests occur in estuarine areas or as dense forests with dissecting tidal channels in relatively protected embayments. These mangrove systems have a generally high proportion of forest to open water, (see Chapter 3, this volume) and waters are usually very turbid (Blaber, 1980; Blaber *et al.*, 1985; Robertson and Duke, 1987a). Within such habitats energy and carbon fixed by mangrove vegetation is likely to be the dominant contributor to food chains. This is borne out by the analysis of $\delta^{13}C$ ratios in a Malaysian mangrove forest (Rodelli *et al.*, 1984), which showed that many consumers in mangrove habitats had a carbon isotope signature close to that of mangrove tissues.

This pattern may not be so clear where mangroves occur as fringing vegetation on coastal lagoons, such as in West Africa and Mexico, or on small islands in relatively clear water, such as in southern Florida, where they co-occur with seagrass meadows. Food chains are likely to be more complex in such systems, with a number of carbon sources supporting consumers (e.g. Zieman *et al.*, 1985). Water column production can also be high in such systems (see Chapter 7, this volume).

In developing food chain models for turbid, estuarine systems in the Indo-West Pacific region, much effort has been expended recently in studies of the fate of mangrove primary production. The role of phytoplankton production has been assumed to be minor. Below, we review recent efforts to budget the fate of mangrove production.

10.2.2 Direct Grazing on Mangrove Tissue

Direct grazing on mangrove leaves, by insects and arboreal crabs, usually accounts for a very small proportion of leaf production (Heald, 1971; Beaver *et al.*, 1979; Onuf *et al.*, 1977;

Johnstone, 1981; Robertson and Duke, 1987b; Farnsworth and Ellison, 1991). However, leaf area loss to grazers is highly variable among species, sites and individual trees.

In tropical Australia a survey of 25 mangrove tree species showed that mean leaf area losses ranged from 0.3 to 35.0% of expanded leaf area, while the mean coefficient of variation for leaf area loss was 266% (Robertson and Duke, 1987b). For the dominant *Rhizophora* forests in the region, Robertson and Duke estimated that 11 $g.m^{-2}.y^{-1}$, or only 2.1% of canopy production, entered the direct grazing pathway. However, some high intertidal forests in the Indo-West Pacific, such as those dominated by *Heritiera littoralis,* which may lose up to 35% of leaf area to insects (Robertson and Duke, 1987b), are likely to have a much greater percentage of canopy production grazed. Using calculations similar to those in Robertson and Duke (1987b), such forests may have ~20% of canopy production consumed directly by insects. Between these two extremes are forests dominated by species such as *Avicennia marina,* which have intermediate levels of leaf area loss (8.8-12.0% of expanded leaf area; Robertson and Duke, 1987b). In addition, although the degree of leaf herbivory did not vary among life-history stages of species of *Rhizophora, Ceriops* and *Avicennia* in tropical Australia (Robertson and Duke, 1987b), insect defoliation of seedlings of *Xylocarpus* spp. is common, and may be a major source of seedling mortality in some areas (Robertson, unpub. data).

Differences in leaf chemistry between species, or within species in different locations, are most often used to explain differences in the degree of herbivory. Onuf *et al.*, (1977) measured significantly higher grazing rates on *Rhizophora* leaves in forests that served as bird rookeries. Leaves from the forest with the guano input had higher tissue nitrogen concentrations, which presumably made them more desirable to insect grazers. However, a comparison of grazing in forest receiving high nitrogen sewage effluent with a pristine forest in New Guinea failed to find any significant difference in grazing rates (Johnstone, 1981).

In Australia, the poisonous latex produced by the leaves of *Excoecaria agallocha* (Ohigaski *et al.*, 1974) is presumably the reason for the very low levels of leaf damage (Robertson, 1991). Similarly, the high tannin and low nitrogen contents of species of the Rhizophoraceae (Robertson, 1988) are probably the cause of low levels of herbivory compared with species such as *Avicennia marina*, which have less tannins and higher nitrogen concentrations (Robertson, 1988).

However, differences in insect communities between locations are also likely to influence patterns of herbivory. Indirect evidence for this comes from a comparison of the ranking of mangrove trees according to leaf area losses to insects in New Guinea (Johnstone, 1981) and tropical Australia (Robertson and Duke, 1987b), which showed no significant correlation between the rankings. Presumably, differences in the herbivore guilds between the two locations had more to do with differences in herbivory than did differences in leaf chemistry.

10.2.3 Turnover of Litter

Direct consumption

The early Florida model of food chains in mangrove forests suggested that the principal flow of energy was along the route, mangrove leaf litter→saprophytic community→detritus consumers (a mixed trophic level composed of detritivores and omnivores)→ lower carnivores→higher carnivores (Odum and Heald, 1975). Most leaf litter was thought to be flushed into mangrove waterways, where microbial decomposition occurred (Heald, 1971; Odum and Heald, 1975).

Recent work in Australia, South-east Asia, Africa, the Caribbean and South America has shown that large proportions of the leaf and other litter reaching the floor of mangrove forests is consumed or hidden underground by crabs (Sasekumar and Loi, 1983; Loke, 1984; Robertson, 1986 and 1991; Robertson and Daniel, 1989a; Lee, 1989; Japar, 1989; Smith *et al.*, 1989; Micheli *et al.*, 1991; Twilley, in press; W. Wiebe, pers. comm.). Consumption and retention of litter within forests by crab populations has profound effects on pathways of energy and carbon flow within forests, the quantities of material available for export from forests and the cycling of nitrogen to support forest primary production (Robertson, 1991).

Macnae (1968) suggested that the absence of leaf litter accumulations in Indo-West Pacific mangrove forests may be due to the feeding activities of crabs of the sub-family Sesarminae, but it was not until the 1980's that quantitative tests of this hypothesis were performed. In north-eastern Australia the proportions of total litter fall consumed or buried by crabs have been measured in four of the major forest types of the region (Robertson, 1986; Robertson and Daniel, 1989a; Robertson, 1991), and the results of these studies are summarized in Table 1.

There is variation in the relative importance of litter turnover by crabs in different mangrove forests (Table 1), which depends more on the frequency of tidal flushing and the species of crabs present, than on differences in the nutritional quality of the litter. In low- to mid- intertidal forests of *Rhizophora* spp., which are flushed twice daily by tides, the crab *Sesarma messa* consumes or buries at least 154 $gDW.m^{-2}.y^{-1}$ or 28 percent of the annual leaf fall of 556 $gDW.m^{-2}.y^{-1}$ (Table 1, and see Figure 5). This figure is likely to be an underestimate, since it includes no estimate of the consumption of other litter components, including flowers and propagules. *Sesarma messa* has been observed to consume both of these components in the field (Smith, 1987; Robertson, pers obs.) as well as anchoring large numbers of propagules in crab holes, thus preventing their export.

The mean instantaneous standing stock of leaf litter on the floor of *Rhizophora* forests is low (Table 1). Because microbial decomposition of leaves is slow on the forest floor (see 10.2.4), the turnover of leaf material due solely to microbial action is low (Table 1). By difference, tidal export of leaf litter from forests to adjacent waterways accounts for the remaining 71 percent of annual leaf fall (Table 1).

Table 1. Litter processing in mangrove forests in tropical Queensland. Units are $g.m^{-2}$ for standing stocks and $g.m^{-2}.y^{-1}$ for fluxes. Figures in brackets are percentages of the litterfall accounted for by various processes. Percentages for all but *Rhizophora* forests do not add up to exactly 100 percent, as each flux was estimated separately. For *Rhizophora* forests export was calculated by difference. Data from Robertson (1986 and 1988) and Robertson and Daniel (1989a).

Parameter	**Forest type**			
	Rhizophora* **(mid-intertidal)**	***Bruguiera/Ceriops*** **(high-intertidal)**	***Ceriops*** **(high-intertidal)**	***Avicennia*** **(high-intertidal)**
Litter fall	556 (100)	1022 (100)	822 (100)	519 (100)
Litter standing stock	2	6	6	84
Litter consumption by crabs	154 (28)	803 (79)	580 (71)	173 (33)
Microbial decay	5 (0.9)	5 (0.5)	5 (0.6)	168 (32)
Export	397 (71)	252 (25)	194 (24)	107 (21)

* data for leaves; does not include reproductive products or stipules

In high intertidal forests dominated by *Bruguiera* or *Ceriops*, crabs remove up to 79 percent of the total annual litter fall to the sediment (Table 1). Only a fraction of this material is consumed immediately; the remainder, particularly leaf fragments, is plastered on the walls of crab burrows, which appears to aid in the leaching of tannins from leaves (Giddins *et al.*, 1986; Micheli, 1992). The mean instantaneous standing stocks of litter are very low in these forests and the flux of litter due solely to microbial decay is also small (Table 1). Twenty- five percent of litter may be exported each year by tides (Table 1). However, only a portion of this exported litter would enter mangrove waterways as some is likely to be consumed in the intervening low- to mid-intertidal forests.

Forests of *Avicennia* in the high-intertidal zone of North Queensland often have crab communities dominated by microphagous ocypodid rather than leaf eating sesarmid crabs (Robertson and Daniel, 1989a). Litter removal rates are lower in these forests than in *Ceriops* or *Bruguiera* forests (Table 1), even though fallen leaves of *Avicennia* have lower C:N ratios and significantly less tannins than those of *Ceriops* or *Bruguiera* (Robertson, 1988). Because litter consumption is low, there is a relatively high standing stock of litter in *Avicennia* forests (Table 1).

Retention and rapid processing of litter within forests influences trophodynamic processes in forest sediments. The very high sediment bacterial productivities in tropical Australian mangrove forests (mean ~ 1 $gC.m^{-2}.d^{-1}$; Alongi, 1988a) are probably facilitated by rapid litter processing. For instance, Robertson and Daniel (1989a) have calculated that in *Bruguiera* forests, sesarmid crabs void ~ 260 $gC.m^{-2}.y^{-1}$ of litter-derived faeces, equivalent to ~ 70% of bacterial production.

Litter removed by crabs also influences potential exports of materials from forests to adjacent habitats (Robertson, 1986; Twilley, in press). An early estimate of tidal export of

Table 2. The nitrogen required for forest primary production and the contribution to nitrogen conservation within various mangrove forests by crabs which consume and/or bury litter within the forest sediment. Primary production data from Boto & Bunt (1982) and Boto *et al.*, (1984).

Forest type	N required for forest 1° production ($gN.m^{-2}.y^{-1}$)	Burial/Consumption of litter N by crabs ($gN.m^{-2}.y^{-1}$)
Bruguiera/Ceriops	7.5	4.8 (64%)
Ceriops	5.1	2.7 (53%)
Avicennia	14.0	1.5 (11%)
Rhizophora	27.4	0.9 (3%)

litter from mangrove (mainly *Rhizophora* spp.) forests to the adjacent waters of Missionary Bay in north Queensland (Boto and Bunt, 1981a) was based on the assumption that all litter fall was available for export. The first estimate of 19.5 kg $DW.ha^{-1}.d^{-1}$ for export (Boto and Bunt, 1981a) was reduced to 15.3 kg $DW.ha^{-1}.d^{-1}$ when the effect of crabs was included, i.e. 22 percent less than the original.

Using estimates of primary production in several forest types (Table 2) it is also possible to appreciate how litter processing and retention by crabs conserves significant amounts of nitrogen within forests. In high intertidal forests in tropical Australia, between 11 and 64% of N requirements for forest primary production is recycled through litter processing by crabs (Table 2). By contrast, litter retention by *Sesarma messa* has little effect on the nitrogen budget in the more regularly flushed *Rhizophora* forests (Table 2).

Estimates of the importance of leaf processing by mangrove crabs elsewhere in the Indo-west Pacific region also point to the importance of members of the sub-Family Sesarminae. Several studies in Malaysia and Hong Kong have estimated that these crabs consume or bury between 9 and 78% of the annual litter fall in mangrove forests (Sasekumar and Loi, 1983; Loke, 1984; Japar, 1989; Lee, 1990; Gong and Ong, 1990). Comparisons of these estimates with data on the pathways of litter processing in some Florida mangrove systems (e.g. Odum and Heald, 1975) led us to hypothesize that there is a fundamental difference in the fate of litter in New and Old World mangrove forests (Robertson, 1986; Robertson and Daniel, 1989a). However, Lugo and Snedaker (1974) suggested that crabs may be important litter consumers in Florida mangrove forests and recent detailed observations in Panama, Ecuador, Jamaica and Kenya indicate that crabs are also important primary consumers of litter in New World and other Old World forests.

Decomposition processes -leaf litter

Decomposition is the sum of three processes: leaching, saprophytic decay and fragmentation. In the previous section we showed how fragmentation of mangrove leaves by crabs can greatly increase their rate of decomposition. Indeed, data in Robertson (1986)

Table 3. Summary of data on field decomposition rates of mangrove leaf litter. Data are the time in days required for loss of half the original mass of leaf material held in litter bags.

Site	*Avicennia* leaves	*Rhizophora* leaves
Intertidal		
temperate	112[1]	
subtropical	42-60[2]	~ 200[6]
tropical	90[3]	226[7]
Subtidal		
subtropical	13-24[4]	45[8]
tropical	11-20[5]	39-40[9]

Sources: 1. Van der Valk and Attiwill, 1984; 2. Albright, 1976; Goulter and Allaway, 1979; Woodroffe, 1982; Twilley *et al.*, 1986; 3. Robertson, 1988; 4. Reice *et al.*, 1984; Angsupanich *et al.*, 1989; 5. Boonruang, 1984; Robertson, 1988; 6. Twilley *et al.*, 1986; 7. Robertson, 1988; 8. Heald, 1971; 9. Boonruang, 1984; Robertson, 1988;

indicate a two orders of magnitude increase in breakdown rates due to crab feeding. However, between 20 and 70 percent of litter fall in most mangrove forests is not processed immediately by macroconsumers, but degrades slowly in forests or adjacent waters. A number of studies (e.g. Heald, 1971; Fell *et al.,* 1975; Goulter and Allaway, 1979; Fell and Masters, 1980; Boonruang, 1984; Flores-Verdugo *et al.,* 1987; Robertson, 1988; Angsupanich *et al.,* 1989) have measured leaf decomposition rates using litter bags in field studies, and is it obvious that absolute decay parameters for mangrove litter are extremely site and species dependent (Table 3). Leaves decompose faster in subtidal regions of mangrove systems than in the intertidal (Table 3), and presumably leaching and saprophytic decay are more effective when leaves are not subjected to drying (Robertson, 1988). In addition, leaves of *Avicennia* species, with lower tannin and higher initial nitrogen concentrations (Cundell et al., 1979; Robertson, 1988) decompose more rapidly than leaves of *Rhizophora* species (Table 3). However, decomposition rates are not always faster in the tropics versus higher latitudes, and this is particularly true for litter in the intertidal zone (Table 3). For instance Robertson (1988) recorded a half-life of 90 days for *Avicennia* litter in the semi-arid tropics in north-eastern Australia, much greater than observed in subtropical Australia and New Zealand (e.g. Albright, 1976; Goulter and Allaway, 1979). The very high temperatures (> 35°C) and low rainfalls experienced in semi-arid tropical mangrove forests obviously retard decay rates.

During the first 10-14 days of decomposition of most mangrove litter, close to 100% of mass and carbon loss is due to leaching of dissolved organic matter (DOM) (Rice and Tenore, 1981; Benner and Hodson, 1985; Camilleri and Ribi, 1986; Robertson, 1988). This initial leaching is not mediated by microbial populations (Camilleri and Ribi, 1986), but subsequent leaching and mineralization of leaf detritus is highly dependent on the action of bacterial and fungal communities (Fell and Masters, 1980; Benner *et al.,* 1988), which develop rapidly on leaves. Benner *et al.,* (1988) have measured bacterial densities up to 3.4×10^8 cells.cm^{-2} of leaf surface area after 6 days of emersion in seawater. During this period approximately 9 x

10^6 cells.cm^{-2} were lost from the leaf material per hour, due to grazing, cell sloughing or both. Estimated rates of bacterial production during the same period were 1.6 to 8.0 x 10^6 cells.cm^{-2}.h^{-1} (Benner *et al.*, 1988). Soluble tannins from leaves make up a major fraction of the DOM in mangrove litter leachate (Benner *et al.*, 1986). However, despite the correlations between loss of tannins from leaves and increases in microbial populations during the early phases of litter bag studies of decomposition (e.g. Cundell *et al.*, 1979; Robertson, 1988), the leachate from mangrove leaves is not inhibitory to microbial decay, except at high (mg.ml^{-1}) concentrations (Benner *et al.*, 1986).

Leachate from decaying litter is both incorporated rapidly into bacterial and fungal mass associated with litter (Fell and Masters, 1980; Benner *et al.*, 1986) and flocculated into flakes either abiotically or through the mediation of free-living microbes (Wiebe and Pomeroy, 1973; Fell and Masters, 1980; Camilleri and Ribi, 1986). In laboratory experiments, Benner *et al.*, (1986) showed that the efficiency of conversion of leachate into microbial biomass was high (64-94%) and that up to 42% of leachate was utilized during 2-12 h of their incubations. Flocculation of DOM into flakes occurs approximately 48 h after litter is emersed in seawater (Camilleri and Ribi, 1986), and flakes are rapidly (hours) colonized by bacteria, fungi, protozoa and meiofauna (Camilleri and Ribi, 1986).

Approximately 30-50% of the organic matter in mangrove leaves is leachable (Cundell *et al.*, 1979) and the remaining fraction consists of plant structural polymers which breakdown relatively slowly during decomposition. This is reflected in the fact that double exponential models often provide the best fit to data on leaf decomposition in subtidal regions (Table 4). In tropical Australia analysis of data from leaves in submerged litter bags showed that the proportion of labile material in leaves increased in the order *Rhizophora, Ceriops, Avicennia.* Within each species, the rate of decay of the more refractory component of the litter was ~ an order of magnitude less than for the labile fraction. In addition, the rates of decay of both labile and refractory components of *Avicennia* litter were more rapid than for the other species (Table 4). Benner and Hodson (1985) have also shown, using labelled material, that rates of microbial mineralization of the more refractory, lignocellulose fraction of *Rhizophora* leaf detritus is an order of magnitude lower than the leachable fraction. In addition, they showed that the polysaccharide component of the lignocellulose was

Table 4. Double exponential decay parameters (A, k_1 k_2) describing decomposition of leaves from three mangrove species in tropical Australia when submerged. Ash free dry weight (AFDW) data were fitted to the model, $X_t/X_0 = Ae^{-k_1 t} + (1-A)e^{-k_2 t}$, where X_t/X_0 is the proportion of initial material remaining at time t. K_1 and K_2 are decay constants, A is the relatively labile proportion of initial leaf litter and (1-A) is the more refractory portion of the initial material. Data from Robertson (1988).

Parameters	**Mangrove species**		
	Rhizophora stylosa	***Ceriops tagal***	***Avicennia marina***
A (%)	40.3	45.3	61.5
k_1 (d^{-1})	0.047	0.109	0.128
k_2 (d^{-1})	0.008	0.005	0.010

mineralized at twice the rate of the lignin component which indicates a progressive lignin enrichment of decomposing litter with time. However, rates of microbial mineralization of *Rhizophora* leaf litter are lower than losses via leaching. Benner *et al.*, (1991) have estimated that approximately 46% of the polysaccharide loss and 74% of the lignin loss from *Rhizophora* leaves results from leaching.

A variety of factors alter the nutritional status of mangrove litter to consumers during the decomposition process. The leaching of soluble tannins from litter increases its palatibility. Nielsen *et al.*, (1986) have shown clearly that flavolins inhibit the consumption of leaves by the crab *Sesarma smithi*: addition of flavolins to already leached material of *Ceriops tagal* resulted in lower consumption rates by crabs relative to controls. During decomposition the concentrations and mass of nitrogen increase in mangrove litter (e.g. Fell et al., 1980; Rice, 1982; Twilley *et al.*, 1986) and it is likely that this makes the detritus more palatable to consumers. Nevertheless the chemical form of the nitrogen may make it inaccessible to consumers. Rice (1982) showed that during the decomposition of *Rhizophora* leaves there was a positive relationship between nitrogen accumulation and the production of non-labile humic nitrogen rather than living protein. Although microbial nitrogen constitutes < 5% of the total nitrogen on mangrove detritus (eg. Robertson, 1988), it is highly likely that the microbial community and its extracellular exudates (Rice, 1982) form the major source of nutrition for grazers on mangrove litter (Benner *et al.*, 1988). In addition, Camilleri and Ribi (1986) have shown clearly that the aggregate formed from the leachate of mangrove litter is common in the field and can be used as a food source by a variety of small invertebrates inhabiting mangrove forests. After colonization by microbes and meiofauna, flakes have a C:N ration of ~ 6, making them highly nutritious foods relative to most particles derived from the fragmentation of litter (Fell and Masters, 1980; Camilleri and Ribi, 1986).

The degree to which litter decomposition by microbes is important in determining carbon turnover within mangrove forests depends on the degree of flushing by tidal waters or floods and the presence or absence of a leaf consuming fauna (Twilley *et al.*, 1986; Robertson, 1986; Flores- Verdugo *et al.*, 1987; Robertson and Daniel, 1989a). In Florida, in basin mangrove forests, which are flooded approximately 150 times per year, only 21 percent of the annual litter fall is exported from the forest and the remainder undergoes leaching and saprophytic decay in situ (Twilley *et al.*, 1986). In tropical Australia, *in situ* microbial decomposition of litter is important only in high intertidal, or basin forests dominated by *Avicennia,* where leaf consuming crabs are absent (Table 1). In those forests, Robertson and Daniel (1989a) estimated that 32 percent of annual litter fall is decomposed initially by microbial communities.

Decomposition of wood

A large proportion of forest primary production in mangrove habitats is in the form of woody components (see Chapter 8, this volume), but the role of dead wood material in mangrove trophodynamics has been largely ignored. Chai (1982) estimated that 21-31% of the mass of mangrove logs was lost after only 445 days in different mangrove forests in

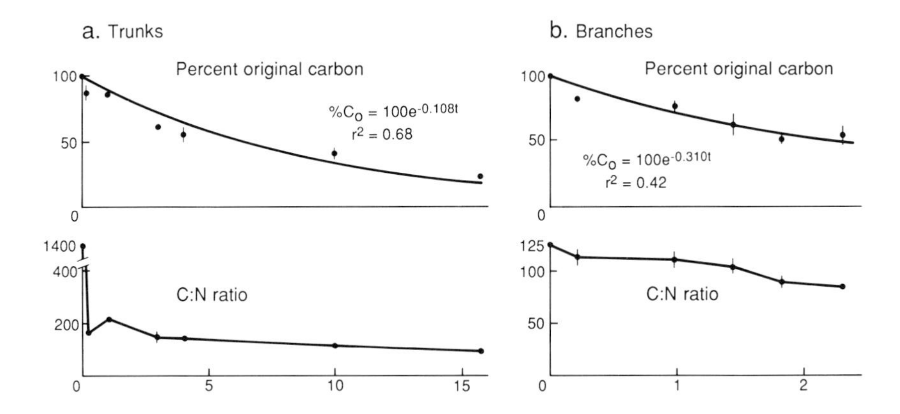

Figure 1. Changes in the percent of original carbon and the C:N ratios during the decomposition of (a) trunks and (b) branches of *Rhizophora* spp. in a tropical Australian mangrove forest. Single order exponential decay equations are given for both wood components. All data from Robertson and Daniel (1989b).

Sarawak, and Ong *et al.*, (1984) observed the break-up of *Rhizophora* stumps, left after logging in Malaysia, after only 2 years. In tropical Australia, Robertson and Daniel (1989b) estimated the standing stocks of fallen wood and the decomposition rates of woody components in order to estimate the annual flux of wood detritus in young and mature riverine *Rhizophora* forests (Figure 1). Trunk wood in mangrove forests decomposes rapidly ($k = 0.108\ y^{-1}$) relative to wood in temperate and tropical terrestrial forests (e.g. Lang and Knight, 1979), but 20 percent of the original carbon in trunks remained after 15.7 y on the forest floor (Figure 1). During the first year of decomposition the C:N ratio of wood dropped from ~ 1400 to ~ 190 but fell slowly thereafter (Figure 1). Branch wood decomposed faster ($k = 0.301\ y^{-1}$; Figure 1) but the C:N ratio decreased slowly during the 2.5 y of the experiment.

Teredinid molluscs were the most important agent of wood breakdown during the decomposition of *Rhizophora* trunk wood (Figure 2). After 15.7 y of decomposition most mangrove trunks consist of a matrix of the tubes produced by teredinids with attached fragments of bark and wood. Although there is rapid leaching of soluble tannins from trunks immediately they hit the forest floor, this process is a very minor component of total weight loss. Direct microbial decomposition also occurs, but Benner and Hodson (1985) have shown that microbial decay of the lignocellulose component of mangrove wood is very slow, particularly under anaerobic conditions such as those experienced on the underside of large tree trunks partly submerged in mud (Robertson and Daniel, 1989b). Teredinid molluscs are capable of rapid consumption of mangrove wood because they possess symbiotic cellulolytic and nitrogen- fixing bacteria which aid in wood breakdown and nutrition of the animals.

By combining estimates of fallen dead wood mass with decomposition rates, Robertson and Daniel (1989a) calculated that detritus from wood breakdown in mature *Rhizophora* forests is as important as leaf litter consumption by crabs in the overall flux of carbon on the floor of mature mangrove forests (Table 5 and see Figure 5). The mass of fallen timber in

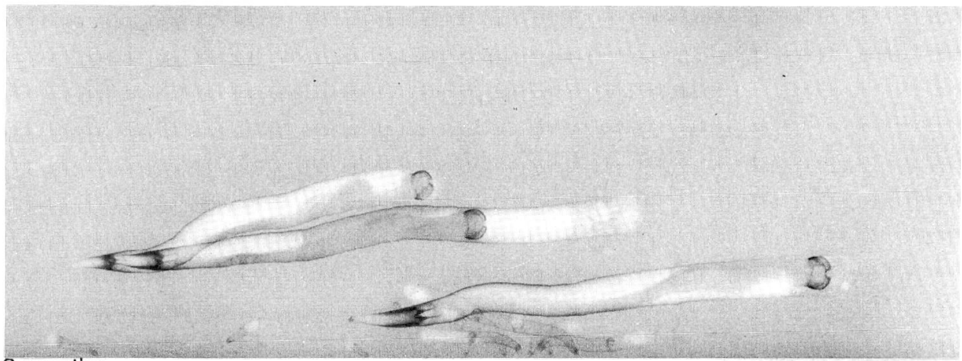

6 months

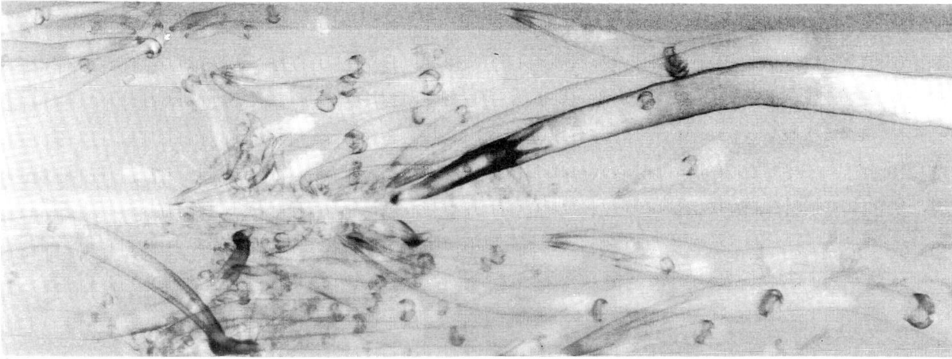

1 year

2 years

Figure 2. Changes in the degree of boring by teredinid molluscs (shipworms) in *Rhizophora* wood without bark during a two-year period of exposure just above the mud surface in the lowest portion of the intertidal gradient of a tropical Australian mangrove forest.

the younger, developing forest they studied was very low, and the contribution of wood to food chains was minor (Table 5).

In the wetter climate of Malaysia, the decomposition rate of mangrove timber is more rapid than that in tropical Australia (Gong and Ong, 1990). Thus in mature forests in the wet tropics, wood decomposition is probably a major contributor to detrital pathways within mangrove systems.

Root decomposition

Although roots are probably a highly productive component of mangrove trees, and mangrove soils often have very high root bio- and necromass (see Chapter 8, this volume), there are only two published studies of mangrove root decomposition and neither are from the tropics. In New Zealand, Albright (1976) showed that roots of *Avicennia marina* (~1mm diameter) buried in the mud lost 30% of their original dry mass in 154 days, while those exposed at the sediment-air surface lost 52% of their original mass during the same period. Working at the southern-most limit of mangrove distribution in Australia where average minimum winter air temperature is 7.2°C, Van der Valk and Attiwill (1984) found that, over 270 days, buried main roots (i.e. roots with a diameter of 1-2 cm) of *Avicennia marina* lost 60% of their initial weight compared to only 15% for smaller, fibrous roots. For comparison, leaves on the mud surface lost 90% of their original weight during the same period.

10.2.4 Role of Bacteria

Several studies during the last two decades point to the central role of sediment bacteria in mangrove trophodynamics. In Florida and Puerto Rico, sediment respiration has been estimated at 197 and 135 $gC.m^{-2}.y^{-1}$, or 62 and 57% of total annual litter fall (Twilley, 1988). In Columbia, Hoppe *et al.*, (1983) found that total community respiration in mangrove sediments was 530 $mg.C.m^{-2}.d^{-1}$ and that the greatest enzymatic potential of particulate matter from various microhabitats in the lagoon they studied was recorded in mangrove sediments. Since enzymatic activity was positively related to bacterial densities, Hoppe *et al.*, (1983) concluded that mangrove sediments were a major site of organic matter decomposition.

Table 5. Estimates of the mass of fallen wood and the fluxes of wood detritus and other mangrove litter components on the floor of mature and young mixed *Rhizophora* spp. forests in Missionary Bay, north Queensland. All fluxes are in $gC.m^{-2}.y^{-1}$; stocks in $gC.m^{-2}$. Data from Robertson and Daniel (1989b).

Parameter	Mature forest	Young forest
Standing stock of dead wood	344	28
Flux of wood detritus	44	4
Leaf consumption by crabs	62	62
Microbial decay of leaves	2	2

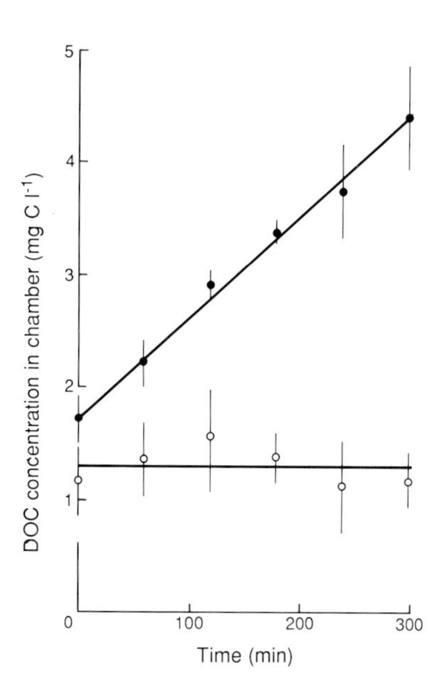

Figure 3. An example of the variations in the mean (±1SD) dissolved organic carbon concentrations in flux chambers overlying untreated (o) and treated-$HgCl_2$ (●) mangrove sediments in tropical Australia. Data from Boto et al., (1989).

Alongi (1988a) measured sediment bacterial biomass and productivity (based on ^3H-thymidine uptake) in several mangrove forests in northern Australia. In the top 2 cm of sediment he measured cell densities ranging from 0.02 to 3.6 (mean, 1.1) x 10^{11} cells.g DW^{-1} and productivities of 0.2-5.1 (mean 1.6) $gC.m^{-2}.d^{-1}$. These standing stocks and production figures are amongst the highest recorded for marine sediments (Alongi, 1990a). Because rates of bacterial production in these sediments are so high and amino acids in sediment pore waters are rapidly and nearly completely used by microbes at the sediment surface (Stanley *et al.,* 1987), Alongi (1989a) argued that sedimentary bacteria act as a carbon 'sink' in tropical mangrove forests (see Figure 5).

This hypothesis has been tested through simultaneous measurements of the fluxes of dissolved organic carbon (DOC) from sediments to overlying waters of a mangrove forest and the productivity and specific growth rates of sediment bacteria (Boto *et al.,* 1989). The study showed that despite the high concentration gradient of DOC between porewaters and the overlying tidal waters, significant efflux of DOC was rarely detected unless sediments were poisoned with mercury (Figure 3). The DOC flux rates from poisoned sediments provided, on average, 35% of the carbon requirements for bacteria at the sediment-water interface. These data suggest that there is little export of DOC from mangrove sediments.

Water-column bacterial production in the Fly River estuary of Papua New Guinea, where there is 800 km^2 of mangrove forest, ranges from 20-498 $mgC.m^{-2}.d^{-1}$ (Robertson *et al.,* 1992). If these figures are indicative of bacterial production in other tropical mangrove waterways, then the potential for bacteria to act as a 'sink' for mangrove carbon is much greater than suggested on the basis of benthic data alone.

10.2.5 Higher Consumers

It is not possible to provide a complete review of the feeding ecology of all mangrove consumer groups in this chapter, and here we concentrate on new information available on trophic processes that are often believed to be important in wetland systems such as mangrove forests (e.g. see Robertson, 1987; Alongi, 1990a).

The role of meiofauna

It is often suggested that most bacterial production in benthic, detritus-based food webs is grazed significantly by meiofauna (e.g. Tenore *et al.,* 1977; Gerlach, 1978; Tietjen, 1980; Findlay and Tenore, 1982). However, recent research in tropical Australia has shown that the meiofauna is likely to play only a minor role in mangrove food chain dynamics.

Alongi (1987a,c) surveyed the densities and taxonomic composition of meiofauna in several northern Australian estuaries and showed that there were very low numbers of animals and very low diversity within all mangrove forests he sampled. Laboratory experiments revealed that one of the factors responsible for low densities of meiofauna was the quantities of soluble tannins derived from decomposing mangrove tissue inside the forests (Alongi, 1987b).

In addition, Tietjen and Alongi (1990) have investigated the ability of nematode species to survive when offered diets of different types of mangrove detritus. Populations of *Monhystera* sp. and *Chromadorina* sp. (both isolated from a nearby mangrove forest) decreased in flasks containing aged detritus; by day 20 of the incubations numbers of *Monhystera* sp. and *Chromadorina* sp. declined by an average of 75 and 46% respectively, when raised on 1 week and 1 month aged detritus of *Rhizophora stylosa* and *Avicennia marina*. Only fresh leaves of *A. marina* sustained increased nematode populations, as a result of their relatively high soluble nitrogen and low tannin content (Tietjen and Alongi, 1990). Bacterial abundance and production were generally not significantly different in flasks containing aged detritus plus nematodes from those in flasks without nematodes (Figure 4). A significant positive correlation between bacterial and nematode abundance occurred in flasks with fresh *A. marina* litter, but not with aged detritus. In general, these experiments indicated that the minimal densities of nematodes necessary to stimulate bacterial production were not achieved on the types of detritus tested by Tietjen and Alongi (1990). The low field densities and inability of nematodes to influence bacterial abundance indicate that the meiofauna may not play a major role in the cycling of organic matter in tropical mangrove forest sediments.

Alongi (1988b) also investigated whether there was any field evidence for interrelationships among microbes (i.e. bacteria, protozoans and microalgae) and meiofauna in mangrove and adjacent mudflat habitats on scales of hours, weeks or months. He found little evidence to suggest that temporal changes in bacterial and microalgal standing stocks and bacterial growth rates were closely linked to temporal variations in protozoan and

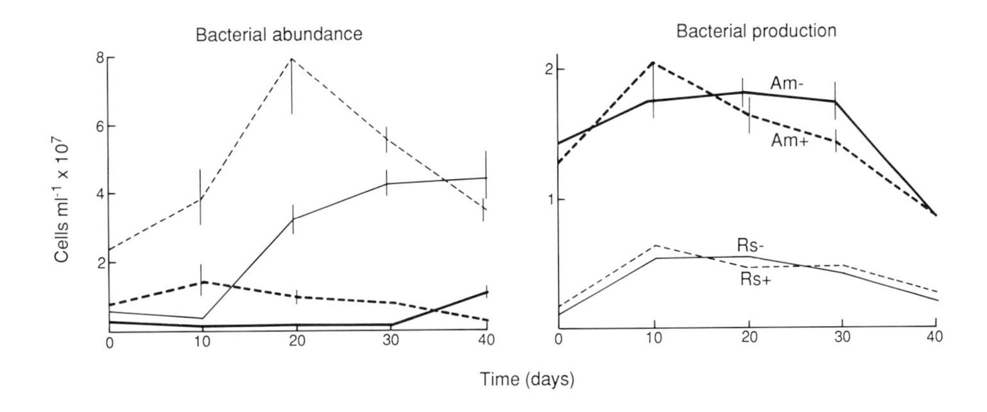

Figure 4. Mean (±1SE) bacterial abundance and production in culture flasks with (+) and without (-) nematodes (*Monhystera* sp.). Bacteria and nematodes were cultured on leaf detritus of the mangroves *Avicennia marina* (Am) and *Rhizophora stylosa* (Rs) that had been aged in seawater for 1 month. From Tietjen and Alongi (1990).

meiofaunal densities. Instead, bacterial population dynamics appear to be controlled more by the degree of tidal flooding and exposure, and daily and seasonal changes in temperature (Alongi 1988a,b).

Diets of benthic macrofauna

Apart from work on leaf-consuming crabs (see 10.2.3), our understanding of the trophic position of most of the benthic macrofauna in mangrove forest systems has not progressed much past that established by Odum and Heald (1975), who described a mixed trophic level composed of fauna consuming detritus, associated microbes and perhaps meiofauna. It is rare that a detailed, quantitative analysis has been made of the diet of a particular group within the macrofauna. Recently, Dye and Lasiak (1986) addressed this deficiency in our understanding by investigating the trophic interactions between bacteria, meiofauna and two species of fiddler crab (*Uca*) which inhabit mudflats immediately adjacent to *Rhizophora* forests in tropical Australia. Despite the large quantities of meiofauna available on the sediment in this region, gut analysis of *Uca vocans* and *U. polita* revealed that microheterotrophs were the major food source for both species. No evidence was found for the ingestion of meiofauna (mainly nematodes) when their densities were compared among natural sediment, and the gut contents and feeding pellets of both crab species. When crabs were excluded from areas of sediment there was a highly significant increase in the density of meiofauna in surface (2 mm depth) sediments. Since no evidence of meiofauna was found in the guts of the fiddler crabs, Dye and Lasiak (1986) interpreted this result as an indication of competition between *Uca* and meiofauna for microbial food resources. In support of this contention, they calculated that during each low tide period populations of *Uca* turned over ~ 43% of the available sediment surface while feeding.

In subtropical Australia, investigations of competition among populations of the gastropod *Bembicium auratum* on a mangrove (*Avicennia*) shore (Branch and Branch, 1980) have shown clearly that on mudflats adjacent to the mangroves, these gastropods consume large quantities of benthic microalgae. When gastropods were excluded from sediments there were rapid and significant increases in chlorophyll *a* in surface sediments, and an order of magnitude increases in the densities of gastropods resulted in similar magnitude decreases in chlorophyll *a*. That populations of gastropods occurring within dense, shaded mangrove forests also have benthic microalgae as a major dietary source appears unlikely, given the very low sediment algal production in such forests (e.g. Kristensen *et al.*, 1988; Alongi, unpub. data).

Although giving no detail of the diet of benthic macrofaunal taxa, the study of isotopic ratios of these consumers in a Malaysian mangrove forest by Rodelli *et al.*, (1984) has shown clearly that macrofaunal taxa differ in their dependance on mangrove primary production as an ultimate food source. For instance, common gastropod species such as *Telescopium telescopium, Cassidula aurisfelis* and *Ellobium aurisjudae* usually had carbon isotope ratios close to that of mangrove detritus. However, some taxa, e.g. *Nerita articulata* and *Cerithidea obtusa* exhibited a wide range of isotope ratios, indicating that they shifted their diets from mangrove to microalgal carbon depending on their location within the forest.

Carbon isotope ratios for species of *Uca* indicated that they were depending on benthic microalgae (Rodelli *et al.*, 1984), in contrast to the findings of Dye and Lasiak (1986) for different species in the same genus. Not surprisingly, leaf-consuming sesarmid crabs have carbon isotope ratios very close to that of mangrove leaves (Rodelli *et al.*, 1984). However, sesarmids and other leaf-eating grapsid crabs in mangrove systems never have isotope ratios that are exactly those of mangrove detritus (Rodelli *et al.*, 1984; Zieman *et al.*, 1985), indicating some input of microalgal carbon to their diets. Field observations on sesarmid crabs in tropical Australia, reveals that they often spend considerable time picking at the mud surface and consuming material other than mangrove leaves (Campbell, 1977).

Species other than grapsid crabs are also direct consumers of mangrove leaf detritus. In tropical Australia, Poovachiranon *et al.*, (1986) showed that aged, intact portions of *Rhizophora* leaves were the preferred food of the amphipod *Parhyale hawaiensis,* and the guts of individuals of this species recovered from the field contained only mangrove leaf detritus. Similar observations have been made for the gastropod *Terebralia palustris* in Okinawa (Nakasone *et al.*, 1985), although this gastropod also feeds directly on the mud surface. The diet of this species appears to shift from deposit feeding to rasping of leaves and bark as individuals grow (Plaziat, 1984).

Nekton

Penaeid prawns are abundant and commercially valuable members of mangrove-associated nekton in most tropical regions, and our understanding of their trophic role in mangrove systems has changed markedly since Odum and Heald (1975) reported that mangrove detritus was a significant component of the gut contents of *Penaeus* spp. in

Florida. Chong and Sasekumar (1981) performed gut analysis on all life-history stages of *Penaeus merguiensis* in Malaysia, and reported that epibenthic postlarvae and juveniles inhabiting mangrove forests and waterways were carnivorous detritivores, consuming large amounts of organic detritus, which they believed to be of mangrove origin. Leh and Sasekumar (1984), who also examined the diet of *P. merguiensis* in Malaysia, concluded that mangrove detritus was the major constituent of the diet of juveniles inhabiting mangrove waterways. Robertson (1988) examined ontogenetic shifts in the diet of *P. merguiensis* in tropical Australia. He found that flocculant detrital material, composed of particulate matter similar to the detrital 'scum' on the surface of the mud in mangrove creeks, was a major dietary item. Such material is composed of organic aggregates formed from dissolved organic material released during the decomposition of mangrove detritus (e.g. Camilleri and Ribi, 1986) and small particles derived from the breakdown of the fecal pellets of crabs, prawns and fish. These organic particles are often bound together in a matrix with diatoms on the mud surface.

Carbon isotope analyses of juvenile *P. merguiensis* in Malaysian mangrove creeks confirms that the detrital material found in the guts of prawns was of mangrove origin (Rodelli *et al.*, 1984). However, the importance of mangrove carbon in supporting *P. merguiensis* decreased markedly as prawns moved offshore to join adult populations (Rodelli *et al.*, 1984).

Despite the earlier claims of the importance of mangrove detritus in the diet of *Penaeus* spp. in Florida, Stoner and Zimmerman (1988) found that the diets of *Penaeus notialis, P. subtilis* and *P. brasiliensis* in a mangrove-fringed lagoon in Puerto Rico were composed of mainly capitellid polychaetes and amphipods, with less than 25% of the diet being detritus. Stable carbon isotope ratios in the penaeids and their food items indicated that the prawns and the majority of sediment dwellers in the lagoon were supported by food chains dependent on benthic algae rather than mangrove detritus.

Although there are a large number of studies on the diets of individual fish species inhabiting mangrove waters (e.g. see Odum *et al.*, 1982 for a review of Florida data), it is rare that the trophic spectrum of a whole mangrove fish community has been studied. On the Pacific coast of Mexico, Yanez-Arancibia (1978) examined the diets of fishes in a large number of mangrove-lined lagoons. His work showed that no one trophic group dominated fish biomass in these systems, nor was there evidence that mangrove detritus was the basis of the food web supporting fish. This is also true for Caribbean lagoonal systems (Yanez-Arancibia *et al.*, 1980). In many of these lagoons phytoplankton and seagrass productivity is as great, or greater than that of mangroves, (Yanez- Arancibia *et al.*, 1980; Day *et al.*, 1982; Flores-Verdugo *et al.*, 1987), so it is not surprising that few of the fishes appear to be wholly dependent on mangrove-based food chains.

In north-eastern Australia the trophic links between juvenile and small adult fish (< 15cm) and mangrove systems have recently been investigated (Robertson, 1988c; Robertson *et al.*, 1988). In terms of both numbers and biomass, zooplankton-feeding fish dominated the fish community of a small mangrove-lined estuary during the late dry season and wet season (Robertson, 1988c), the major recruitment period of most fish into the system (Robertson and

Duke, 1990). During this recruitment period, crab zoea dominated in the diets of most juvenile fish captured during spring tide periods, and crab zoea were two orders of magnitude more abundant in the mangrove waterways than in adjacent nearshore habitats during the same periods (Robertson *et al.*, 1988). Most crab zoea were of the sub- family Sesarminae (P. Dixon, pers. comm.) indicating that the larvae of mangrove leaf-eating crabs may be a major trophic resource of fish recruiting to mangrove forests during the late dry and early wet seasons. Further detailed study of this trophic interaction are necessary to ascertain its general importance, since it is well known that mangrove crabs release most larvae on spring tides (see Chapter 7, this volume). During the remainder of the year copepods were the major dietary item of the zooplanktivores in the fish community (Robertson *et al.*, 1988).

Several commercially important, large carnivorous fish species inhabiting tropical Australian waters appear to have a clear trophic connection to mangrove primary production. Barramundi, *Lates calcarifer,* consume the prawn *Penaeus merguiensis* as a major portion (~ 30% by volume) of their diet (Dunstan, 1958; Davis, 1985; Robertson, 1988b). Since this prawn occurs only in mangrove habitats as a juvenile (see Chapter 7, this volume), and appears to be dependent on mangrove carbon as a food source (see above) there is evidence of a trophic connection between the fish and mangroves. The mangrove jack, *Lutjanus argentimaculatus*, consumes mainly leaf- consuming sesarmid crabs when feeding in mangrove habitats (Robertson and Duke, 1990). Similarly, Sasekumar *et al.*, (1984) have shown that the majority of carnivorous fish which feed in mangrove forests at high tide in Malaysia, feed on grapsid crabs.

In north-western Australia, mangrove fish communities appear to have a high proportion of piscivorous species (Blaber *et al.*, 1985; see Chapter 7 this volume). An analysis of the feeding selectivity of the dominant sharks, carangids and scombrids showed that their most preferred prey were species of Atherindae, *Sillago* spp. and *Harengula* sp., although species of Ambassidae were the numerically dominant small fishes in the mangrove creeks where the work was done.

10.3 Export From Mangrove Habitats

The magnitude and direction of material fluxes between wetlands and adjacent marine habitats appears to be relatively site specific and depends on the geomorphology and tidal regimes of different regions (e.g. Mann, 1975; Odum *et al.*, 1979; Twilley, 1988).

Despite early indications of the importance of material exports from mangrove forests to adjacent waters (e.g. Heald, 1971; Carter *et al.*, 1973; Lugo and Snedaker, 1974), there have been few thorough and quantitative estimates for tropical mangrove forests. These are reviewed below, as is the influence of exported material on adjacent benthic systems.

10.3.1 Particulate Carbon

Golley *et al.*, (1962) and Lugo and Snedaker (1973) estimated the net flux of mangrove particulate matter from different Caribbean mangrove forests, but both data sets were based on sampling over very few tidal cycles. Given the ratio of net fluxes to the total quantities of detritus present in mangrove creeks and the often episodic nature of export from wetlands (Nixon, 1980; Twilley, 1985), it is unlikely that these data provide reliable estimates of export across the mangrove- nearshore boundary.

In Coral Creek, one of eight tidal creeks in Missionary Bay on Hinchinbrook Island in tropical Australia, Boto and Bunt (1981) used a hydrodynamic model developed for the system by Wolanski *et al.*, (1980), together with a detailed topographic survey of the creek, to estimate the forest area flooded daily by tides (~ 400 ha). Three years of data on litterfall in different parts of the catchment were used with the hydrodynamic data to estimate the daily export of litter. Initially it was assumed that all litter was available for export. However, as discussed earlier (section 10.2.3) Boto and Bunt's (1981) original estimate was revised by Robertson (1986) to account for litter consumption and burial by sesarmid crabs. The revised estimate of 7.5 kg $C.ha^{-1}.d^{-1}$ still represents a substantial export. The export of microparticulate organic matter (MPOM) from Coral Creek was estimated to be 1.6 kg $C.ha^{-1}.d^{-1}$ (Boto and Bunt, 1981). Thus total particulate export was 9.1 kg $C.ha^{-1}.d^{-1}$ or 3322 kgC. $ha^{-1}.y^{-1}$ (see Figure 5).

Recently, in a small (11.5 km^2) mangrove forest in Thailand, Wattayakorn *et al.*, (1990) used a hydrodynamic model together with measurements of suspended litter mass on eight tidal cycles, including neap and spring tide periods, to estimate particulate export. They estimated export of between 0.03 and 0.12 kgC. $ha^{-1}.d^{-1}$, two orders of magnitude lower than observed in Coral Creek. It is likely that the one net used to trap mangrove litter in this study did not provide an adequate sampling design for estimating litter standing stocks.

Following subsequent work in Coral Creek, other creeks in Missionary Bay, and survey work in the bay itself (Robertson *et al.*, 1989) it was possible to model the fate of litter exported from the 42.5 km^2 of mangrove forests. Despite the ~ 12,600 tonnes C of litter that enters the bay each year from the mangrove forests (Boto and Bunt, 1981; Robertson, 1986), the mean instantaneous standing stock of litter on the sea floor in most microhabitats in Missionary Bay is extremely low. Robertson *et al.*, (1989) estimated that ~ 70 tonnes C (of exported litter) was processed annually within the Bay, the remaining ~ 12,500 tonnes C was exported to the adjacent Coral Sea (Figure 5).

In two high-intertidal (basin) forests in Florida, Twilley (1985) showed that the monthly net export of carbon was proportional to the cumulative tidal amplitude within the forest. He estimated that export of particulate material from these forests was only ~ 16 $gC.m^{-2}.y^{-1}$.

Gomez (1988) investigated the export of organic matter from a disturbed (logged) and an undisturbed mangrove forest in the Philippines. She found significantly greater concentrations of particulate matter on ebb tides than on flood tides, greater export in the wet

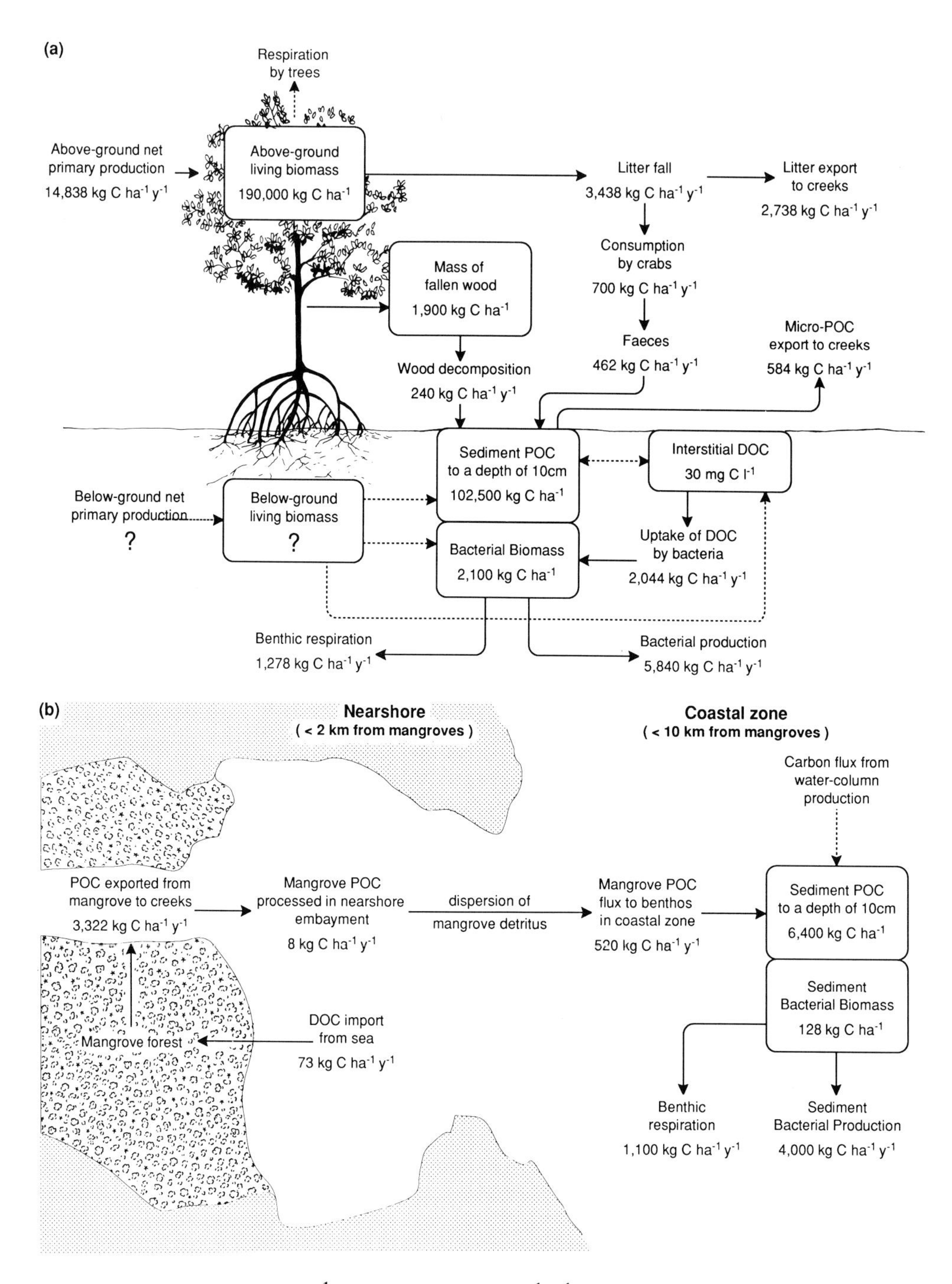

Figure 5. Major stocks (kg C.ha^{-1}) and fluxes (kg C.ha^{-1}y^{-1}) of carbon (a) in mangrove forests of Missionary Bay, northeastern Australia, and (b) between the mangrove forests and the shallow shelf region of the Coral Sea (<10km from the mangroves). Question marks and dashed lines indicate unquantified, but probably important stocks and fluxes. See text for sources of data.

season than the dry season and greater export from disturbed sites only during the wet season. Unfortunately, the sizes of her study sites were not given, so estimates of export per area of forest are not available.

10.3.2 Dissolved Organic Carbon

In the basin forests of southern Florida, Twilley (1985) showed that when flooded, dissolved organic carbon (DOC) concentrations in waters ebbing from the forests were twice as high as in receiving embayment waters. Annual net exports of DOC for two forests, based on the sum of tidal and rainfall data, were 36.7 $gC.m^{-2}.y^{-1}$ (73% of DOC + POC) and 44.3 $gC.m^{-2}.y^{-1}$ (79% of DOC + POC). Seasonal variations in DOC export reflected the frequency of tidal inundation and rainfall events (Twilley, 1985).

The only other long-term (> 12 mo.) data set on the fluxes of dissolved material in mangrove systems comes from the tidally-dominated mangrove forests on Coral Creek, a part of Missionary Bay (Boto and Wellington, 1988). DOC concentrations were typically lower in Coral Creek waters than those reported in other mangrove and non-mangrove estuarine systems (e.g. Valiela *et al.*, 1978; Twilley, 1985; Balasubramanian and Venugopalan, 1984), and most of the dissolved organic matter appeared to be highly refractory (Boto and Wellington, 1988). Although there were statistically significant fluxes of DOC during individual tidal cycles, there was a small annual net import of DOC from adjacent waters of $73kgC.ha^{-1}.y^{-1}$ (Boto and Wellington, 1988). This represents only a small fraction of total carbon flux across the system boundaries (Figure 5).

Recently, Moran *et al.*, (1991) have used the natural fluorescence of vascular plant-derived lignin phenols to track the influence of mangrove forests on dissolved organic material (DOM) in coastal waters, 1 km from the nearest mangrove forest at their study site in the Bahamas. Using simple mixing models they estimated that 10 percent of the DOM at their coastal station was derived from the nearby mangroves.

These studies do not allow us to generalize about the factors which control DOC flux in mangrove systems. However, the fact that there is a small import of DOC to the tidally dominated (no freshwater input) Coral Creek site, indicates that much of the DOC export observed in the one-day study of Nixon *et al.*, (1984) in Malaysian mangrove estuaries was likely to have originated from upland, rather than mangrove habitats. However, there was no freshwater influence at the Bahaman field site studied by Moran *et al.*, (1991), indicating that non-estuarine mangroves on open coasts can export significant DOC.

10.3.3 Influence of Mangrove Carbon on Adjacent Systems

Several approaches have been used to assess the influence that materials exported from mangrove forests may have an adjacent marine habitats. The most integrated approach has used comparisons of the element isotope ratios of mangrove plant material and consumers

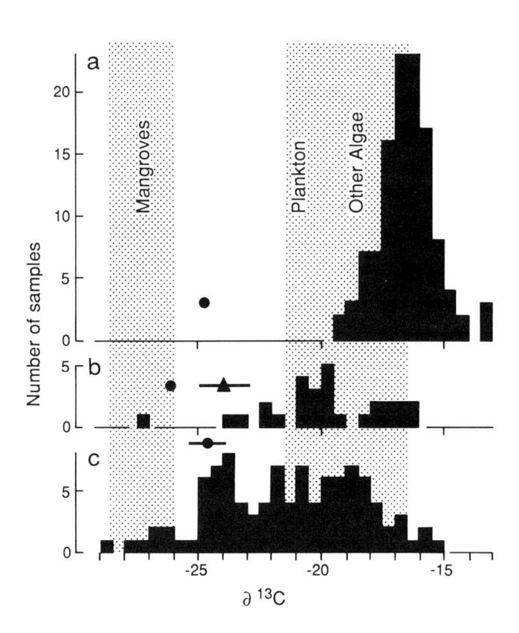

Figure 6. $\delta^{13}C$ isotope concentrations in live mangrove, phytoplankton and other algae in Malaysian waters and frequency distributions of $\delta^{13}C$ values for animals collected in (a) offshore waters (2-18 km from coast), (b) coastal inlets (<2km from coast) and (c) in mangrove forests and their associated mudflats. Closed circles show the ˇ13C values for sediments in the different habitats and the triangle (and SD bars) gives the value for suspended detrital material in the inlet habitat. All data from Rodelli *et al.*, 1984

(Rodelli *et al.*, 1984; Zieman *et al.*, 1984; Fleming *et al.*, 1990). Rodelli *et al.* (1984) showed that the carbon isotope ratios of organisms in coastal inlets adjacent to mangrove forests and in offshore (18 km from coast) regions on the Malaysian coastal showed little or no evidence of mangrove-derived carbon in their tissues (Figure 6). However, δ13C values for sediments and particulate matter in offshore habitats showed strong evidence of export of mangrove carbon. Exported mangrove particulate matter was thus considered to be highly refractory and had very little influence on coastal food chains. The exceptions were some detritus-feeding prawns and fishes in inlets close to mangrove forests, which appeared to derive a component of their nutrition from mangrove detritus (Rodelli *et al.*, 1984).

Two studies in Florida have shown that while consumers in mangrove waterways are dependent on mangrove detritus, there is also a sharp decline in the importance of mangrove carbon to benthic food chains with distance offshore (Zieman *et al.*, 1984; Fleming *et al.*, 1990).

In tropical Australia several studies centered on Missionary Bay have investigated the potential contribution of exported mangrove detritus to nearshore sediments and food chains. Using a mass balance approach Robertson *et al.*, (1989) estimated the total amount of mangrove particulate detritus exported from Missionary Bay and the nearby Hinchinbrook Channel mangrove forests (Wolanski *et al.*, 1990) to the nearshore (<10 km from the coast) waters of the Great Barrier Reef lagoon to be ~ 25,000 tonnes $C.y^{-1}$. Combining data on current speeds and directions in the GBR lagoon (Wolanski and Pickard, 1985) and the sinking rate of leaf litter it was estimated that deposition of mangrove detritus in the lagoon occurred at a mean rate of $520kgC.ha^{-1}.y^{-1}$ (Figure 5). Sediment bacterial production (measured as ^3H- thymidine uptake) in this region of the nearshore averaged $4000kgC.ha^{-1}.y^{-1}$; exported mangrove detritus could therefore potentially support ~ 13% of benthic bacterial production (Figure 5). As Robertson *et al.* (1989) pointed out, these estimates are likely to be the upper limit of the influence of mangrove export on nearshore benthic food chains, because this region has one of the largest concentrations of mangrove forest on the tropical east coast of Australia; in many regions the influence of exported mangrove particulates is likely to minimal, except in localized areas (see below) because of the high ratio of open water to mangrove forest area.

Torgerson and Chivas (1985) analysed the sediments in Missionary Bay to assess the relative influence of terrestrial (= mangrove) and marine carbon. They estimated rates of accumulation of from $440kgC.ha^{-1}.y^{-1}$ in sites 1-2 km. from the mangroves to $10\text{-}40kgC.ha^{-1}.y^{-1}$ 10 km from shore. These figures fit reasonably well with the mass balance estimates of Robertson et al. (1989), but imply rapid breakdown of leaf litter once it reaches the sediment surface. Total benthic metabolism in the region averages $110kgC.ha^{-1}.y^{-1}$ (Figure 5).

Subsequent work on bacterial productivity and DOC fluxes in nearshore sediments in the same region (Alongi *et al.*, 1989) has shown that additions of exported mangrove detritus to intact boxcore samples of sediment did not affect significantly bacterial densities and production rates, or DOC fluxes to overlying waters (Table 6). These results suggested that much of the mangrove litter exported to the nearshore habitats in the region is highly refractory and of poor nutritional quality. This conclusion was borne out by further work with several benthic ciliates from nearshore regions, whose population growth rates were not increased after additions of exported mangrove detritus to experimental microcosms (Alongi, 1990b; and see Chapter 6 this volume). However, sufficient quantities of litter are deposited

Table 6. The effect on (N) bacterial densities (cells x $10^{-9}.g^{-1}$ sediment), (P) bacterial production ($gC.m^{-2}.d^{-1}$) and DOC efflux from the sediment ($gC.m^{-2}.d^{-1}$) of adding powdered mangrove litter detritus to incubated sediment in boxcores taken 10 km offshore from the nearest mangrove forests in Missionary Bay, north Queensland. Controls received no detritus; low ration was equivalent to 135 mg $DW.m^{-2}.d^{-1}$ and the high ration 675 mg $DW.m^{-2}.d^{-1}$. Added litter had a C:N ratio of 52:1. Values are means ±SD. Data from Alongi *et al.*, (1989).

Day	**Control**		**Low ration**		**High ration**	
	N	**P**	**N**	**P**	**N**	**P**
0	3.6±0.5	1.4±0.2	3.6±0.4	1.2±0.3	3.4±0.6	1.0±0.3
1	2.8±0.3	1.2±0.2	2.8±1.1	1.2±0.4	2.1±0.4	2.0±0.8
3	3.4±0.7	1.1±0.1	3.9±1.0	1.1±0.4	3.4±1.1	1.4±0.3
5	3.4±0.6	1.2±0.2	2.6±1.2	1.1±0.3	5.2±1.6	1.2±0.3
10	3.1±0.6	0.8±0.2	3.8±0.6	0.6±0.1	4.6±1.7	0.7±0.2
DOC efflux after 10 days	0.67±0.37		0.78±0.27		0.69±0.24	

in nearshore habitats to influence sediment C:N ratios and POC-tannin concentrations (Alongi *et al.*, 1989; and see also Rodelli *et al.*, 1984 for similar conclusions on coastal Malaysian sediments).

Moran *et al.*, (1991) showed a positive correlation between rates of bacterioplankton secondary production and fluorescence due to mangrove lignin phenols in coastal waters of the Bahamas. Since ~10 percent of the DOM at this coastal site was derived from mangrove forests, there was some evidence of a positive influence of mangrove DOM on coastal pelagic food webs.

It appears from the available information that exported mangrove particulate material has limited influence on offshore (~5-10 km from mangrove sources) food chains in many regions, owing to its wide dispersal and refractory nature. One mechanism which may lead to concentrations of detritus, and significant influences on food chain processes in inshore areas (<5 km from mangrove shores) is the development of coastal boundary layers in highly seasonal areas of the tropics (Wolanski and Ridd, 1990; and see Chapter 3 this volume), which inhibit the mixing between estuarine and offshore waters. This phenomenon has been shown to trap coastal waters for days to months in tropical Australia, depending on coastal topography (Wolanski and Ridd, 1990; Wolanski *et al.*, 1990). Water enriched with mangrove particulate matter is often ejected in regions of offshore waters as tidal jets peeling off from capes and headlands (Wolanski and Ridd, 1990).

10.4 Major gaps in knowledge

10.4.1 Below-ground processes

Next to nothing is known about root biomass, production and decomposition or the quantities and fate of root exudates in mangrove forests. Research in terrestrial forests, and in the temperate analogue of mangrove swamps, saltmarshes, has shown that an understanding of below-ground processes has led to major modifications of models of organic matter processing (e.g. Gosz *et al.*, 1976; McClaugherty *et al.*, 1982; Morris *et al.*, 1984). At present, even estimates of root biomass in mangrove ecosystems are rare, and some are of dubious quality (see Chapter 8, this volume).

Boto *et al.*, (1989) have demonstrated that DOC derived from sediments can support a large proportion of the high bacterial production at the sediment water interface in tropical mangrove forests. The source of this DOC is unknown. Is it derived from root decomposition or from the exudates of living roots? Similarly, to what degree does anaerobic decomposition in the sediments support food chains based on methanogenic or sulphate reducing bacteria (see Odum *et al.*, 1982). What is required to answer such questions is detailed work on production and remineralization processes in mangrove sediments that encompass different forest types, particle sizes and degrees of inundation. Until this is done, carbon and other nutrient budgets for tropical mangrove forests are preliminary at best.

10.4.2 Sedimentation and rates of burial of organic matter

The possible confusion about whether mangrove forests are net exporters or importers of particulate matter, highlighted by Woodroffe (Chapter 2, this volume), results in part from observations at different time scales. In the present chapter we have reviewed ecological data on exports of particulate carbon which indicate that there can be substantial losses from mangrove systems. However, mangroves in many geomorphic settings receive and trap significant quantities of sediment and its associated organic matter over long time periods. To provide comprehensive budgets of materials in mangrove systems requires measurements of sedimentation rates. Such data are extremely rare for mangrove forests (see Chapter 2, this volume).

10.4.3 Food chains supporting higher consumers

A large proportion of the juvenile fish which use mangroves as nursery habitats (see Chapter 7, this volume) are zooplanktivores, yet almost nothing is known about the trophic links supporting mangrove-associated zooplankton. Water column food webs in mangrove systems are likely to be complex, involving interactions between mangrove-derived DOC and POC, bacteria, phytoplankton and their exudates and protozoa and other small heterotrophs. A key question is whether mangrove carbon is the most important carbon source for zooplankton and how the relative importance of mangrove or other sources of carbon supporting zooplankton changes with location?

10.4.4 Trophic interactions and plant population processes

Trophic interactions between plants and animals are important contributors to plant population processes in mangrove systems (see Robertson, 1991 for review). Although direct herbivory on the leaves of mature mangrove trees is usually of minor importance in overall carbon budgets (see 10.2.2), insects appear to defoliate the seedlings of many mangrove tree species. The role of herbivory in controlling survival and growth of juvenile life history stages of mangroves appears to be an important area for research. Similarly, while insects and crabs are known to be important in post-dispersal predation on mangrove seeds, their role in controlling seed set in mangrove trees via pre-dispersal predation is unknown, and likely to be a productive area of research.

10.5 References

Albright, L.J., 1976. *In situ* degradation of mangrove tissues. *New Zealand Journal of Marine and Freshwater Research* **10**:385-389.

Alongi, D.M., 1987a. Intertidal zonation and seasonality of meiobenthos in tropical mangrove estuaries. *Marine Biology* **95**:447-458.

Alongi, D.M., 1987b. The influence of mangrove-derived tannins on intertidal meiobenthos in tropical estuaries. *Oecologia* **71**:537-540.

Alongi, D.M., 1987c. Inter-estuary variation and intertidal zonation of free-living nematode communities in tropical mangrove systems. *Marine Ecology Progress Series* **40**:103-114.

Alongi, D.M., 1988a. Bacterial productivity and microbial biomass in tropical mangrove sediments. *Microbial Ecology* **15**:59-79.

Alongi, D.M., 1988b. Microbial-meiofaunal interrelationships in a tropical mangrove and sandflat ecosystem. *Journal of Marine Research* **46**:349-365.

Alongi, D.M., 1989a. The role of soft bottom benthic communities in tropical mangrove and coral reef ecosystems. *Reviews in Aquatic Sciences* **1**:243-280.

Alongi, D.M., 1990a. The ecology of tropical soft bottom benthic ecosystems. *Oceanography and Marine Biology: Annual Review* **28**:381-496.

Alongi, D.M., 1990b. Abundances of benthic microfauna in relation to outwelling of mangrove detritus in a tropical coastal region. *Marine Ecology Progress Series* **63**:53-63.

Alongi, D.M., 1990c. Effects of mangrove detrital outwelling on nutrient regeneration and oxygen fluxes in coastal sediments of the central Great Barrier lagoon. *Estuarine, Coastal and Shelf Science* **31**:581-598.

Alongi, D.M., Boto, K.G., and Tirendi, F., 1989. Effect of exported mangrove litter on bacterial productivity and dissolved organic carbon fluxes in adjacent tropical nearshore sediments. *Marine Ecology Progress Series* **58**:133-144.

Angsupanich, S., Miyoshi, H., and Hata Y., 1989. Degradation of mangrove leaves immersed in the estuary of Nakama River, Okinawa. *Nippon Suisan Gakkaishi* **55**:147-151.

Balasubramanian, T., and Venugopalan, V.K., 1984. Dissolved organic matter in Pitchavaran mangrove environment, Tamil Nadu, South India. In: Soepadmo, E., Rao, A.N., and Macintosh, D.J. (Eds.), *Proceedings of the Asian Symposium on mangrove environment: research and management*, pp. 496-513, University of Malaya. Kuala Lumpur.

Beever, J.W. III, Simberloff, D., and King, L.L., 1979. Herbivory and predation by the mangrove tree crab *Aratus pisonii*. *Oecologia* **43**:317-328.

Benner, R., and Hodson, R.E., 1985. Microbial degradation of the leachable and lignocellulosic components of leaves and wood from *Rhizophora mangle* in a tropical mangrove swamp. *Marine Ecology Progress Series* **23**:221-230.

Benner, R., Hodson, R.E., and Kirchman, D., 1988. Bacterial abundance and production on mangrove leaves during initial stages of leaching and biodegradation. *Archiv fuer Hydrobiologie* **31**:19-26.

Benner, R., Peele, R., and Hodson, R.E., 1986. Microbial utilization of dissolved organic matter from leaves of the red mangrove, *Rhizophora mangle*, in the Fresh Creek estuary, Bahamas. *Estuarine, Coastal and Shelf Science* **23**:607-620.

Benner, R., Weliky, K., and Hedges, J.I., 1991. Early diagenesis of mangrove leaves in a tropical estuary: molecular-level analyses of neutral sugars and lignin-derived phenols. *Geochimica and Cosmochimica Acta* **54**:1991-2001.

Blaber, S.J.M., 1980. Fish of the Trinity inlet system of North Queensland with notes on the ecology of tropical Indo-Pacific estuaries. *Australian Journal of Marine and Freshwater Research* **31**:137-146.

Blaber, S.J.M., 1986. Feeding selectivity of a guild of piscivorous fish in mangrove areas of north-west Australia. *Australian Journal of Marine and Freshwater Research* **37**:329-336.

Blaber, S.J.M., Young, J.W., and Dunning, M.C., 1985. Community structure and zoogeographic affinities of the coastal fishes of the Dampier region of North-western Australia. *Australian Journal of Marine and Freshwater Research* **36**:247-266.

Boonruang, P., 1984. The rate of degradation of mangrove leaves, *Rhizophora apiculata* B.L. and *Avicennia marina* (Forsk) Vierh. at Phuket Island, Western Peninsula of Thailand. In: Soepadmo, E., Rao, A.N., and Macintosh, D.J., (Eds.), *Proceedings of the Asian Symposium on Mangrove Environment: Research and Management,* pp. 200-208,University of Malaysia and UNESCO, Kuala Lumpur.

Boto, K.G., Alongi, D.M., and Nott, A.L.J., 1989. Dissolved organic carbon-bacteria interactions at the sediment water interface in a tropical mangrove system. *Marine Ecology Progress Series* **51**:243-251.

Boto, K.G., and Bunt, J.S., 1981. Tidal export of particulate organic matter from a northern Australian mangrove system. *Estuarine, Coastal and Shelf Science* **13**:247-255.

Boto, K.G., Bunt, J.S. and Wellington, J.T. 1984. Variation in mangrove forest productivity in northern Australia and Papua New Guinea. *Estuarine, Coastal and Shelf Science* **19**:321-329.

Boto, K.G., Robertson, A.I., and Alongi, D.M., 1991. Mangrove and nearshore connections - a status report from the Australian perspective. In: Alcala, A., (Ed.), *Proceedings ASEAN-Australia Coastal Living Resources Program,* pp. 459-467, University of Philippines, Manila.

Boto, K.G., and Wellington, J.T., 1988. Seasonal variations in concentrations and fluxes of dissolved organic and inorganic materials in a tropical tidally-dominated mangrove waterway. *Marine Ecology Progress Series* **50**:151-160.

Branch, G.M., and Branch, A., 1980. Competition in *Bembecium auratum* (Gastropoda) and its effects on microalgal standing stock in mangrove mud. *Oecologia* **46**:106-114

Bunt, J.S., Boto, K.G., and Boto, G., 1979. A survey method for estimating potential levels of mangrove forest primary production. *Marine Biology* **52**:123-128.

Camilleri, J.C., and Ribi, G., 1986. Leaching of dissolved organic carbon (DOC) from dead leaves, formation of flakes of DOC, and feeding on flakes by crustaceans in mangroves. *Marine Biology* **91**:337-344.

Campbell, G.R., 1977. A comparative study of the distribution, physiology, and behaviour of five Sesarminae species occurring along the Ross River estuary, Townsville. M.Sc. thesis, James Cook University of North Queensland, Australia, 127pp.

Carter, M.R., Burns, L.A., Cavinder, T.R., Dugger, K.R., Fore, P.L., Hicks, D.B., Revells, H.L., and Schmidt, T.W., 1973. Ecosystems analysis of the Big Cypress Swamp and estuaries. U.S. EPA Region IV, South Florida Ecology Study.

Chai, P.P.K., 1982. Ecological studies of mangrove forests in Sarawak. Ph. D. Thesis, University of Malaya, Malaysia, 424 pp.

Cundell, A.M., Brown, M.S., Stanford, R., and Mitchell, R., 1979. Microbial degradation of *Rhizopora mangle* leaves immersed in the sea. *Estuarine, Coastal and Marine Science* **9**:281-286.

Davis, T.L.O., 1985. The food of barramundi, *Lates calcarifer* (Bloch), in coastal and inland waters of Van Diemans Gulf and the Gulf of Carpentaria, Australia. *Journal of Fish Biology* **26**:669-682.

Day, J.W., Day, R.H., Barreiro, M.T., Ley-Lou, F., and Madden, C.J., 1982. Primary production in the Laguna de Terminos, a tropical estuary in the southern Gulf of Mexico. *Oceanologica Acta* **5** (suppl.), 269-276.

Dunstan, D.J., 1959. The barramundi in Queensland waters. CSIRO Division of Fisheries and Oceanography Technical Paper No 5.

Dye, A.H., and Lasiak, T.A., 1986. Microbenthos, meiobenthos and fiddler crabs: trophic interactions in a tropical mangrove sediment. *Marine Ecology Progress Series* **32**:259-264.

Farnsworth, E.J., and Ellison, A.M., 1991. Patterns of herbivory in Belizeau mangrove swamps. *Biotropica* **23**:555-567.

Fell, J.W., Cefalu, R.C., Master, I.M., and Tallman, A.S., 1975. Microbial activities in the mangrove (*Rhizophora mangle*) leaf detrital system. In: Walsh, G.E., Snedater, S.C., and Tens, H.J., (Eds.), *Proceedings of the international symposium on biology and management of mangroves*, pp. 661-679, University of Florida, Gainesville.

Fell, J.W., and Masters, I.M., 1980. The association and potential role of fungi in mangrove detrital systems. *Botanica Marina* **23**:257-263.

Fell, J.W., Master, I.M., and Newell, S.Y., 1980. Laboratory model of the potential role of fungi in the decomposition of red mangrove (*Rhizophora mangle*) leaf litter. In: Tenore, K.R., and Coull, B.C., (Eds.), *Marine Benthic Dynamics*, pp. 359-372, University of South Carolina Press, Columbia, SC.

Findlay, S.E.G., and Tenore, K.R., 1982. Effects of a free-living marine nematode (*Diplolaimella chitwoodi*) on detrital carbon mineralization. *Marine Ecology Progress Series* **8**:161-166.

Fleming, M., Lin, G., and L das S., Sternberg, L., 1990. Influence of mangrove detritus in an estuarine ecosystem. *Bulletin of Marine Science* **47**:663-669.

Flores-Verdugo, F.J., Day, Jr. J.W. and Briseno-Duenas R., 1987. Structure, litter fall, decomposition and detritus dynamics of mangroves in a Mexican coastal lagoon with an ephemeral inlet. *Marine Ecology Progress Series* **35**:83-90.

Gerlach, S.A., 1978. Food-chain relationships in subtidal silty sand marine sediments and the role of meiofauna in stimulating bacterial productivity. *Oecologia* **33**:55-69.

Giddins, R.L., Lucas, J.S., Neilson, M.J., and Richards, G.N., 1986. Feeding ecology of the mangrove crab *Neosarmatium smithi* (Crustancea: Decapoda: Sesarmidae). *Marine Ecology Progress Series* **33**:147-155.

Gomez, P.L., 1988. Production and transport of organic matter in mangrove-dominated estuaries. In: Jansson, B.O., (Ed.), *Coastal-Offshore Ecosystem Interactions*, pp. 181-187, Springer- Verlag, New York.

Gong, W.-K., and Ong, J.-E., 1990. Plant biomass and nutrient flux in a managed mangrove forest in Malaysia. *Estuarine, Coastal and Shelf Science* **31**:519-530.

Goulter, P.F.E., and Allaway, W.G., 1979. Litter fall and decomposition in a mangrove stand, *Avicennia marina* (Forsk) Vierh. in Middle Harbour, Sydney. *Australian Journal of Marine and Freshwater Research* **30**:541-6.

Gosz, J.R., Likins, G.E., and Bormann, F.H., 1976. Organic matter and nutrient dynamics of the forest and forest floor in the Hubbard Brook Forest. *Oecologia* **22**:305-320

Haines, E.B., 1977. The origins of detritus in Georgia salt marsh estuaries. *Oikos* **29**:254-260.

Heald, E.J., 1971. The production of organic detritus in a south Florida estuary. Sea Grant Technical Bulletin No 6, University of Miami Sea Grant Program (Living Resources), Miami, Florida 110 pp.

Hoffman, W.E., and Dawes, C.J., 1980. Photosynthetic rates and primary production by two Florida benthic red algal species from a salt marsh and a mangrove community. *Bulletin of Marine Science* **30**:358-364.

Hoppe, H.G., Gocke, K., Zamorano, D., and Zimmerman, R., 1983. Degradation of macromolecular organic compounds in a tropical lagoon (Cienaga Grande, Columbia) and its ecological significance. *Internationale Revue der gesamten Hydrobiologie* **68**:811-824.

Japar, S.B., 1989. Studies on leaf litter decomposition of the mangrove *Rhizophora apiculata* Bl. PhD. Thesis, Universiti Sains Malaysia, Malaysia, 322 pp.

Johnstone, I.M., 1981. Consumption of leaves by herbivores in mixed mangrove stands. *Biotropica* **13**:252-259.

Kristensen, E., Andersen, F.O., and Kofoed, L.H., 1988. Preliminary assessment of benthic community metabolism in a south-east Asian mangrove swamp. *Marine Ecology Progress Series* **48**:137-145.

Lang, G.E., and Knight, D.H., 1979. Decay rates for the boles of tropical trees in Panama. *Biotropica* **11**:316-317

Lee, S.Y., 1989. Litter production and turnover of the mangrove *Kandelia candel* (L.) Druce in a Hong Kong tidal pond. *Estuarine, Coastal and Shelf Science* **29**:75-87.

Loke, Y.M., 1984. Energetics of leaf litter production and its pathway through the sesarmid crabs in a mangrove ecosystem. M. Sc. Thesis, Universiti Sains Malaysia, Malaysia, 140 pp.

Lugo, A.E., and Snedaker, S.C., 1974. The ecology of mangroves. *Annual Review of Ecology and Systematics* **5**:39-64.

Lugo, A.E., Evink, G., Brinson, M., Broce, A., and Snedaker, S.C., 1975. Diurnal rates of photosynthesis, respiration and transpiration in mangrove forests of south Florida. In: Golley, F.B., and Medina, E., (Eds.), *Tropical ecological systems,* Springer, New York.

Macnae, W., 1968. A general account of the fauna and flora of mangrove swamps and forests in the Indo-West Pacific region. *Advances in Marine Biology* **6**:73-270.

McClaugherty, C.A., Aber, J.D., and Melillo, J.M., 1982. The role of fine roots in the organic matter and nitrogen budgets of two forested ecosystems. *Ecology* **63**:1481-1490.

Mann, K.H., 1975. Relationship between morphometry and biological functioning in three coastal inlets of Nova Scotia. In: Cronin, L.E., (Ed.), *Estuarine Research,* pp. 634-644, Academic Press Inc., New York.

Micheli, F., 1992. Feeding ecology of mangrove crabs (*Sesarma messa* and *Sesarma smithi*) in north eastern Australia. Submitted to *Journal of Experimental Marine Biology and Ecology.*

Micheli, F., Gherardi, F., and Vannini, M., 1991. Feeding and burrowing ecology of two East African mangrove crabs. *Marine Biology* **111**:247-254.

Moran, M.A., Wicks, R.J., and Hodson, R.E., 1991. Export of dissolved organic matter from a mangrove swamp ecosystem: evidence from natural fluoresence, dissolved lignin phenols, and bacterial secondary production. *Marine Ecology Progress Series* **76**:175-184.

Morris, J.T., Houghton, R.A., and Botkin, D.B., 1984. Theoretical limits of belowground production by *Spartina alterniflora*: an analysis through modelling. *Ecological Modelling* **26**:155-176.

Nielson, M.J., Giddins, R.L., and Richards, G.N., 1986. Effects of tannins on the palatability of mangrove leaves to the tropical sesarmid crab *Neosarmatium smithi. Marine Ecology Progress Series* **34**:185-186.

Nixon, S.W., Furnas, B.N., Lee, V., Marshall, N., Ong, J-E., Wong, C-H., Gong, W-K., and Sasekumar, A., 1984. The role of mangroves in the carbon and nutrient dynamics of Malaysian estuaries. In: Soepadmo, E., Rao, A.N., and MacIntosh, D.J. (Eds.), *Proceedings of the Asian symposium on mangrove environment: research and management,* pp. 534-544, University of Malaya and UNESCO, Kuala Lumpur.

Odum, W.E., Fisher, J.S., and Pickral, J.C., 1979. Factors controlling the flux of particulate organic carbon from estuarine wetlands. In: Livingstone, R.J., (Ed.), *Ecological processes in coastal and marine ecosystems,* pp. 69-80, Plenum Press, New York.

Odum, W.E., and Heald, E.J., 1975. The detritus-based food web of an estuarine mangrove community. In: Cronin, L.E., (Ed.), *Estuarine Research,* pp. 265-286, Academic Press Inc., New York.

Odum, W.E., McIvor, C.C., and Smith III, T.J., 1982. The ecology of the mangroves of south Florida: a community profile. U.S. Fish and Wildlife Service, Office of Biological Services, Washington D.C., FWS/OBS-81/24, 144 pp.

Ohigashi, H., Katsumata, H., Kawazu, K., Koshimizu, K., and Mitsui, T., 1974. A piscicidal constituent of *Excoecaria agallocha. Agricultural Biological Chemistry* **38**:1093-1095.

Ong, J-E., Gong, W-K., Wong, C-H., and Dhanarajan, G., 1984. Contributions of aquatic productivity in a managed mangrove ecosystem in Malaysia. In: Soepadmo, E., Rao, A.N., and MacIntosh, D.J., (Eds.), *Proceedings of the Asian symposium on mangrove environment: research and management*, pp. 209-215, University of Malaya and UNESCO, Kuala Lumpur.

Onuf, C.P., Teal J.M., and Valiela I., 1977. Interactions of nutrients, plant growth and herbivory in a mangrove ecosystem. *Ecology* **58**:514-526.

Plaziat, J.C., 1984. Mollusk distribution in the mangal. In: Por, F.D., and Dor, I., (Eds.), *Hydrobiology of the mangal*, pp. 111-144, Dr. W. Junk, The Hague.

Reice, S.R., Spira, Y., and Por, F.D., 1984. Decomposition in the mangal of Sinai: the effects of spatial heterogeneity. In: Por, F.D., and Dor, I., (Eds.), *Hydrobiology of the mangal*, pp. 193-200, Dr. W. Junk, The Hague.

Rice, D.L., 1982. The detritus nitrogen problem: new observations and perspectives from organic geochemistry. *Marine Ecology Progress Series* **9**:153-162.

Rice, D.L., and Tenore, K.R., 1981. Dynamics of carbon and nitrogen during the decomposition of detritus derived from estuarine macrophytes. *Estuarine, Coastal and Shelf Science* **13**: 681-690.

Robertson, A.I., 1986. Leaf-burying crabs: their influence on energy flow and export from mixed mangrove forests (*Rhizophora* spp.) in north eastern Australia. *Journal of Experimental Marine Biology and Ecology* **102**:237-248.

Robertson, A.I., 1987. The determination of trophic relationships in mangrove-dominated systems: areas of darkness. In: Field, C.D., and Dartnall, A.J., (Eds.), *Mangrove ecosystems of Asia and the Pacific: status, exploitation and management*, pp. 292-304, Australian Institute of Marine Science, Townsville.

Robertson, A.I., 1988a. Decomposition of mangrove leaf litter in tropical Australia. *Journal of Experimental Marine Biology and Ecology* **116**:235-247.

Robertson, A.I., 1988b. Abundance, diet and predators of juvenile banana prawns *Penaeus merguiensis* in a tropical mangrove estuary. *Australian Journal of Marine and Freshwater Research* **39**:467-478.

Robertson, A.I., 1988c. Food chains in tropical Australian mangrove habitats: a review of recent research. In: Field, C.D., and Vannucci, M., (Eds.), *Symposium on new perspectives in research and management of mangrove ecosystems*, pp. 23-36, UNDP-UNESCO, New Delhi.

Robertson, A.I., 1991. Plant-animal interactions and the structure and function of mangrove forest ecosystems. *Australian Journal of Ecology* **16**:433-443.

Robertson, A.I., Alongi, D.M., Daniel, P.A., and Boto, K.G., 1989. How much mangrove detritus enters the Great Barrier Reef Lagoon. *Proceedings of the 6th International Coral Reef Symposium* **2**:601-606.

Robertson, A.I., and Daniel, P.A., 1989b. Decomposition and the annual flux of detritus from fallen timber in tropical mangrove forests. *Limnology and Oceanography* **34**:640-646.

Robertson, A.I., and Daniel, P.A., 1989a. The influence of crabs on litter processing in high intertidal mangrove forests in tropical Australia. *Oecologia* **78**:191-198.

Robertson, A.I., Daniel, P.A., Dixon, P., and Alongi, D.M., 1992 Pelagic biological processes along a salinity gradient in the Fly delta and river plume (Papua New Guinea). *Continental Shelf Research.*

Robertson, A.I., and Duke, N.C., 1987a. Insect herbivory on mangrove leaves in north Queensland. *Australian Journal of Ecology* **12**:1-7.

Robertson, A.I., and Duke, N.C., 1987b. Mangroves as nursery sites: comparisons of the abundance and species composition of fish and crustaceans in mangroves and other nearshore habitats in tropical Australia. *Marine Biology* **96**:193-205.

Robertson, A.I., and Duke, N.C., 1990. Recruitment, growth and residence time of fishes in a tropical Australian mangrove system. *Estuarine, Coastal and Shelf Science* **31**:723-743.

Rodelli, M.R., Gearing, J.N., Gearing, P.J., Marshall, N., and Sasekumar, A., 1984. Stable isotope ratio as a tracer of mangrove carbon in Malaysian ecosystems. *Oecologia* **61**:326-333.

Saenger, P., Hegerl, E.J., and Davie, J.D.S., 1983. Global status of mangrove ecosystems. *The Environmentalist* **3**:1-88.

Sasekumar, A., and Loi, J.J., 1983. Litter production in three mangrove forest zones in the Malay Peninsular. *Aquatic Botany* **17**:283-290.

Sasekumar, A., Ong, T-L., and Thong, K-L., 1984. Predation of mangrove fauna by marine fishes. In: Soepadmo, A., Rao, A.N., and McIntosh, D. (Eds.), *Proceedings Asian Symposium Mangrove Environment: Research and Management,* pp. 378-384, University of Malaya and UNESCO, Kuala Lumpur.

Smith, T.J., III. 1987. Seed predation in relation to tree dominance and distribution in mangrove forests. *Ecology* **68**:266-273.

Smith, T.J., III, Chan, H.T., McIvor, C.C., and Robblee, M.B., 1989. Comparisons of seed predation in tropical, tidal forests from three continents. *Ecology* **70**:146-151.

Stanley, S.O., Boto, K.G., Alongi, D.M., and Gillan, F.T., 1987. Composition and bacterial utilization of free amino acids in tropical mangrove sediments. *Marine Chemistry* **22**:13-30.

Stoner, A.W., and Zimmerman, R.J., 1988. Food pathways associated with penaeid shrimps in a mangrove-fringed estuary. *U.S. Fishery Bulletin* **86**:543-551.

Tenore, K.R., Tietjen, J.H., and Lee, J.J., 1977. Effects of meiofauna on incorporation of aged eelgrass by the polychaete *Nephthys incisa*. *Journal of the Fisheries Research Board of Canada* **34**:563-567.

Tietjen, J.H., 1980. Microbial-meiofaunal interrelationships: a review. In: Colwell, R.R., and Foster, J., (Eds.), *Aquatic microbial ecology*, pp. 130-140, University of Maryland Press, College Park.

Tietjen, J.H., and Alongi, D.M., 1990. Population growth and effects of nematodes on nutrient regeneration and bacteria associated with mangrove detritus from north eastern Queensland (Australia). *Marine Ecology Progress Series* **68**:169-179.

Torgerson, T., and Chivas, A.R., 1985. Terrestrial organic carbon in marine sediment: a preliminary balance for a mangrove environment derived from ^{13}C. *Chemical Geology (Isotope Geoscience Section)* **53**:379-390.

Twilley, R.R., 1985. The exchange of organic carbon in basin mangrove forests in a southwest Florida estuary. *Estuarine, Coastal and Shelf Science* **20**:543-557.

Twilley, R.R., Lugo, A.E., and Patterson-Zucca, C., 1986. Litter production and turnover in basin mangrove forests in southwest Florida. *Ecology* **7**:670-683.

Twilley, R.R., 1988. Coupling of mangroves to the productivity of estuarine and coastal waters. In: Jansson, B.O., (Ed.), *Coastal-Offshore Ecosystem: Interactions*, pp. 155-180, Springer-Verlag Berlin Heidelberg.

Twilley, R.R., in press. Properties of mangrove ecosystems related to the energy signature of coastal environments. In: Hall, C.A.S., (Ed.), *Maximum Power.*

Valiela, I., Teal, J., Volkmann, S., Shafer, D., and Carpenter, E.J., 1978. Nutrient and particulate fluxes in a salt marsh ecosystem: tidal exchanges and inputs by precipitation and groundwater. *Limnology and Oceanography* **23**:798-812.

Van der Valk, A.G., and Attiwill, P.M., 1984. Decomposition of leaf and root litter of *Avicennia marina* at Westernport Bay, Victoria, Australia. *Aquatic Botany* **18**:205-221.

Wolanski, E., Jones, M., and Bunt, J.S., 1980. Hydrodynamics of a tidal creek-mangrove swamp system. *Australian Journal of Marine and Freshwater Research* **31**:431-50.

Wolanski, E., Mazda, Y., King, B., and Gay, S., 1990. Dynamics, flushing and trapping in Hinchinbrook Channel, a giant mangrove swamp, Australia. *Estuarine, Coastal and Shelf Science* **31**:555-580.

Wolanski, E., and Pickard, G.D., 1985. Long-term observations of currents on the central Great Barrier Reef continental shelf. *Coral Reefs* **4**:47-57.

Wolanski, E., and Ridd, P., 1990. Mixing and trapping in Australian tropical coastal waters. In: Cheng, R.T., (Ed.), *Residual currents and long-term transport,* pp. 165-183, Springer-Verlag, New York.

Woodroffe, C.D., 1982. Litter production and decomposition in the New Zealand mangrove, *Avicennia marina* var. *resinifera*. New Zealand *Journal of Marine and Freshwater Research* **16**:179-188.

Yanez-Arancibia, A., 1978. Taxonomia, ecologia y estructura de las communidades de peces en lagunas costeras con bocas efimeras del Pacifico de Mexico. *Publicaciones especial del Centro de Ciencias del Mar y Limnologia Universidad del Nacional Autonoma de Mexico* **2**:1-306.

Yanez-Arancibia, A., Linares, F.A., and Day, Jr., J.W., 1980. Fish community structure and function in Terminos Lagoon, a tropical estuary in the southern Gulf of Mexico. In; Kennedy, V.S., (Ed.), *Estuarine Perspectives*, pp. 465-482, Academic Press, New York.

Zieman, J.C., Macko, S.A., and Mills, A.L., 1984. Role of seagrasses and mangroves in estuarine food webs: temporal and spatial changes in stable isotope composition and amino acid content during decomposition. *Bulletin of Marine Science* **35**:380-392.

11

Concluding remarks: research and mangrove conservation

A.I. Robertson

The contributions to this book show that our understanding of the factors controlling the structure and function of tropical mangrove ecosystems has advanced significantly in the last decade. Whole areas of knowledge, such as plant-animal interactions and their effect on forest structure, and the details of nutrient cycles and food webs have been revealed by recent research, particularly in the long term studies undertaken by the Australian Institute of Marine Science in tropical Australia.

However, there is little cause for congratulations among scientists working on mangrove ecosystems. During the same period that has seen a blossoming of research activity, losses of mangrove habitat in many tropical nations have exceeded 1% of mangrove forest area per year (e.g. Umali *et al.*, 1987; Hatcher *et al.*, 1989). As is the case with many other tropical habitats, new research findings do not appear to have influenced the conservation of mangrove resources.

Sound scientific knowledge is essential to the management of tropical mangrove forests. For instance, early in this century research on the growth of mangroves in Malaysia was used to establish silvicultural techniques that are the basis for the sustained yield management for timber and charcoal of the Matang mangrove forest reserve (Watson, 1928; Haron, 1981). However, I can identify several (non-political) reasons why scientific knowledge has not advanced the cause of sustainable management of most mangrove forests.

Many mangrove scientists are not in a position to provide the information required by managers. Firstly, despite the advances during the last decade, there remain huge gaps in our knowledge of the resource (see most chapters in this book). For instance, as yet there is no single study that has measured total net forest primary production in any mangrove system. Nor do we understand much about what controls the within-habitat distribution, survival or growth of individual mangrove tree species or the response(s) of mangrove communities to natural disturbances. Secondly, the usual rationale for research on mangroves is that better understanding of patterns and processes in these systems will benefit sustainable development (see Chapter 1, this volume). However, much of the research is still curiousity-driven, and few scientists attempt to interpret their results in a management framework.

An answer to these dilemmas is a more pragmatic structuring of research priorities. Research programs need to be guided by real management issues. In this way the necessary fundamental research can be done, while providing appropriate advice to managers. Recently, the rapid loss of natural resources worldwide has made scientists aware that research needs to be driven by, and at the same time provide advice on the management of natural habitats (e.g. Lubchenco *et al.*, 1991; Risser *et al.*, 1991; Hubbell and Foster, 1992). Some particularly urgent problems facing mangrove habitats world-wide are their likely responses to sea-level change, their capacity to act as sinks for pollutants, their ability to survive in the face of harvest for wood products and what their loss means to fishery harvests. The resolution of these problems requires that scientists not only do the appropriate research, but also provide appropriate recommendations to managers.

A further challenge to mangrove scientists is the proper communication of their research findings. Publication of scientific findings on mangrove ecosystems in specialist journals should not be the final step in the scientific process, but the beginning of the flow of information to managers and the public.

The developing nations of the world have the greatest need for reliable scientific information on mangrove ecosystems, but are least able to afford the specialist journals where most new information is published. Information exchange networks developed by international agencies such as UNESCO have greatly increased the transfer of research findings. However, scientists in developed countries need to establish closer ties with scientists and managers in developing nations. Only in this way can scientific advice be provided with sensitivity to local cultural and political frameworks.

Finally, at every opportunity, all mangrove scientists must be prepared to educate politicians and the public about the value of mangrove forests. Funding for research must increase if the major gaps in our knowledge of mangrove ecosystems are to be filled. This will only occur if governments are convinced of the importance of the resource. Heightened public awareness leads to a more informed national and international debate on resource conservation. At thc local or village level understanding the function of mangroves can generate small scale action to prevent mangrove destruction or to re-establish mangrove forests (e.g. Kemf, 1988).

Human populations in tropical countries with large coastlines and mangrove resources are increasing at an annual rate of between 1.3 and 3.8% per annum (United Nations, 1990). Increased pressure on mangrove forests is, therefore, inevitable. During the next decade scientists must play a more active role in conserving, managing and restoring these forests and their associated coastal waters or we may lose them forever.

References

Haron, H.A.H., 1981. A working plan for the second 30 year rotation of the Matang Mangrove Forest Reserve, Perak. Perak State Forestry Department, Ipoh, Malaysia, 115 pp.

Hatcher, B.G., Johannes, R.E., and Robertson, A.I., 1989. Review of research relevant to conservation of shallow tropical marine ecosystems. *Oceanography and Marine Biology: An Annual Review* **27**:337-414.

Hubbell, S.P. and Foster, R.B., 1992. Short-term dynamics of a neotropical forest: why ecological research matters to tropical conservation and management. *Oikos* **63**:48-61.

Kemf, E., 1988. The re-greening of Vietnam. *New Scientist* **188**(1618):53-57.

Lubchenco, J., Olsen, A.M., Brubaker, L.B., Carpenter, S.R., Holland, M.M., Hubbell, S.P., Levin, S.A., McMahon, J.A. Matson, P.A., Melillo, J.M., Mooney, H.A., Peterson, C.H., Pulliam, H.R., Real, L.A., Regal, P.J., and Risser, P.G., 1991. The Sustainable Biosphere Initiative: an ecological research agenda. *Ecology* **72**:371-412.

Risser, P.G., Lubchenco, J., and Levin, S.A., 1991. Biological research priorities - a sustainable biosphere. *Bioscience* **41**:625-627.

Umali, R.M., Zamora, P.M., Gatera, R.R., Jara, R.S., Camacho, A.S., and Vannucci, M., 1987. Mangroves of Asia and the Pacific: status and management. Technical Report of the UNDP/UNESCO Research and Training Pilot Programme on mangrove ecosystems in Asia and the Pacific (RAS/79/002), Quezon City, The Philippines, 538pp.

United Nations, 1990. World population prospects 1990. Population Studies No. 120, ST/ESA/SER.A/120, United Nations, New York, 607pp.

Watson, J.G., 1928. Mangrove forests of the Malay Peninsula. *Malayan Forest Records* **6**:1-275.

Coastal and Estuarine Sciences

1. Coastal Upwelling, Francis A. Richards (Ed.) 529 pages. 1981.
2. Oceanography of the Southwestern U.S. Continental Shelf, L. P. Atkinson, D. W. Menzel, and K. A. Bush (Eds.) 156 pages. 1985.
3. Baroclinic Processes on Continental Shelves, Christopher N. K. Mooers (Ed.) 130 pages. 1986.
4. Three-Dimensional Coastal Ocean Models, Norman S. Heaps (Ed.) 208 pages. 1987.

Coastal and Estuarine Series

(formerly Lecture Notes on
Coastal and Estuarine Studies)

Vol. 1: J. Sündermann, K.-P. Holz (Eds.), Mathematical Modelling of Estuarine Physics. Proceedings, 1978. 265 pages. 1980.

Vol. 2: D.P. Finn, Managing the Ocean Resources of the United States: The Role of the Federal Marine Sanctuaries Program. 193 pages. 1982.

Vol. 3: M. Tomczak Jr., W. Cuff (Eds.), Synthesis and Modelling of Intermittent Estuaries. 302 pages. 1983.

Vol. 4: H.R. Gordon, A.Y. Morel, Remote Assessment of Ocean Color for Interpretation of Satellite Visible Imagery. 114 pages. 1983.

Vol. 5: D.C.L. Lam, C.R. Murthy, R.B. Simpson, Effluent Transport and Diffusion Models for the Coastal Zone. 168 pages. 1984.

Vol. 6: M.J. Kennish, R.A. Lutz (Eds.), Ecology of Barnegat Bay, New Jersey. 396 pages. 1984.

Vol. 7: W.R. Edeson, J.-F. Pulvenis, The Legal Regime of Fisheries in the Caribbean Region. 204 pages. 1983.

Vol. 8: O. Holm-Hansen, L. Bolis, R. Gilles (Eds.), Marine Phytoplankton and Productivity. 175 pages. 1984.

Vol. 9: A. Pequeux, R. Gilles, L. Bolis (Eds.), Osmoregulation in Estuarine and Marine Animals. 221 pages. 1984.

Vol. 10: J.L. McHugh, Fishery Management. 207 pages. 1984.

Vol. 11: J.D. Davis, D. Merriman (Eds.), Observations on the Ecology and Biology of Western Cape Cod Bay, Massachusetts. 289 pages. 1984.

Vol. 12: P.P.G. Dyke, A.O. Moscardini, E.H. Robson (Eds.), Offshore and Coastal Modelling. 399 pages. 1985.

Vol. 13: J. Rumohr, E. Walger, B. Zeitzschel (Eds.), Seawater-Sediment Interactions in Coastal Waters. An Interdisciplinary Approach. 338 pages. 1987.

Vol. 14: A.J. Mehta (Ed.), Estuarine Cohesive Sediment Dynamics. 473 pages. 1986.

Vol. 15: R.W. Eppley (Ed.), Plankton Dynamics of the Southern California Bight. 373 pages. 1986.

Vol. 16: J. van de Kreeke (Ed.), Physics of Shallow Estuaries and Bays. 280 pages. 1986.

Vol. 17: M.J. Bowman, C.M. Yentsch, W.T. Peterson (Eds.), Tidal Mixing and Plankton Dynamics. 502 pages. 1986.

Vol. 18: F. Bo Pedersen, Environmental Hydraulics: Stratified Flows. 278 pages. 1986.

Vol. 19: K.N. Fedorov, The Physical Nature and Structure of Oceanic Fronts. 333 pages.1986.

Vol. 20: A. Rieser, J. Spiller, D. VanderZwaag (Eds.), Environmental Decisionmaking in a Transboundary Region. 209 pages. 1986.

Vol. 21: Th. Stocker, K. Hutter, Topographic Waves in Channels and Lakes on the f-Plane. 176 pages. 1987.

Vol. 22: B.-O. Jansson (Ed.), Coastal Offshore Ecosystem Interactions. 367 pages. 1988.

Vol. 23: K. Heck, Jr. (Ed.), Ecological Studies in the Middle Reach of Chesapeake Bay. 287 pages. 1987.

Vol. 24: D.G. Shaw, M.J. Hameedi (Eds.), Environmental Studies in Port Valdez, Alaska. 423 pages. 1988.

Vol. 25: C.M. Yentsch, F.C. Mague, P.K. Horan (Eds.), Immunochemical Approaches to Coastal, Estuarine and Oceanographic Questions. 399 pages. 1988.

Vol. 26: E.H. Schumann (Ed.), Coastal Ocean Studies off Natal, South Africa. 271 pages. 1988.

Vol. 27: E. Gold (Ed.), A Law of the Sea for the Caribbean: An Examination of Marine Law and Policy Issues in the Lesser Antilles. 507 pages. 1988.

Vol. 28: W.S. Wooster (Ed.), Fishery Science and Management. 339 pages. 1988.

Vol. 29: D.G. Aubrey, L. Weishar (Eds.), Hydrodynamics and Sediment Dynamics of Tidal Inlets. 456 pages. 1988.

Vol. 30: P.B. Crean, T.S. Murty, J.A. Stronach, Mathematical Modelling of Tides and Estuarine Circulation. 471 pages. 1988.

Vol. 31: G. Lopez, G. Taghon, J. Levinton (Eds.), Ecology of Marine Deposit Feeders. 322 pages. 1989.

Vol. 32: F. Wulff, J.G. Field, K.H. Mann (Eds.), Network Analysis in Marine Ecology. 284 pages. 1989.

Vol. 33: M.L. Khandekar, Operational Analysis and Prediction of Ocean Wind Waves. 214 pages. 1989.

Vol. 34: S.J. Neshyba, Ch.N.K. Mooers, R.L. Smith, R.T. Barber (Eds.), Poleward Flows Along Eastern Ocean Boundaries. 374 pages. 1989.

Vol. 35: E.M. Cosper, V.M. Bricelj, E.J. Carpenter (Eds.), Novel Phytoplankton Blooms. 799 pages. 1989.

Vol. 36: W. Michaelis (Ed.), Estuarine Water Quality Management. 478 pages. 1990.

Vol. 37: C.M. Lalli (Ed.), Enclosed Experimental Marine Ecosystems: A Review and Recommendations. X, 218 pages. 1990.

Vol. 38: R.T. Cheng (Ed.), Residual Currents and Long-term Transport. XI, 544 pages. 1990.

Vol. 39: M.I. El-Sabh, N. Silverberg (Eds.), Oceanography of a Large-Scale Estuarine System. X, 434 pages. 1990.

Vol. 40: D. Prandle (Ed.), Dynamics and Exchanges in Estuaries and the Coastal Zone. 648 pages. 1992.